250 Years of Industrial Consumption and Transformation of Nature: Impacts on Global Ecosystems and Life

Authored by

Hubert Engelbrecht

Environmental Geology, Munich, Germany

General:

1. Any dispute or claim arising out of or in connection with this License Agreement or the Work (including non-contractual disputes or claims) will be governed by and construed in accordance with the laws of the U.A.E. as applied in the Emirate of Dubai. Each party agrees that the courts of the Emirate of Dubai shall have exclusive jurisdiction to settle any dispute or claim arising out of or in connection with this License Agreement or the Work (including non-contractual disputes or claims).
2. Your rights under this License Agreement will automatically terminate without notice and without the need for a court order if at any point you breach any terms of this License Agreement. In no event will any delay or failure by Bentham Science Publishers in enforcing your compliance with this License Agreement constitute a waiver of any of its rights.
3. You acknowledge that you have read this License Agreement, and agree to be bound by its terms and conditions. To the extent that any other terms and conditions presented on any website of Bentham Science Publishers conflict with, or are inconsistent with, the terms and conditions set out in this License Agreement, you acknowledge that the terms and conditions set out in this License Agreement shall prevail.

Bentham Science Publishers Ltd.
Executive Suite Y - 2
PO Box 7917, Saif Zone
Sharjah, U.A.E.
Email: subscriptions@benthamscience.org

BENTHAM SCIENCE

CONTENTS

FOREWORD

This is a very ambitious book. The author, Hubert Engelbrecht takes a holistic view on geology, minerals extraction, and the hugely interlinked webs of civilizational use of the resources. He looks into the processes and consequences of 250 years of industrialization for minerals, ecosystems and life. This concentrated synopsis, which attempts to consider all kinds of interactions between humans and nature, opens the eyes for the enormous sizes and momenta of transformations generated.

The author chooses to introduce his subject by describing how natural cycles of matter and energy were modulated by geological and extra-terrestrial processes as well as by evolving life. He distinguishes two kinds of life, one evolving seemingly in a deterministic manner and generating the oxygenated atmosphere; and the other culminating, in a non-deterministic manner, in intelligent organisms capable of escaping from the constraints of photosynthetic energy: the miracle of the emergence of humankind creating innumerable artificial niches in the technosphere.

After a short history of industrialization, the author brings the reader down to the basics: how core and lifeblood of industry work and how its output grew with time: mining, processing and further refinement of immense amounts of mineral commodities, used to meet the demands of 7.5 billion humans.

But nothing came without side-effects. Extraction and processing of minerals and the industrial use of biological resources caused multiple and serious environmental impacts - *e.g.* atmospheric warming, acidification of oceans, eutrophication of lakes, pollution, deforestation, and an accelerated loss of biodiversity. In a subchapter on atmospheric turnovers, natural and artificial processes are compared.

For a strategic approach to reduce unwanted damage, the author observes the gap between the amount of resources extracted and processed and the amounts safely discarded or recycled. It is the large number of open loops that impair our ecosystems and indeed our health.

The author finishes with listing up decisive reasons, which caused the actual ecological crisis and mentions ideologies, economic systems as well as innate and learned behavior of humans. This book is recommended for reading for students and professionals of engineering, chiefly mining and civil engineering, of business economics, and indeed of political sciences and humanities.

Ernst Ulrich von Weizsäcker
Former Co-Chair of the International Resource Panel,
Emmendingen,
Germany

PREFACE

The purpose of this book is to make public discussions on the part of the extractive industries' play: they are primary economic drivers and carry a considerable part of responsibility to promote sustainable practices and environmental protection, as well as controlling speed and direction of the transformations. Their products are indispensable for the economy and they provide positive contributions - creation, knowledge, culture, and life, - but also risk, conflict, hazard and destruction. The current, irreversible transformations on planetary scale are part of the most complex, singular and enormous experiment ever conducted. This book offers a responsible, generalist view on rising global ecological problems caused by transformations resulting from long-term industrial extraction of mineral raw materials, their manifold utilizations, and rising number of individuals consuming them in ever increasing amounts. Climate change is only one problem among many that were caused by significant alterations of natural cycles, especially the carbon, nitrogen, and water cycles. Other problems, all characterized by violations of precaution, prudence, mindfulness and restraint, are unpredictable in the long-term. The unexpected side-effects, developmental speeds, trends, and extents of these problems are our moral responsibility to solve, though some may be irreversible. Many improvements were made during industrialization. But a grave cultural crisis arouse at the same time. Because it is unreasonable, insensitive, and at high risk, to cut and interfere on huge economic scale into a natural system: merely a few hundred years of techno-scientifical development cannot have resulted in sufficient knowledge of that system, which evolved and diversified infinitely in a time-abyss of ca. 4.5 Ga. Mankind overstrained itself by economizing the best from ingenious human brains (inventions, ideas) and from the earth (mineral commodities), resulting in incipient loss of control. It is unwise to damage ecosystems, which prospered on a planet orbiting in one of the very rare habitable zones in space. Engineering assessment and advice from the human and ecological disciplines should be better included, because the fire of Prometheus entailed the serious consequence of the opening of Pandora's Box. This study intends to connect, as postulated by physicist H.-P. Dürr, profound expert knowledge with the broad and deep ranges of environmental geology and ecology, which are constituents of the science of consumption, conversion, and transformation of nature by humans. This interdisciplinary approach, which attempts to fully describe the complex human transformation of nature, occurred mentally from outside: *i.e.* virtually from space onto the globe, in order to obtain a general view on all areas affected by industrial development and to realistically represent magnitude, speed and momentum of the resulting changes.

Hubert Engelbrecht
Environmental Geology,
Munich,
Germany

Anthology of Important Mottos, Statements and Stimuli

Immanuel Kant: *Act only according to that maxim whereby you can at the same time will that it should become a universal law without contradiction.*

Fjodor Dostojewskij: *Each one is responsible for all exclusively.*

Viktor Frankl: *Being a human means self-knowledge and responsibility.*

Lao Te: *You are responsible for Your concerns and for Your omissions.*

Arthur Schopenhauer: *We are responsible-le for our actions and for all what we tolerate without contradiction.*

Leon Trotzkij: *Responsibility originates by contradicting not punctually.*

White Rose: *Each person is responsible for all what it tolerates.*

Albert Einstein: *Problems never can be solved by persisting in the same mentality, which caused their origination.*

J. Robert Oppenheimer: *We knew that the world would not be the same.*

Odo Marquard: *It is a rational position to avoid the state of emergency.*

Friedrich Dürrenmatt: *The world resembles a petrol station without ban on smoking.*

Philippus T. Paracelsus: *Quantum facit velenum.*

Karl R. Popper: *Our mental view on future time must be: we are now responsible for that, what will occur in future.*

Marquise de Pompadour: *Aprés nous, le dèluge.*

Georg C. Lichtenberg: *Living together is a combination of boundless unconcern and the same pharisaism.*

Demosthenes: *Wells run dry, if too often and too much water is abstracted.*

Ernest H. Shackleton: *Difficulties are just things to overcome after all.*

Rachel Carson: *I am afraid it is true that, since to beginning of time, man has been a mostly untidy animal.*

Georgius Agricola: *...there is a greater detriment from mining than the values of the metals ... produced.*

Albert Schweitzer: *We have invented many things ... we live in a frightening age.*

Grace M. Hopper: *Humans are allergic to change.*

Heraclitus of Ephesus: *No person steps into the same stream twice.*

Isaak Newton: *Actio est reactio.*

Aristoteles: *There are so many utility goods and luxury I don't need.*

Imre Kertész: *Life is a piece of art: it must be carefully developed. ....getting completely absorbed into and assimilated by the system in order to lose personality and self.*

Francisco J. de Goya y Lucientes: *Absence of common sense generates monsters.*

Friedrich Hölderlin: *But where hazard threatens, rescue also grows.*

Elias Canetti: *Organizations specialized in growth, multiplication and production are the most successful and most momentous structures ever created...; ... excessive production began to suppress other sectors of life.*

Fritz Schumacher: *Small is beautiful.*

Theodor W. Adorno: *Industries' global interactions with humans are restricted to their properties as customer and employee.*

Bernd Guggenberger: *The vanishing of reality.*

Eugene F. Kranz: *Failure is not an option.*

Alexander Gerst: *We have only one earth.*

Jane Goodall: *Reasons for hope.*

CHAPTER 1

Ecosystem Transformations - Natural and Anthropogenic Forcings

Abstract: The evolution of ecosystems and life embedded in the spheres of Earth, as well as the theoretical background - entropic dissipation of energy and matter causing rising complexity - is outlined. This is followed by a description of how natural - terrestrial and extra-terrestrial - forcings (*e.g.* discrete volcanic eruptions, large igneous provinces, silicate weathering, continental drift, albedo, bioevolution; Milankovitch cycles, solar irradiance, galactic tides, impactors, gamma ray bursts) changed the ecosystems in the geological past and how life adapted to and transformed these ecosystems. Examples for the latter case are the Precambrian Great Oxygenation Event, the colonization of continents by plants since the Silurian and the spread of mammals since the early Tertiary. Brain growth, the invention of tools, the controlled use of fire, cooking, and creative organization of niches enabled hominines to escape natural habitats to innumerable artificial habitats. The didactic value of fossilized remnants of geological climate indicators and of ecosystem changes (preserved in geotopes) is emphasized.

Keywords: Albedo, Anthropocene, Bioevolution, Carbon dioxide, Continental drift, Dissipation, Earth history, Ecosystem transformation, Extra-terrestrial forcings, Geogenic forcings, Hominins, Life's origin, Natural forcings, Oxygenic photosynthesis, Silicate weathering, Turnovers, Volcanism.

Since ca. 4.4 Ga, geological processes have resulted in diverse spheres [1] - *e.g.* the hydro- and lithospheres and separate spatialities like oceans and continents forming the earth's surface. Of these the oceans were probably the first colonized by life since ca. 4.1 Ga ago [2]. Whether the origin of life, in the form of the Last Common Universal Ancestral Cell from inorganic molecules, occurred by chemoevolution in Darwin's warm little pond; in fissure brines present in the deep subsurface; in a hot "primordial soup" (Oparin); or in a gas mixture (Miller & Urey) stimulated by intense UV-radiation from the young sun, by electric energy from lightning, or by redox-reactions on the surfaces of Fe- and Zn-sulphides transferring energy to synthesize organic molecules in hot aquatic environment; or whether life was imported by deorbiting extra-terrestrial bodies containing organic molecules [3], cannot be precisely answered at present [4 - 7]. One type of early ecosystem[1] began to develop *via* catalytic organic synthesis probably at alkaline hydrothermal vent systems in anoxic Hadean oceans [8].

It is a hallmark of Earth history that geological processes, extra-terrestrial factors and events, as well as evolving life itself, endlessly modified or even destroyed ecosystems (see below). The lithified relics of the matter processed, converted and cycled through these ecosystems are present in km-thick piles of sediment, which are sub-dividable into geological formations[2]. These products filled sedimentary basins, which were integrated *via* subsidence and subduction into the geological cycle. Later on, due to tectonic forces, they were exhumed and exposed in metamorphosed and deformed states in mountain belts of the earth's surface [9]. Geophysical, palaeontological, sedimentological, stratigraphical, geochrono-logical, geochemical, and mineralogical analytics of these formations often result in rather precise reconstructions of changing ecosystems (palaeoecology, palaeoclimatology) as well as the distribution of oceans and continents (paleogeography) [9, 10].

Natural changes in ecosystems depend, beside the geological processes described above, primarily on the energy fluxes on and above the earth's surface, which is controlled by global solar irradiance, by scattered and reflected terrestrial infrared irradiation, as well as by the heat and matter transfer between the earth's surface, oceans' interiors and the atmosphere [11]. The absorbed and transferred heat circulating in the ecosystems embedded in the spheres of Earth depends predominantly on the efficacy of the planetary albedo and the varying quantities of greenhouse gases and aerosols present in the atmosphere [11, 12].

Matter and energy, circulating in the earth's spheres since ca. 4.5 Ga ago [9] irreversibly dissipate[3] according to the theory of synergetic self-organisation[4], which is controlled by entropy-transfer and the maximum entropy producing principle [13 - 15] in open systems. The asymmetry of time[5] and the directedness/irreversibility of evolution are founded in the second law of thermodynamics and in the fact that entropy from physical and chemical processes increases with time [16]. Evolution is interwoven into that directed process, which even more efficiently dissipates energy and matter by generating self-organized structures that evolve into more and more complexity [17].

Species, which, since 4.0-4.1 Ga ago [2], have been subjected to casual mutation and natural selection, attempt to survive *via* competition, niche construction, innovation, creative adaptation, cooperation [18], and symbiosis. These activities, as well as the necessities to cope with resource limitations and to maintain vital functions by finding and metabolizing food and energy-carriers, caused deterministic transformations of the ecosystems they live in [9]. The most important instances of bioevolutionary ecosystem transformations are the Great Oxygenation Event, the Cambrian Radiation, the Silurian colonization of land by primitive plants, the radiation and spread of flowering plants in the Cretaceous, and the mammals domination since the Cretaceous-Tertiary boundary [9, 19]; see below. Hominines diverged from apes 5-7 Ma ago and diversified into several lineages [20]. Reorganization, expansion of neocortex and frontal lobe, improvement of interconnectedness of neuronal networks, and adaptive change in shape of the hominine brain since ca. 1.9 Ma ago enabled *Homo habilis* and its phylogenetic successors - the population of *Homo sapiens* diverged probably between 350,000 and 260,000 a [21] - to intelligently improve advanced cognitive and physical functions and skills: *e.g.* articulating, planning, communicating, social cooperation, problem solving. In consequence, it willingly (indeterministically) escaped the limitations of its natural habitats *via* inventing tools, hunting, controlled use of fire, practicing agriculture, deforestation, cooking, producing higher nutrition food, applying artificial selection (breeding, domestication), and practicing arts [22 - 24]. As consequence, atmospheric levels of CO_2, N_2O and CH_4 began to rise since the middle Holocene [25]. Several severe late and latest Holocene epi- and pandemics effected synchronous atmospheric cooling, because forests regrowing on plague-abandoned farms assimilated tens of GT of atmospheric CO_2 [25]. According to Malthusian-Darwinian dynamics [26], the cultural developments - farming, trade, transportation, technology, and science - of modern man led to the emergence, construction, and maintenance of innumerable niches and domesticated ecosystems [27, 28], as well as in the provocation of innumerable bioevolutionary reactions [29 - 32]. Since 1971, genetic engineering has opened new fields in artificial selection, as well as experimental evolution [33]. Creation of huge geomorphic transformations [34] profoundly and irreversibly modified many ecosystems. Reliable reconstructions of the geogenic carbon-release rates during the Paleocene-Eocene thermal maximum[6] 56 Ma ago indicate that even during that extreme period, the 2014 input rate of anthropogenic carbon into the atmosphere has no geological analogue and must be seen as unprecedented and unparalleled during the last 66 Ma [35].

The above mentioned anthropogenic impacts occur in addition to variations in and turnovers of ecosystems brought about by natural events and processes; see above [3, 36]. In the following, it is necessary to detail the types of natural events and processes that modulate ecosystems, because 1) the impacts of a few of them can be seen as proxies and analogues to some anthropogenic effects on ecosystems, and 2) the information preserved in fossilized deposits can be deciphered by means of chemical and physical analytical methods and inform us about the causes and effects of ecosystem transformations and highlight the procedures and durations of biotic responses, *i.e.* the processes of adapting to and recovering from ecosystem transformations. This knowledge enables us to better determine the intensities and reaches, and to more reliably predict, the effects of anthropogenic impacts on ecosystems [37]. Recognitions from palaeobiological responses analyzed from fossilized ecosystem turnovers is recommended to be applied as guiding principles in actual conservation biology paradigms to foster biodiversity and its resilience to rising anthropogenic pressure [38].

Among the first to identify fossil climate indicators - *e.g.* halite, redbeds, ventifacts, loess, tillites, striated pebbles, coal - and to have attempted to reconstruct the development of the corresponding palaeoecosystems in the geological past, were the meteorologists Wladimir Köppen and Alfred Wegener [39].

Volcanic eruptions: Ashes, gases and aerosols ejected into the stratosphere by brief eruptions decrease atmospheric transparency and increase the albedo effect. Depending on residence time and the amount of the suspended particulate matter and secondary aerosols (*e.g.* sulphuric acid aerosols formed from SO_2), brief atmospheric cooling will result [40]. Emitted volcanic gases like H_2O, CO_2, H_2S, CO, *etc.*, however, contribute to the natural greenhouse effect and HCl and HF decompose the ozone layer [41, 42]. Examples of major eruptions that occurred in historic time are Santorin (1650 BC), Krakatoa (535 AD, 1883 AD), Laki (1783-1784 AD), Tambora (1815 AD), Mt. St. Helens (1980 AD), Pinatubo (1991 AD), and Eyjafjallajökull (1821-1823 AD, 2010 AD). The largest volcanic eruptions caused severe global economic stress, migrations of people, sociocultural problems, famines, plagues, and even extinctions, *e.g.* the demise of the Minoan culture and the Sasanian and East Turkish Empires [43 - 47]. It took about 70 years for the scientific community to realize that the Year Without a Summer, 1816, was caused by a volcanic super eruption and to begin to understand the global impacts of such events [48].

Rapid effusion of very large volumes of volcanics ($>> 0.1$ Mkm^3 within 1-5 Ma) [49], as occurs with the origination of large marine and terrestrial igneous provinces [50], changes the physicochemical state of the hydro-, cryo-, pedo-, and atmospheres substantially on global scale. Large quantities of volcanic gases are also emitted, which cause a shift towards warm-humid climate; ocean-acidification, warming, eutrophication, and deoxygenation; diminution of the equator-pole temperature gradient, more pronounced ocean water stratification; melting of polar ice caps; sea level rise; possible dissociation of submarine CH_4- and CO_2 clathrates; and sometimes selective mass extinctions of species [51 - 53]. Examples of large igneous provinces are the Siberian flood basalts (252 Ma ago) and eruptions shortly afterwards [54], the Central Atlantic (Newark) rift basalt effusions (202-199 Ma ago), formation of the Shatzky Rise (147 Ma ago), Parana-

Etendeka (132-130 Ma ago) Traps, the Karoo Traps (180 Ma ago), the mid-Cretaceous Manihiki (123 Ma ago), Ontong-Java (122 Ma ago, 90 Ma ago), and Kerguelen (110 Ma ago) events, as well as the Deccan Traps (66 Ma ago) and Afar event (31-29 Ma ago). These are often linked to contemporaneous ecosystem-turnovers, *e.g.* the demise of reef systems and carbonate platforms [55, 56], as well as oceanic anoxic events, the latter originating as a consequence of intensified near-surface bioproductivity and its degradation products, which cause deoxygenation in deeper water layers and the deposition of sapropelites, in which CH_4 and H_2S originate at concentrations above tipping points [53]. The volatile emissions of the Siberian flood basalts, estimated at ca. 170 teratonnes within 60,000 a, caused the extinction of 80-90% of terrestrial and marine species [55, 57, 58]. Rapid injections of vast amounts of CO_2 and SO_2 into the atmosphere caused major biotic catastrophes [59].

Silicate weathering: Chemical decomposition of large volumes of flood basalts and traps, exposed in the tropics, effectively absorbs atmospheric CO_2 and diminishes the natural greenhouse effect[7] [60]. This also applies to subaerially exposed continental terrain in general [61 - 64]. Silicate weathering of the low latitude landmasses of the supercontinent Rodinia contributed to the origination of several late Proterozoic glaciations (850-635 Ma ago) [65].

Continental drift/plate tectonics: landmasses at high latitudes are affected by cooling of their surfaces and subsequent glaciation, which can reinforce itself by the snow albedo effect, if enough moisture is transported poleward by marine and atmospheric circulations [12]. Such an arrangement enhanced icehouse conditions during the late Ordovician to early Silurian (the Andean-Saharan glaciation 450-420 Ma ago), late Devonian to late Permian (the Karoo glaciation 360-260 Ma ago) and the Quarternary glaciation (2.58-0.01 Ma ago) [9].

Planetary albedo: The diminished albedo of subaerially exposed, deglaciated land contributes to climate warming: Models of the middle Pliocene global warming suggest that it resulted from the considerably reduced extent of high-latitude terrestrial ice sheets and sea ice cover, which increased the heat-uptake through absorption of solar irradiation [66].

Bioevolution: In the beginning, significant palaeobiological evolutionary steps consisted in the development from fermentation below the Pasteur-point[8] to aerobic metabolism, improving the effectiveness of the transfer of chemical energy in organisms by ca. 18 times and completion of the evolution from procaryota *via* endosymbiosis to eucaryota ca. 3.2 Ga ago [67 - 69]. Unicellular life consists of functional subunits like nuclei, vacuoles, chloroplasts (containing chlorophyll[9]), and mitochondria, as well as other components, enabling these phytogenic cells, *via* oxygenic photosynthesis, to assimilate CO_2 by forming carbonates, carbohydrates, lipids, sterines, saccharines, *etc.*, as well as free O_2 [67].

Since ca. 3.7 - 2.7 Ga ago, palaeobiological events - *e.g.* the development of mats of photosynthesizing cyanophyceae in littoral zones - affected global oxygenation substantially [70, 71]. Subsequent to the oxygenation of the upper layers of the ocean 2.3 Ga ago, O_2 began to accumulate in the atmosphere and the ozone layer began to form [53, 72]. The resulting Great Oxygenation Event consisted in an inversion of the volume-relations of the gas-constituents of the atmosphere and in a fundamental modification of the carbon-cycle: O_2 - a trace gas constituent (0.0001-0.001%) of the Archaean (4.0-2.5 Ga ago) atmosphere - evolved to a main constituent of the Proterozoic (2.5-0.545 Ga ago) atmosphere [73]. CO_2 underwent reverse development, whereby many PT of assimilated carbon as integrated *via* sedimentation into the geological cycle, being transformed and buried for millions of years in the lithosphere in form of limestone, dolostone, marlstone, coal, tar and hydrocarbons[10]. Consumption of methane - one of the main constituents of the Archaean atmosphere - was brought about by bacterial metabolism [74]. The substantial diminution of atmospheric CO_2 and CH_4 caused a decrease in the greenhouse effect, compensating for the modest irradiation arriving from the early, relatively faint sun, which was responsible for a cool climate that hosted several Precambrian and early Palaeozoic glaciation events [12] that temporarily diminished the progress of bioevolution.

Free oxygen proliferated beginning ca. 1.1 Ga the chain of respiration in animal unicellulars, in which important carriers of chemical energy *via* oxidative phosphorylation are formed to spur metabolism[11]. During the development of multicellular life (ca. 0.8-0.6 Ga ago) and more highly evolved aerobes, physiological transport of oxygen was made possible by the protein haemoglobin [9, 67, 75]. Carnivory has been linked to higher levels of oxygenation in the atmosphere, hydrosphere, and deep marine waters, and is supposed to have originated in the late Ediacaran (580-540 Ma ago); the evolution of sense organs is also discussed in this context [76].

Colonized since ca. 3.5 Ga ago by microbial communities [68], terrestrial environments were invaded ca. 465 Ma ago by flora species (algae, lichens, and bryo-, psilo-, lycopodo-, and pteridophyta of the Palaeophyticum) [77]. This important palaeobiological evolutionary step fostered atmospheric CO_2 consumption *via* photosynthesis, resulting in maximum atmospheric O_2 (ca. 35% by volume) in the latest Carboniferous and the formation of coal, while the drop in atmospheric CO_2 was enhanced by Ca-Mg silicate weathering, induced by the rhizomes of vascular plants; both effects contributed to the Karoo glaciations [9, 78].

Adaptation to declining atmospheric CO_2 concentration - *e.g.* the development of megaphyll leaves and C4-photosynthesis - during the Cretaceous caused the radiation and taxonomic diversification of angiosperms, as well as their competitive replacement of gymnosperms and pteridophytes [19].

During early Tertiary, mammals radiatively adapted to environmental habitats previously occupied by predatory dinosaurs and large reptiles. Further diversification, migration, exchange, and dispersion, but also selective extinctions, occurred in middle and late Tertiary due to climatic cooling, sea level drop, origination of land- and filter bridges, reduced moisture, as well as changed food availability [9, 79, 80].

Orbital forcing: During the Cenozoic era, periodic variations in the spatial and temporal amount of solar irradiation arriving at the earth's surface occur in cycles of ca. 100 ka, 40 ka and 22 ka. These are controlled by eccentricity of the earth's orbit, obliquity of the ecliptic, as well as nutation of the rotational axis (Milankovitch 1930 in [12]). The superpositioning of these cycles causes *e.g.* the division of ice ages into stages - glacials and interglacials - and monsoon variations. As a result, cyclostratigraphic sedimentary deposits accumulated in the sinks of ecosystems exposed to orbital forcings [81].

Repetitive patterns of sedimentation have been recognized *e.g.* in the epicontinental Vocontian Basin (NW Tethyan margin) of the earliest Cretaceous [82], in the lacustrine environment of the Germanic Triassic [83], in the carbonate shelf environment of latest Permian [84], and in Mesoproterozoic back-arc basin deposits [85].

A critical functional relationship between the orbitally controlled intensity of boreal summer insolation and global atmospheric CO_2-concentration explains the onset of the eight Quaternary glacial-interglacial cycles [86].

Extra-terrestrial forcing: Solar forcing of a few minor, but rapid, changes, as well as high-frequency oscillations of the Holocene palaeoclimate are recorded in isotopic signatures (variations of ^{14}C, ^{10}Be) measured in ice-cores and lake sediments [87 - 89], which indicate variations of intensity of solar irradiance, as well as the influence of the solar cycle on sedimentation. Detection and attribution of solar forced climate change in the 20th century is, however, complicated by similarity and degeneracy of signal-patterns from anthropogenic greenhouse gases and from solar irradiance [90].

The effects of comet and asteroid impacts on the transformation of ecosystems depend on their mass, heat content, chemical composition, transferred kinetic energy, and the geochemical composition of the target area [9]. Major extinctions, first order unconformities, and stratigraphic turnovers characterizing *e.g.* the early Carnian, Triassic-Jurassic and Cretaceous-Tertiary boundaries are most probably influenced by the effects of Saint Martin (220 ± 32 Ma ago), Rochecouart (201 ± 2 Ma ago) - and Chicxulub 65 Ma ago) - astroblemes [91, 92]. Formation of astroblemes on planet Earth by bodies deorbiting the scattered disc, the Kuiper-belt, or the Oort-Cloud at and beyond the periphery of the solar system may consist of at least three different processes: - Periodic cratering, where galactic tidal forces move objects from their orbits and shift them to the inner solar system [93]; - the solar system orbiting within 250 Ma the galactic centre and oscillates in a period of 35 Ma up and down the galaxy's plane, which probably consists of a dark matter disc capable of gravitationally perturbing objects [94]; - random cratering, which may be caused by transient gravitational waves that exert weak gravitational wave strain on objects, destabilizing their orbits at the periphery of solar system [95].

Some major palaeoecosystem changes and turnovers associated with diversification and selective mass extinctions, like the Cambrian Radiation, Kellwasser crisis, and Hangenberg event have been described in Earth history but lacked clear explanations, because several factors probably interacted in a complex manner to cause them [96 - 98]. Alternatively, an intragalactic, large, and long lasting gamma ray burst with a source-distance < 5000 light years may have hit planet Earth with ca. 90% probability and affected one of the major extinction events of yet unknown cause, like the terminal Ordovician crisis [99].

It is supposed that the Cambrian Radiation was probably linked to the crossing of a threshold value of bioavailable oxygen and to behavioral innovations (onset of ecosystem engineering, *e.g.* increase of bioturbation intensity) and evolution of sense organs [76, 100].

Detailed analysis of the last deglaciation, which started ca. 19 ka ago, found it was initiated by orbital forcing, causing warming of the Arctic area. Mixing of huge amounts of melt water with saline boreal water decelerated sinking of cold North Atlantic Deep Water and the Atlantic Meridional Overturning Circulation system, driven by thermohaline density gradients, slowed down. This caused warming of the deep waters of the southern hemisphere oceans. Because less CO_2 is soluble in warmer seawater, a surplus of CO_2 degassed and accumulated, with a delay of ca. 800-1000 a, in the atmosphere and contributed to later global greenhouse warming characterizing the postglacial era [101 - 103]. Late glacial (ca. 12,000 an ago) dissociation of gas hydrates present in high latitude hydrocarbon provinces and subsequent release of CH_4 into the atmosphere contributed to warming [104].

ENSO[12]-related short term displacements of the inner tropical convergence zone during the latest Holocene caused alternation of the hydrological cycle of the tropical rain belt. Multi-year (< 10 a) droughts in the Southern Caribbean and monsoon changes in Eastern Asia chronologically correspond to the demise and collapse of classic Mayan civilization and Tang dynasty between 750-910 AD [105, 106]. Recent ENSO-forced climate variations were identified in time series measurements (covering 65 a) from the Amazon Basin that was compiled from *in situ* river gauges and satellite-based gravimeters [107, 108].

NOTES

[1] Ecosystem: etymology: oikós [Classical Greek] = house; sýstema [Classical Greek] = compound. Definition: biologist Sir Arthur Tansley (1871-1955) coined in the year 1935, the term ecosystem: dynamic interactions between life-forms - biocoenosis consisting of producers, consumers, and destruents - and their randomly selectable living-spaces in the Earth's biogeosphere. Ecosystems are complex, self-organizing, open, and scale-invariant, keeping material structures and energetic states in dynamic equilibrium. Ecosystems are driven by thermodynamic gradients and by energetic and material resources transferred into them.

[2] Geological formation: it forms a rock unit unambiguously distinguishable according to its chemical, mineralogical, and lithological composition, as well as to its fabric. Sedimentary formations often contain distinctive marks and fossils; both originated in ecosystems having successfully existed over geological times. Specific marks, fossils, biomarkers, and chemical compounds indicate the type of ecosystem from which the geological formation originated.

[3] Dissipation: thermodynamically non-equilibrated structures, which exchange matter and energy with adjacent structures in open systems. According to Ilya Prigogine, dissipative structures are characterized by stability facing minor disorders and by the allocation of their state far from thermodynamic equilibrium. Their stability is founded on the balance between non-linearity and dissipation of energy. They are fundamental for the formation of new structures and patterns in the inanimate nature and of their evolution [13].

[4] It occurs in an open and dynamic system far from thermodynamic equilibrium, if energy is transferred into it and subsystems spontaneously begin to from irreversibly new, stable and regular structures [13].

[5] This concept, the arrow of time, is valid in the macrocosmos and was developed in the year 1927 by the British astronomer Arthur Eddington.

[6] This interval, situated at the geological boundary between Paleocene and Eocene, is characterized by two phases of significant input - in sum ca. 3 teratonnes - of Carbon into the atmosphere, causing perturbation of the carbon cycle, global warming, as well as sedimentary and biotic turnovers [109].

[7] This effect has successfully been applied at the CarbFix site on Island, where 95% of hydrolyzed and injected CO_2 mineralized within just two years, thus opening an option to safely and quickly immobilize and store anthropogenic CO_2 long-term as environmentally benign carbonate minerals [110].

[8] Level of oxygen containing ca. 1% of the atmosphere's O_2-content, above which fermenting organisms adapt to aerobic respiration.

[9] It probably developed in thermophiles bacteria locating and utilizing infrared radiation of submarine hydrothermal systems [53].

[10] At present, carbon is distributed in the geosphere as follows: the lithosphere contains 50×10^6 GT carbon, of which ca. 75% is bound in an organic carbonates and ca. 25% in organogenic sediment (coal, oil shale, methane-hydrate, *etc.*), as well as natural gas. The small rest (0,08%) - 40×10^3 GT carbon - is stored at the earth's surface in biomass (ca. 21%), in the hydrosphere (70%) and in the atmosphere (<1%) [111].

[11] However, new geochemical proxies of oxygen content of mesoproterozoic bottom sea waters indicate that there was sufficient O_2 to fuel animal respiration and that oxygen itself did not limit phylogenesis, spread and radiation of animal life [112].

[12] ENSO, El Nino Southern Oscillations: A climatic anomaly occurring with decadal frequency in the equatorial Pacific. The physical processes occurring in the ocean-atmosphere system is represented by the Walker circulation, which is convectively driven by a temperature-gradient between warmer Indonesian sea waters in the west and cooler South-American sea surface water in the east, where upwelling of deep sea water occurs. The latter is forced by westbound trade winds, which also affect uplift of the thermocline beneath the sea surface adjacent

to the South-American coast. Southern Oscillation - defined by physicist and meteorologist Sir Gilbert Walker - describes cyclic fluctuations in the exchange of these air masses. The state of this system is expressed by the Southern Oscillation Index SOI, derived from the differences in air-pressure measured on Tahiti and at Darwin: $SOI=10(dP-dP_{avg}): S_{dP}$. dP: mean of air pressure measured over 1 month; dP_{avg}: mean of dP over many years; S_{dP}: standard deviation of that average value over many years. Persistent negative/positive SOI values indicate slow/fast Walker circulation, surface sea water warming/cooling and forcing of an El Nino (1) La Nina (2) event. (1): Shift to slow westbound, or even reverse of, trade winds, which affect uplift of the thermocline, cessation of upwelling, and ingression of warm surface waters off the South-American coast. (2): Transition to intensified westbound trade winds, causing further uplift and steepening of the thermocline, as well as amplification of upwelling of cold deep waters off the South-American coast. Causal connections between (1) and (2) and climatic anomalies affecting distant regions exist [13, 113].

CONFLICT OF INTEREST

The author (editor) declares no conflict of interest, financial or otherwise.

ACKNOWLEDGEMENTS

Declare none.

REFERENCES

[1]   Wilde, S.A.; Valley, J.W.; Peck, W.H.; Graham, C.M. Evidence from detrital zircons for the existence of continental crust and oceans on the Earth 4.4 Gyr ago. *Nature,* **2001**, *409*(6817), 175-178. [http://dx.doi.org/10.1038/35051550] [PMID: 11196637]

[2]   Bell, E.A.; Boehnke, P.; Harrison, T.M.; Mao, W.L. Potentially biogenic carbon preserved in a 4.1 billion-year-old zircon. *Proc. Natl. Acad. Sci. USA,* **2015**, *112*(47), 14518-14521. [http://dx.doi.org/10.1073/pnas.1517557112] [PMID: 26483481]

[3]   Ward, P.; Kirschvink, J. *A New History of Life: The Radical New Discoveries about the Origins and Evolution of Life on Earth*; Bloomsbury, **2015**.

[4]   Follmann, H.; Brownson, C. Darwin's warm little pond revisited: from molecules to the origin of life. *Naturwissenschaften,* **2009**, *96*(11), 1265-1292. [http://dx.doi.org/10.1007/s00114-009-0602-1] [PMID: 19760276]

[5]   de Marcellus, P.; Meinert, C.; Myrgorodska, I.; Nahon, L.; Buhse, T.; d'Hendecourt, Ll.S.; Meierhenrich, U.J. Aldehydes and sugars from evolved precometary ice analogs: importance of ices in astrochemical and prebiotic evolution. *Proc. Natl. Acad. Sci. USA,* **2015**, *112*(4), 965-970. [http://dx.doi.org/10.1073/pnas.1418602112] [PMID: 25583475]

[6]   Altwegg, K.; Balsiger, H.; Bar-Nun, A.; Berthelier, J.J.; Bieler, A.; Bochsler, P.; Briois, C.; Calmonte, U.; Combi, M.R.; Cottin, H.; De Keyser, J.; Dhooghe, F.; Fiethe, B.; Fuselier, S.A.; Gasc, S.; Gombosi, T.I.; Hansen, K.C.; Haessig, M.; Jäckel, A.; Kopp, E.; Korth, A.; Le Roy, L.; Mall, U.; Marty, B.; Mousis, O.; Owen, T.; Rème, H.; Rubin, M.; Sémon, T.; Tzou, C.Y.; Hunter Waite, J.; Wurz, P. Prebiotic chemicals-amino acid and phosphorus-in the coma of comet 67P/Churyumov-Gerasimenko. *Sci. Adv.,* **2016**, *2*(5), e1600285.

[http://dx.doi.org/10.1126/sciadv.1600285] [PMID: 27386550]

[7] Borgonie, G.; Linage-Alvarez, B.; Ojo, A.O.; Mundle, S.O.; Freese, L.B.; Van Rooyen, C.; Kuloyo, O.; Albertyn, J.; Pohl, C.; Cason, E.D.; Vermeulen, J.; Pienaar, C.; Litthauer, D.; Van Niekerk, H.; Van Eeden, J.; Sherwood Lollar, B.; Onstott, T.C.; Van Heerden, E. Eukaryotic opportunists dominate the deep-subsurface biosphere in South Africa. *Nat. Commun.,* **2015**, *6*, 8952.
[http://dx.doi.org/10.1038/ncomms9952] [PMID: 26597082]

[8] Herschy, B.; Whicher, A.; Camprubi, E.; Watson, C.; Dartnell, L.; Ward, J.; Evans, J.R.; Lane, N. An origin-of-life reactor to simulate alkaline hydrothermal vents. *J. Mol. Evol.,* **2014**, *79*(5-6), 213-227.
[http://dx.doi.org/10.1007/s00239-014-9658-4] [PMID: 25428684]

[8] Dodd, M.S.; Papineau, D.; Grenne, T.; Slack, J.F.; Rittner, M.; Pirajno, F.; O'Neil, J.; Little, C.T. Evidence for early life in Earth's oldest hydrothermal vent precipitates. *Nature,* **2017**, *543*(7643), 60-64.
[http://dx.doi.org/10.1038/nature21377] [PMID: 28252057]

[9] Condie, K.C.; Sloan, R.E. *Origin and Evolution of Earth: Principles of Historical Geology*; Prentice Hall: New York, **1998**.

[10] Posamentier, H.W.; Walker, R.G. Facies models revisited, SEPM Special Publication 84: Tulsa, Oklahoma, **2006**.
[http://dx.doi.org/10.2110/pec.06.84]

[11] Hartmann, D.L. *Global Physical Climatology, International Geophysics Series 56*; Dmowska, R.; Holten, J.R., Eds.; Academic Press: San Diego, **1994**.

[12] Klostermann, J. Das Klima im Eiszeitalter, Schweizerbart: Stuttgart, **2009**.

[13] Greulich, W. *Lexikon der Physik*; Vol. 1-5, Spektrum Akademischer Verlag GmbH: Heidelberg, **1999**.

[14] Haken, H.; Wunderlin, A. *Die Selbststrukturierung der Materie: Synergetik in der unbelebten Welt*; Vieweg and Teubner Verlag: Braunschweig, **1991**.
[http://dx.doi.org/10.1007/978-3-322-83602-1]

[15] Kleidon, A.; Malhi, Y.; Cox, P. M. Maximum entropy production in environmental and ecological systems. *Philos. T. Roy. Soc. B,* **2010**, *365*(1545), 1297-1302.

[16] Zeh, H.D. *The Physical Basis of The Direction of Time*; Springer: Berlin, Heidelberg, **2007**.

[17] Kováč, L. *Closing Human Evolution: Life in the Ultimate Age*; Springer: Berlin, Heidelberg, **2015**.
[http://dx.doi.org/10.1007/978-3-319-20660-8]

[18] Nowak, M.; Highfield, R. *SuperCooperators: Altruism, Evolution, and Why We Need Each Other to Succeed*; Free Press: New York, **2012**.

[19] McElwain, J.C.; Willis, K.J.; Lupia, R. Ehleringer, J.R.; Cerling, T.E.; Dearing, M.D. *A History of Atmospheric CO_2 and its Effects on Plants, Animals and Ecosystems*; Springer: New York, **2005**.

[20] Hawks, J. *The Princeton Guide to Evolution*; Losos, J.B.; Baum, D.A.; Futuyma, D.J., Eds.; Princeton University Press: Princeton, **2013**, pp. 183-188.

[21] Schlebusch, C. M.; Malmström, H.; Günther, T. Southern African ancient genomes estimate modern human divergence to 350,000 to 260,000 years ago. *Science,* **2017**. in press

[22] Beaumont, P.B. The Edge: More on Fire-Making by about 1.7 Million Years Ago at Wonderwerk Cave in South Africa. *Curr. Anthropol.,* **2011**, *52*(4), 585-595.
[http://dx.doi.org/10.1086/660919]

[23] Lieberman, D.E.; Pilbeam, D.R.; Wrangham, R.W. Transitions in prehistory: essays in honor of Ofer Bar-Yosef. In: *American School of Prehistoric Research Monograph Series*; Cambridge, USA, **2009**.

[24] Herculano-Houzel, S. *Human Advantage: A New Understanding of How our Brain Became Remarkable*; MIT Press: Cambridge, USA, **2016**.
[http://dx.doi.org/10.7551/mitpress/9780262034258.001.0001]

[25]   Ruddiman, W.F. The anthropogenic Greenhouse Era began thousands of years ago. *Clim. Change,* **2003**, *61*(3), 261-293.
[http://dx.doi.org/10.1023/B:CLIM.0000004577.17928.fa]

[26]   Nekola, J.C.; Allen, C.D.; Brown, J.H.; Burger, J.R.; Davidson, A.D.; Fristoe, T.S.; Hamilton, M.J.; Hammond, S.T.; Kodric-Brown, A.; Mercado-Silva, N.; Okie, J.G. The Malthusian-Darwinian dynamic and the trajectory of civilization. *Trends Ecol. Evol. (Amst.),* **2013**, *28*(3), 127-130.
[http://dx.doi.org/10.1016/j.tree.2012.12.001] [PMID: 23290501]

[27]   Palumbi, S.R. Humans as the world's greatest evolutionary force. *Science,* **2001**, *293*(5536), 1786-1790.
[http://dx.doi.org/10.1126/science.293.5536.1786] [PMID: 11546863]

[28]   Ertsen, M. *Human Niche Construction, Conference Report, Rachel Carson Center: Munich,* **2015**.

[29]   Vandvik, V.; Töpper, J.P.; Cook, Z.; Daws, M.I.; Heegaard, E.; Måren, I.E.; Velle, L.G. Management-driven evolution in a domesticated ecosystem. *Biol. Lett.,* **2014**, *10*(2), 20131082.
[http://dx.doi.org/10.1098/rsbl.2013.1082] [PMID: 24522633]

[30]   Russell, E. *Evolutionary History: Uniting History and Biology to Understand Life on Earth*; Cambridge Univ. Press: New York, **2011**.
[http://dx.doi.org/10.1017/CBO9780511974267]

[31]   Sultan, S.E. *Organism and Environment: Ecological Development, Niche Construction, and Adaptation*; Oxford University Press: UK, **2015**.
[http://dx.doi.org/10.1093/acprof:oso/9780199587070.001.0001]

[32]   Kerner, S.; Dann, R.; Bangsgaard, P. *Climate and Ancient Societies*; Tusculanum Press: University of Copenhagen, **2015**.

[33]   Garland, T., Jr; Rose, M.R. *Experimental Evolution: Concepts, Methods, and Applications of Selection Experiments*; University of California Press: Berkeley, California, USA, **2009**.

[34]   Hooke, R. LeB. On the history of humans as geomorphic agents. *Geology,* **2000**, *28*(9), 843-846.
[http://dx.doi.org/10.1130/0091-7613(2000)28<843:OTHOHA>2.0.CO;2]

[35]   Zeebe, R.E.; Ridgwell, A.; Zachos, J.C. Anthropogenic carbon release rate unprecedented during the past 66 million years. *Nat. Geosci.,* **2016**, *9*, 325-329.
[http://dx.doi.org/10.1038/ngeo2681]

[36]   Hallam, T. *Catastrophes and Lesser Calamities - The Causes of Mass Extinctions*; Oxford, University Press: UK, **2003**.

[37]   Payne, J.L.; Clapham, M.E. End-Permian Mass Extinction in the Oceans: An Ancient Analog for the Twenty-First Century? *Annu. Rev. Earth Planet. Sci.,* **2012**, *40*, 89-111.
[http://dx.doi.org/10.1146/annurev-earth-042711-105329]

[38]   Barnosky, A.D.; Hadly, E.A.; Gonzalez, P.; Head, J.; Polly, P.D.; Lawing, A.M.; Eronen, J.T.; Ackerly, D.D.; Alex, K.; Biber, E.; Blois, J.; Brashares, J.; Ceballos, G.; Davis, E.; Dietl, G.P.; Dirzo, R.; Doremus, H.; Fortelius, M.; Greene, H.W.; Hellmann, J.; Hickler, T.; Jackson, S.T.; Kemp, M.; Koch, P.L.; Kremen, C.; Lindsey, E.L.; Looy, C.; Marshall, C.R.; Mendenhall, C.; Mulch, A.; Mychajliw, A.M.; Nowak, C.; Ramakrishnan, U.; Schnitzler, J.; Das Shrestha, K.; Solari, K.; Stegner, L.; Stegner, M.A.; Stenseth, N.C.; Wake, M.H.; Zhang, Z. Merging paleobiology with conservation biology to guide the future of terrestrial ecosystems. *Science,* **2017**, *355*(6325), eaah4787.
[http://dx.doi.org/10.1126/science.aah4787] [PMID: 28183912]

[39]   Köppen, W.; Wegener, A. *Die Klimate der Geologischen Vorzeit / The Climates of the Geological Past; Reproduction by Thiede J*; Lochte, K.; Dummermuth, A., Eds.; Borntraeger Science Publishers: Stuttgart, **2015**.

[40]   Robock, A. Volcanic eruptions and climate. *Rev. Geophys.,* **2000**, *38*, 191-219.
[http://dx.doi.org/10.1029/1998RG000054]

[41] Wohletz, K. **2000**.http://www.ees.lanl.gov/geodynamics/Wohletz/Krakatau_6th_Century.pdf

[42] Symonds, R.B. *Compositions, Origins, Emission Rates and Atmospheric Impacts of Volcanic Gases,* **1998**. http://pubs.usgs.gov/of/1998/of98-776/of98-776.pdf

[43] Sivertsen, B.J. *The Parting of the Sea: How Volcanoes, Earthquakes, and Plagues Shaped the Story of Exodus,* **2009**. http://press.princeton.edu/chapters/s8883.html
[http://dx.doi.org/10.1515/9781400829958]

[44] Mauch, C.; Pfister, C., Eds. *Natural Disasters, Cultural Responses, Case Studies Towards a Global Environmental History*; International Environmental History Publication of the German Historical Institute, Lexington Books, Rowman & Littlefield Publishers: Lanham, **2009**.

[45] Donald, H.J. *Forces of Nature and Cultural Responses*; Pfeifer, K.; Pfeifer, N., Eds.; Springer: Heidelberg, Berlin, **2013**, pp. 111-137.

[46] Behringer, W. *Tambora und das Jahr ohne Sommer: Wie ein Vulkan die Welt in die Krise stürzte,* C.H.Beck: München; , **2015**.
[http://dx.doi.org/10.17104/9783406676161]

[47] Büntgen, U.; Myglan, V.S.; Charpentier Ljungqvist, F. Cooling and societal change during the Late Antique Little Ice Age from 536 to around 660 AD. *Nat. Geosci.,* **2016**, *9*, 231-236.
[http://dx.doi.org/10.1038/ngeo2652]

[48] Bodenmann, T.; Brönnimann, S.; Hirsch Hadorn, G. Perceiving, explaining, and observing climatic changes: A historical case study of the "year without a summer" 1816. *Meteorol. Z.,* **2011**, *20*(6), 577-587.
[http://dx.doi.org/10.1127/0941-2948/2011/0288]

[49] Bryan, S.E.; Ernst, R.E. Revised definition of Large Igneous Provinces (LIPs). *Earth Sci. Rev.,* **2008**, *86*, 175-202.
[http://dx.doi.org/10.1016/j.earscirev.2007.08.008]

[50] Bryan, S.E.; Ukstins Peate, I.; Peate, D.W. The largest volcanic eruptions on Earth. *Earth Sci. Rev.,* **2010**, *102*(3-4), 207-229.
[http://dx.doi.org/10.1016/j.earscirev.2010.07.001]

[51] Erba, E.; Duncan, R.A.; Bottini, C. *The origin, evolution and environmental impact of oceanic Large Igneous Provinces*; Neal, C.R.; Sager, W.W.; Sano, T.; Erba, E., Eds.; The Geological Society of America 2015, Special Paper, **2015**, 511, pp. 271-303.

[52] Wignall, P.B. Large igneous provinces and mass extinctions. *Earth Sci. Rev.,* **2001**, *53*(1-2), 1-33.
[http://dx.doi.org/10.1016/S0012-8252(00)00037-4]

[53] Treude, T. *Anoxia - Evidence for Eucaryote Survival and Paleontological Strategies. Cellular Origin, Life in Extreme Habitats and Astrobiology*; Altenbach, A.V.; Bernhard, J.M.; Seckbach, J., Eds.; Springer: Heidelberg, Berlin, **2012**, Vol. 21, pp. 17-38.

[54] Hochuli, P.A.; Sanson-Barrera, A.; Schneebeli-Hermann, E.; Bucher, H. Severest crisis overlooked-Worst disruption of terrestrial environments postdates the Permian-Triassic mass extinction. *Sci. Rep.,* **2016**, *6*, 28372.
[http://dx.doi.org/10.1038/srep28372] [PMID: 27340926]

[55] Sobolev, S.V.; Sobolev, A.V.; Kuzmin, D.V.; Krivolutskaya, N.A.; Petrunin, A.G.; Arndt, N.T.; Radko, V.A.; Vasiliev, Y.R. Linking mantle plumes, large igneous provinces and environmental catastrophes. *Nature,* **2011**, *477*(7364), 312-316.
[http://dx.doi.org/10.1038/nature10385] [PMID: 21921914]

[56] Föllmi, K.B. Early Cretaceous life, climate and anoxia. *Cretac. Res.,* **2012**, *35*, 230-257.
[http://dx.doi.org/10.1016/j.cretres.2011.12.005]

[57] Burgess, S.D.; Bowring, S.; Shen, S-Z. High-precision timeline for Earth's most severe extinction. *Proc. Natl. Acad. Sci. USA,* **2014**, *111*(9), 3316-3321.

[http://dx.doi.org/10.1073/pnas.1317692111] [PMID: 24516148]

[58]   Sobolev, A.V.; Arndt, N.T.; Sobolev, S.V. *Origin and Environmental impact of Large Igneous Provinces, Research proposal*; European Research Council, **2012**.

[59]   Bond, D.P.; Wignall, P.B. *Volcanism, Impacts and Mass Extinctions: Causes and Effects*; Keller, G.; Kerr, A.C., Eds.; The Geological Society of America, Special Paper, **2014**, 505, pp. 29-55.

[60]   Kent, D.V.; Muttoni, G. Modulation of Late Cretaceous and Cenozoic climate by variable drawdown of atmospheric p CO_2 from weathering of basaltic provinces on continents drifting through the equatorial humid belt. *Clim. Past*, **2013**, *9*, 525-546.
[http://dx.doi.org/10.5194/cp-9-525-2013]

[61]   Dallanave, E.; Tauxe, L.; Muttoni, G. Silicate weathering machine at work: Rock magnetic data from the late Paleocene–early Eocene Cicogna section, Italy. *G3 (Bethesda)*, **2011**, *11*(7), Q07008.

[62]   Sheldon, N.D.; Tabor, N.J. New Frontiers in Paleopedology and Terrestrial Paleoclimatology SEPM Special Publication: Tulsa, Oklahoma, **2013**; 104, pp. 71-78.

[63]   Waldbauer, J.R.; Chamberlain, C.P. A history of atmospheric CO2 and its effects on plants, animals and ecosystems.*Ecological Studies*; Ehleringer, J.R.; Cerling, T.E.; Dearing, M.D., Eds.; Springer: New York, **2005**.

[64]   Gaillardet, J.; Dupré, B.; Louvat, P. Global silicate weathering and CO_2 consumption rates deduced from the chemistry of large rivers. *Chem. Geol.*, **1999**, *159*(1-4), 3-30.
[http://dx.doi.org/10.1016/S0009-2541(99)00031-5]

[65]   Hoffmann, P.F.; Schrag, D.P. The snowball Earth hypothesis: testing the limits *of global change. Terra Nova*, **2002**, *14*, 129-155.
[http://dx.doi.org/10.1046/j.1365-3121.2002.00408.x]

[66]   Haywood, A.M.; Valdes, P.J. Modelling Pliocene warmth: contribution of atmosphere, oceans and cryosphere. *Earth Planet. Sci. Lett.*, **2004**, *218*(3-4), 363-377.
[http://dx.doi.org/10.1016/S0012-821X(03)00685-X]

[67]   Fallert-Müller, A. *Lexikon der Biochemie*; Spektrum Akademischer Verlag, Elsevier GmbH: München, **1999/2000**.

[68]   Beraldi-Campesi, H. Early life on land and the first terrestrial ecosystems. *Ecol. Process.*, **2013**, *2*, 1.
[http://dx.doi.org/10.1186/2192-1709-2-1]

[69]   Lyons, T.W.; Reinhard, C.T.; Planavsky, N.J. The rise of oxygen in Earth's early ocean and atmosphere. *Nature*, **2014**, *506*(7488), 307-315.
[http://dx.doi.org/10.1038/nature13068] [PMID: 24553238]

[70]   Nutman, A.P.; Bennett, V.C.; Friend, C.R.; Van Kranendonk, M.J.; Chivas, A.R. Rapid emergence of life shown by discovery of 3,700-million-year-old microbial structures. *Nature*, **2016**, *537*(7621), 535-538.
[http://dx.doi.org/10.1038/nature19355] [PMID: 27580034]

[71]   Frimmel, H.E.; Hennigh, Q. First whiffs of atmospheric oxygen triggered onset of crustal gold cycle. *Miner. Depos.*, **2015**, *50*(1), 5-23.
[http://dx.doi.org/10.1007/s00126-014-0574-8]

[72]   Holland, H.D. The oxygenation of the atmosphere and oceans. *Philos. Trans. R. Soc. Lond. B Biol. Sci.*, **2006**, *361*(1470), 903-915.
[http://dx.doi.org/10.1098/rstb.2006.1838] [PMID: 16754606]

[73]   Canfield, D.E. *Oxygen - A Four Billion Year History*; University Press: Princeton, USA, **2014**.

[74]   Sundquist, E.T.; Visser, K. *Biogeochemistry: Treatise on Geochemistry*; Schlesinger, W.H., Ed.; Elsevier Ltd.: Amsterdam, **2005**, Vol. 8, pp. 425-472.

[75]   Fenchel, T. *The Origin and Early Evolution of Life*; Oxford University Press: New York, **2002**.

[76] Sperling, E.A.; Frieder, C.A.; Raman, A.V.; Girguis, P.R.; Levin, L.A.; Knoll, A.H. Oxygen, ecology, and the Cambrian radiation of animals. *Proc. Natl. Acad. Sci. USA,* **2013**, *110*(33), 13446-13451. [http://dx.doi.org/10.1073/pnas.1312778110] [PMID: 23898193]

[77] Beerling, D.J. *A history of atmospheric CO$_2$ and its effects on plants, animals and ecosystems*; Ehleringer, J.R.; Cerling, T.E.; Dearing, M.D. Ecological Studies, Springer: New York, **2005**, Vol. 177, pp. 114-132.

[78] Berner, R.A. *A history of atmospheric CO$_2$ and its effects on plants, animals and ecosystems*; Ehleringer, J.R.; Cerling, T.E.; Dearing, M.D. Ecological Studies, Springer: New York, **2005**, 177, pp. 1-7.

[79] Kemp, T.S. *The Origin and Evolution of Mammals*; Oxford University Press, **2005**.

[80] Janis, C.J. Tertiary Mammal Evolution in the Context of Changing Climates, Vegetation, and Tectonic Events. *Annu. Rev. Ecol. Syst.,* **1993**, *Vol. 24*, 467-500. [http://dx.doi.org/10.1146/annurev.es.24.110193.002343]

[81] DeBoer, P.L.; Smith, D.G. *Orbital forcing and cyclic sequences*; DeBoer, P.L.; Smith, D.G., Eds.; Blackwell: Spec. Publ. Int. Assoc. Sediment.: Oxford, UK, **1994**, 19, pp. 1-14.

[82] Charbonnier, G.; Duchamp-Alphonse, S.; Adatte, T. Eccentricity paced monsoon-like system along the northwestern Tethyan margin during the Valanginian (Early Cretaceous): New insights from detrital and nutrient fluxes into the Vocontian Basin (SE France). *Palaeogeogr. Palaeocl.,* **2016**, *443*, 145-155. [http://dx.doi.org/10.1016/j.palaeo.2015.11.027]

[83] Kozur, H. W.; Bachmann, G. H. Updated correlation of the Germanic Triassic with the Tethyan scale and assigned numeric ages. *Berichte Geol. B.-A.,* **2008**, *76*, 53-58.

[84] Wu, H.; Zhang, S.; Hinnov, L.A.; Jiang, G.; Feng, Q.; Li, H.; Yang, T. Time-calibrated Milankovitch cycles for the late Permian. *Nat. Commun.,* **2013**, *4*, 2452. [http://dx.doi.org/10.1038/ncomms3452] [PMID: 24030138]

[85] Zhang, S.; Wang, X.; Hammarlund, E.U.; Wang, H.; Costa, M.M.; Bjerrum, C.J.; Connelly, J.N.; Zhang, B.; Bian, L.; Canfield, D.E. Orbital forcing of climate 1.4 billion years ago. *Proc. Natl. Acad. Sci. USA,* **2015**, *112*(12), E1406-E1413. [PMID: 25775605]

[86] Ganopolski, A.; Winkelmann, R.; Schellnhuber, H.J. Critical insolation-CO$_2$ relation for diagnosing past and future glacial inception. *Nature,* **2016**, *529*(7585), 200-203. [http://dx.doi.org/10.1038/nature16494] [PMID: 26762457]

[87] Mayewski, P.A.; Rohling, E.E.; Stager, J.C. Holocene climate variability. *Quat. Res.,* **2004**, *62*, 243-255. [http://dx.doi.org/10.1016/j.yqres.2004.07.001]

[88] Stager, J.C.; Ryves, D.B.; Cumming, B.F. Solar variability and the levels of Lake Victoria, East Africa, during the last millenium. *J. Paleolimnol.,* **2005**, *33*(2), 243-251. [http://dx.doi.org/10.1007/s10933-004-4227-2]

[89] Beer, J.; Vonmoos, M.; Muscheler, R. *Solar variability and planetary climates, Space Climate Series of ISSI*; Calisesi, Y.; Bonnet, R.M.; Gray, L., Eds.; Springer: Heidelberg, Berlin, **2007**, 23, pp. 67-79.

[90] Ingram, W.J. *Solar variability and planetary climates, Space Climate Series of ISSI*; Calisesi, Y.; Bonnet, R.M.; Gray, L., Eds.; Springer: Heidelberg, Berlin, **2007**, 23, pp. 67-79.

[91] *Large Meteorite Impacts and Planetary Evolution V*; Osinski, G.R.; Kring, D.A., Eds. Geological Society of America Special Papers, **2015**, Vol. 518. [http://dx.doi.org/10.1130/2015.2518(v)]

[92] Keller, G.; Kerr, A.C., Eds. *Volcanism, Impacts, and Mass Extinctions: Causes and Effects*; Geological Society of America Special Papers, **2014**, Vol. 505.

[93] Nurmi, P.; Valtonen, M.J.; Zheng, J.Q. Periodic variation of Oort Cloud flux and cometary impacts on the Earth and Jupiter. *Mon. Not. R. Astron. Soc.,* **2001**, *327*(4), 1367-1376.
[http://dx.doi.org/10.1046/j.1365-8711.2001.04854.x]

[94] Randall, L.; Reece, M. Dark matter as a trigger for periodic comet impacts. *Phys. Rev. Lett.,* **2014**, *112*(16), 161301.
[http://dx.doi.org/10.1103/PhysRevLett.112.161301] [PMID: 24815633]

[95] Abbott, B.P.; Abbott, R.; Abbott, T.D.; Abernathy, M.R.; Acernese, F.; Ackley, K.; Adams, C.; Adams, T.; Addesso, P.; Adhikari, R.X.; Adya, V.B.; Affeldt, C.; Agathos, M.; Agatsuma, K.; Aggarwal, N.; Aguiar, O.D.; Aiello, L.; Ain, A.; Ajith, P.; Allen, B.; Allocca, A.; Altin, P.A.; Anderson, S.B.; Anderson, W.G.; Arai, K.; Arain, M.A.; Araya, M.C.; Arceneaux, C.C.; Areeda, J.S.; Arnaud, N.; Arun, K.G.; Ascenzi, S.; Ashton, G.; Ast, M.; Aston, S.M.; Astone, P.; Aufmuth, P.; Aulbert, C.; Babak, S.; Bacon, P.; Bader, M.K.; Baker, P.T.; Baldaccini, F.; Ballardin, G.; Ballmer, S.W.; Barayoga, J.C.; Barclay, S.E.; Barish, B.C.; Barker, D.; Barone, F.; Barr, B.; Barsotti, L.; Barsuglia, M.; Barta, D.; Bartlett, J.; Barton, M.A.; Bartos, I.; Bassiri, R.; Basti, A.; Batch, J.C.; Baune, C.; Bavigadda, V.; Bazzan, M.; Behnke, B.; Bejger, M.; Belczynski, C.; Bell, A.S.; Bell, C.J.; Berger, B.K.; Bergman, J.; Bergmann, G.; Berry, C.P.; Bersanetti, D.; Bertolini, A.; Betzwieser, J.; Bhagwat, S.; Bhandare, R.; Bilenko, I.A.; Billingsley, G.; Birch, J.; Birney, R.; Birnholtz, O.; Biscans, S.; Bisht, A.; Bitossi, M.; Biwer, C.; Bizouard, M.A.; Blackburn, J.K.; Blair, C.D.; Blair, D.G.; Blair, R.M.; Bloemen, S.; Bock, O.; Bodiya, T.P.; Boer, M.; Bogaert, G.; Bogan, C.; Bohe, A.; Bojtos, P.; Bond, C.; Bondu, F.; Bonnand, R.; Boom, B.A.; Bork, R.; Boschi, V.; Bose, S.; Bouffanais, Y.; Bozzi, A.; Bradaschia, C.; Brady, P.R.; Braginsky, V.B.; Branchesi, M.; Brau, J.E.; Briant, T.; Brillet, A.; Brinkmann, M.; Brisson, V.; Brockill, P.; Brooks, A.F.; Brown, D.A.; Brown, D.D.; Brown, N.M.; Buchanan, C.C.; Buikema, A.; Bulik, T.; Bulten, H.J.; Buonanno, A.; Buskulic, D.; Buy, C.; Byer, R.L.; Cabero, M.; Cadonati, L.; Cagnoli, G.; Cahillane, C.; Calderón Bustillo, J.; Callister, T.; Calloni, E.; Camp, J.B.; Cannon, K.C.; Cao, J.; Capano, C.D.; Capocasa, E.; Carbognani, F.; Caride, S.; Casanueva Diaz, J.; Casentini, C.; Caudill, S.; Cavaglià, M.; Cavalier, F.; Cavalieri, R.; Cella, G.; Cepeda, C.B.; Cerboni Baiardi, L.; Cerretani, G.; Cesarini, E.; Chakraborty, R.; Chalermsongsak, T.; Chamberlin, S.J.; Chan, M.; Chao, S.; Charlton, P.; Chassande-Mottin, E.; Chen, H.Y.; Chen, Y.; Cheng, C.; Chincarini, A.; Chiummo, A.; Cho, H.S.; Cho, M.; Chow, J.H.; Christensen, N.; Chu, Q.; Chua, S.; Chung, S.; Ciani, G.; Clara, F.; Clark, J.A.; Cleva, F.; Coccia, E.; Cohadon, P.F.; Colla, A.; Collette, C.G.; Cominsky, L.; Constancio, M., Jr; Conte, A.; Conti, L.; Cook, D.; Corbitt, T.R.; Cornish, N.; Corsi, A.; Cortese, S.; Costa, C.A.; Coughlin, M.W.; Coughlin, S.B.; Coulon, J.P.; Countryman, S.T.; Couvares, P.; Cowan, E.E.; Coward, D.M.; Cowart, M.J.; Coyne, D.C.; Coyne, R.; Craig, K.; Creighton, J.D.; Creighton, T.D.; Cripe, J.; Crowder, S.G.; Cruise, A.M.; Cumming, A.; Cunningham, L.; Cuoco, E.; Dal Canton, T.; Danilishin, S.L.; D'Antonio, S.; Danzmann, K.; Darman, N.S.; Da Silva Costa, C.F.; Dattilo, V.; Dave, I.; Daveloza, H.P.; Davier, M.; Davies, G.S.; Daw, E.J.; Day, R.; De, S.; DeBra, D.; Debreczeni, G.; Degallaix, J.; De Laurentis, M.; Deléglise, S.; Del Pozzo, W.; Denker, T.; Dent, T.; Dereli, H.; Dergachev, V.; DeRosa, R.T.; De Rosa, R.; DeSalvo, R.; Dhurandhar, S.; Díaz, M.C.; Di Fiore, L.; Di Giovanni, M.; Di Lieto, A.; Di Pace, S.; Di Palma, I.; Di Virgilio, A.; Dojcinoski, G.; Dolique, V.; Donovan, F.; Dooley, K.L.; Doravari, S.; Douglas, R.; Downes, T.P.; Drago, M.; Drever, R.W.; Driggers, J.C.; Du, Z.; Ducrot, M.; Dwyer, S.E.; Edo, T.B.; Edwards, M.C.; Effler, A.; Eggenstein, H.B.; Ehrens, P.; Eichholz, J.; Eikenberry, S.S.; Engels, W.; Essick, R.C.; Etzel, T.; Evans, M.; Evans, T.M.; Everett, R.; Factourovich, M.; Fafone, V.; Fair, H.; Fairhurst, S.; Fan, X.; Fang, Q.; Farinon, S.; Farr, B.; Farr, W.M.; Favata, M.; Fays, M.; Fehrmann, H.; Fejer, M.M.; Feldbaum, D.; Ferrante, I.; Ferreira, E.C.; Ferrini, F.; Fidecaro, F.; Finn, L.S.; Fiori, I.; Fiorucci, D.; Fisher, R.P.; Flaminio, R.; Fletcher, M.; Fong, H.; Fournier, J.D.; Franco, S.; Frasca, S.; Frasconi, F.; Frede, M.; Frei, Z.; Freise, A.; Frey, R.; Frey, V.; Fricke, T.T.; Fritschel, P.; Frolov, V.V.; Fulda, P.; Fyffe, M.; Gabbard, H.A.; Gair, J.R.; Gammaitoni, L.; Gaonkar, S.G.; Garufi, F.; Gatto, A.; Gaur, G.; Gehrels, N.; Gemme, G.; Gendre, B.; Genin, E.; Gennai, A.; George, J.; Gergely, L.; Germain, V.; Ghosh, A.; Ghosh, A.; Ghosh, S.; Giaime, J.A.; Giardina, K.D.; Giazotto, A.; Gill, K.; Glaefke, A.; Gleason, J.R.; Goetz, E.; Goetz, R.; Gondan, L.; González, G.; Gonzalez Castro, J.M.; Gopakumar, A.; Gordon, N.A.; Gorodetsky, M.L.; Gossan, S.E.; Gosselin, M.; Gouaty, R.; Graef, C.; Graff, P.B.; Granata, M.; Grant, A.; Gras, S.; Gray, C.; Greco, G.; Green, A.C.; Greenhalgh, R.J.; Groot, P.; Grote, H.; Grunewald, S.; Guidi, G.M.; Guo, X.; Gupta, A.; Gupta, M.K.; Gushwa, K.E.;

Gustafson, E.K.; Gustafson, R.; Hacker, J.J.; Hall, B.R.; Hall, E.D.; Hammond, G.; Haney, M.; Hanke, M.M.; Hanks, J.; Hanna, C.; Hannam, M.D.; Hanson, J.; Hardwick, T.; Harms, J.; Harry, G.M.; Harry, I.W.; Hart, M.J.; Hartman, M.T.; Haster, C.J.; Haughian, K.; Healy, J.; Heefner, J.; Heidmann, A.; Heintze, M.C.; Heinzel, G.; Heitmann, H.; Hello, P.; Hemming, G.; Hendry, M.; Heng, I.S.; Hennig, J.; Heptonstall, A.W.; Heurs, M.; Hild, S.; Hoak, D.; Hodge, K.A.; Hofman, D.; Hollitt, S.E.; Holt, K.; Holz, D.E.; Hopkins, P.; Hosken, D.J.; Hough, J.; Houston, E.A.; Howell, E.J.; Hu, Y.M.; Huang, S.; Huerta, E.A.; Huet, D.; Hughey, B.; Husa, S.; Huttner, S.H.; Huynh-Dinh, T.; Idrisy, A.; Indik, N.; Ingram, D.R.; Inta, R.; Isa, H.N.; Isac, J.M.; Isi, M.; Islas, G.; Isogai, T.; Iyer, B.R.; Izumi, K.; Jacobson, M.B.; Jacqmin, T.; Jang, H.; Jani, K.; Jaranowski, P.; Jawahar, S.; Jiménez-Forteza, F.; Johnson, W.W.; Johnson-McDaniel, N.K.; Jones, D.I.; Jones, R.; Jonker, R.J.; Ju, L.; Haris, K.; Kalaghatgi, C.V.; Kalogera, V.; Kandhasamy, S.; Kang, G.; Kanner, J.B.; Karki, S.; Kasprzack, M.; Katsavounidis, E.; Katzman, W.; Kaufer, S.; Kaur, T.; Kawabe, K.; Kawazoe, F.; Kéfélian, F.; Kehl, M.S.; Keitel, D.; Kelley, D.B.; Kells, W.; Kennedy, R.; Keppel, D.G.; Key, J.S.; Khalaidovski, A.; Khalili, F.Y.; Khan, I.; Khan, S.; Khan, Z.; Khazanov, E.A.; Kijbunchoo, N.; Kim, C.; Kim, J.; Kim, K.; Kim, N.G.; Kim, N.; Kim, Y.M.; King, E.J.; King, P.J.; Kinzel, D.L.; Kissel, J.S.; Kleybolte, L.; Klimenko, S.; Koehlenbeck, S.M.; Kokeyama, K.; Koley, S.; Kondrashov, V.; Kontos, A.; Koranda, S.; Korobko, M.; Korth, W.Z.; Kowalska, I.; Kozak, D.B.; Kringel, V.; Krishnan, B.; Królak, A.; Krueger, C.; Kuehn, G.; Kumar, P.; Kumar, R.; Kuo, L.; Kutynia, A.; Kwee, P.; Lackey, B.D.; Landry, M.; Lange, J.; Lantz, B.; Lasky, P.D.; Lazzarini, A.; Lazzaro, C.; Leaci, P.; Leavey, S.; Lebigot, E.O.; Lee, C.H.; Lee, H.K.; Lee, H.M.; Lee, K.; Lenon, A.; Leonardi, M.; Leong, J.R.; Leroy, N.; Letendre, N.; Levin, Y.; Levine, B.M.; Li, T.G.; Libson, A.; Littenberg, T.B.; Lockerbie, N.A.; Logue, J.; Lombardi, A.L.; London, L.T.; Lord, J.E.; Lorenzini, M.; Loriette, V.; Lormand, M.; Losurdo, G.; Lough, J.D.; Lousto, C.O.; Lovelace, G.; Lück, H.; Lundgren, A.P.; Luo, J.; Lynch, R.; Ma, Y.; MacDonald, T.; Machenschalk, B.; MacInnis, M.; Macleod, D.M.; Magaña-Sandoval, F.; Magee, R.M.; Mageswaran, M.; Majorana, E.; Maksimovic, I.; Malvezzi, V.; Man, N.; Mandel, I.; Mandic, V.; Mangano, V.; Mansell, G.L.; Manske, M.; Mantovani, M.; Marchesoni, F.; Marion, F.; Márka, S.; Márka, Z.; Markosyan, A.S.; Maros, E.; Martelli, F.; Martellini, L.; Martin, I.W.; Martin, R.M.; Martynov, D.V.; Marx, J.N.; Mason, K.; Masserot, A.; Massinger, T.J.; Masso-Reid, M.; Matichard, F.; Matone, L.; Mavalvala, N.; Mazumder, N.; Mazzolo, G.; McCarthy, R.; McClelland, D.E.; McCormick, S.; McGuire, S.C.; McIntyre, G.; McIver, J.; McManus, D.J.; McWilliams, S.T.; Meacher, D.; Meadors, G.D.; Meidam, J.; Melatos, A.; Mendell, G.; Mendoza-Gandara, D.; Mercer, R.A.; Merilh, E.; Merzougui, M.; Meshkov, S.; Messenger, C.; Messick, C.; Meyers, P.M.; Mezzani, F.; Miao, H.; Michel, C.; Middleton, H.; Mikhailov, E.E.; Milano, L.; Miller, J.; Millhouse, M.; Minenkov, Y.; Ming, J.; Mirshekari, S.; Mishra, C.; Mitra, S.; Mitrofanov, V.P.; Mitselmakher, G.; Mittleman, R.; Moggi, A.; Mohan, M.; Mohapatra, S.R.; Montani, M.; Moore, B.C.; Moore, C.J.; Moraru, D.; Moreno, G.; Morriss, S.R.; Mossavi, K.; Mours, B.; Mow-Lowry, C.M.; Mueller, C.L.; Mueller, G.; Muir, A.W.; Mukherjee, A.; Mukherjee, D.; Mukherjee, S.; Mukund, N.; Mullavey, A.; Munch, J.; Murphy, D.J.; Murray, P.G.; Mytidis, A.; Nardecchia, I.; Naticchioni, L.; Nayak, R.K.; Necula, V.; Nedkova, K.; Nelemans, G.; Neri, M.; Neunzert, A.; Newton, G.; Nguyen, T.T.; Nielsen, A.B.; Nissanke, S.; Nitz, A.; Nocera, F.; Nolting, D.; Normandin, M.E.; Nuttall, L.K.; Oberling, J.; Ochsner, E.; O'Dell, J.; Oelker, E.; Ogin, G.H.; Oh, J.J.; Oh, S.H.; Ohme, F.; Oliver, M.; Oppermann, P.; Oram, R.J.; O'Reilly, B.; O'Shaughnessy, R.; Ott, C.D.; Ottaway, D.J.; Ottens, R.S.; Overmier, H.; Owen, B.J.; Pai, A.; Pai, S.A.; Palamos, J.R.; Palashov, O.; Palomba, C.; Pal-Singh, A.; Pan, H.; Pan, Y.; Pankow, C.; Pannarale, F.; Pant, B.C.; Paoletti, F.; Paoli, A.; Papa, M.A.; Paris, H.R.; Parker, W.; Pascucci, D.; Pasqualetti, A.; Passaquieti, R.; Passuello, D.; Patricelli, B.; Patrick, Z.; Pearlstone, B.L.; Pedraza, M.; Pedurand, R.; Pekowsky, L.; Pele, A.; Penn, S.; Perreca, A.; Pfeiffer, H.P.; Phelps, M.; Piccinni, O.; Pichot, M.; Pickenpack, M.; Piergiovanni, F.; Pierro, V.; Pillant, G.; Pinard, L.; Pinto, I.M.; Pitkin, M.; Poeld, J.H.; Poggiani, R.; Popolizio, P.; Post, A.; Powell, J.; Prasad, J.; Predoi, V.; Premachandra, S.S.; Prestegard, T.; Price, L.R.; Prijatelj, M.; Principe, M.; Privitera, S.; Prix, R.; Prodi, G.A.; Prokhorov, L.; Puncken, O.; Punturo, M.; Puppo, P.; Pürrer, M.; Qi, H.; Qin, J.; Quetschke, V.; Quintero, E.A.; Quitzow-James, R.; Raab, F.J.; Rabeling, D.S.; Radkins, H.; Raffai, P.; Raja, S.; Rakhmanov, M.; Ramet, C.R.; Rapagnani, P.; Raymond, V.; Razzano, M.; Re, V.; Read, J.; Reed, C.M.; Regimbau, T.; Rei, L.; Reid, S.; Reitze, D.H.; Rew, H.; Reyes, S.D.; Ricci, F.; Riles, K.; Robertson, N.A.; Robie, R.; Robinet, F.; Rocchi, A.; Rolland, L.; Rollins, J.G.; Roma, V.J.; Romano, J.D.; Romano, R.; Romanov, G.; Romie, J.H.; Rosińska, D.; Rowan, S.; Rüdiger, A.; Ruggi, P.; Ryan,

K.; Sachdev, S.; Sadecki, T.; Sadeghian, L.; Salconi, L.; Saleem, M.; Salemi, F.; Samajdar, A.; Sammut, L.; Sampson, L.M.; Sanchez, E.J.; Sandberg, V.; Sandeen, B.; Sanders, G.H.; Sanders, J.R.; Sassolas, B.; Sathyaprakash, B.S.; Saulson, P.R.; Sauter, O.; Savage, R.L.; Sawadsky, A.; Schale, P.; Schilling, R.; Schmidt, J.; Schmidt, P.; Schnabel, R.; Schofield, R.M.; Schönbeck, A.; Schreiber, E.; Schuette, D.; Schutz, B.F.; Scott, J.; Scott, S.M.; Sellers, D.; Sengupta, A.S.; Sentenac, D.; Sequino, V.; Sergeev, A.; Serna, G.; Setyawati, Y.; Sevigny, A.; Shaddock, D.A.; Shaffer, T.; Shah, S.; Shahriar, M.S.; Shaltev, M.; Shao, Z.; Shapiro, B.; Shawhan, P.; Sheperd, A.; Shoemaker, D.H.; Shoemaker, D.M.; Siellez, K.; Siemens, X.; Sigg, D.; Silva, A.D.; Simakov, D.; Singer, A.; Singer, L.P.; Singh, A.; Singh, R.; Singhal, A.; Sintes, A.M.; Slagmolen, B.J.; Smith, J.R.; Smith, M.R.; Smith, N.D.; Smith, R.J.; Son, E.J.; Sorazu, B.; Sorrentino, F.; Souradeep, T.; Srivastava, A.K.; Staley, A.; Steinke, M.; Steinlechner, J.; Steinlechner, S.; Steinmeyer, D.; Stephens, B.C.; Stevenson, S.P.; Stone, R.; Strain, K.A.; Straniero, N.; Stratta, G.; Strauss, N.A.; Strigin, S.; Sturani, R.; Stuver, A.L.; Summerscales, T.Z.; Sun, L.; Sutton, P.J.; Swinkels, B.L.; Szczepańczyk, M.J.; Tacca, M.; Talukder, D.; Tanner, D.B.; Tápai, M.; Tarabrin, S.P.; Taracchini, A.; Taylor, R.; Theeg, T.; Thirugnanasambandam, M.P.; Thomas, E.G.; Thomas, M.; Thomas, P.; Thorne, K.A.; Thorne, K.S.; Thrane, E.; Tiwari, S.; Tiwari, V.; Tokmakov, K.V.; Tomlinson, C.; Tonelli, M.; Torres, C.V.; Torrie, C.I.; Töyrä, D.; Travasso, F.; Traylor, G.; Trifirò, D.; Tringali, M.C.; Trozzo, L.; Tse, M.; Turconi, M.; Tuyenbayev, D.; Ugolini, D.; Unnikrishnan, C.S.; Urban, A.L.; Usman, S.A.; Vahlbruch, H.; Vajente, G.; Valdes, G.; Vallisneri, M.; van Bakel, N.; van Beuzekom, M.; van den Brand, J.F.; Van Den Broeck, C.; Vander-Hyde, D.C.; van der Schaaf, L.; van Heijningen, J.V.; van Veggel, A.A.; Vardaro, M.; Vass, S.; Vasúth, M.; Vaulin, R.; Vecchio, A.; Vedovato, G.; Veitch, J.; Veitch, P.J.; Venkateswara, K.; Verkindt, D.; Vetrano, F.; Viceré, A.; Vinciguerra, S.; Vine, D.J.; Vinet, J.Y.; Vitale, S.; Vo, T.; Vocca, H.; Vorvick, C.; Voss, D.; Vousden, W.D.; Vyatchanin, S.P.; Wade, A.R.; Wade, L.E.; Wade, M.; Waldman, S.J.; Walker, M.; Wallace, L.; Walsh, S.; Wang, G.; Wang, H.; Wang, M.; Wang, X.; Wang, Y.; Ward, H.; Ward, R.L.; Warner, J.; Was, M.; Weaver, B.; Wei, L.W.; Weinert, M.; Weinstein, A.J.; Weiss, R.; Welborn, T.; Wen, L.; Weßels, P.; Westphal, T.; Wette, K.; Whelan, J.T.; Whitcomb, S.E.; White, D.J.; Whiting, B.F.; Wiesner, K.; Wilkinson, C.; Willems, P.A.; Williams, L.; Williams, R.D.; Williamson, A.R.; Willis, J.L.; Willke, B.; Wimmer, M.H.; Winkelmann, L.; Winkler, W.; Wipf, C.C.; Wiseman, A.G.; Wittel, H.; Woan, G.; Worden, J.; Wright, J.L.; Wu, G.; Yablon, J.; Yakushin, I.; Yam, W.; Yamamoto, H.; Yancey, C.C.; Yap, M.J.; Yu, H.; Yvert, M.; Zadrożny, A.; Zangrando, L.; Zanolin, M.; Zendri, J.P.; Zevin, M.; Zhang, F.; Zhang, L.; Zhang, M.; Zhang, Y.; Zhao, C.; Zhou, M.; Zhou, Z.; Zhu, X.J.; Zucker, M.E.; Zuraw, S.E.; Zweizig, J. LIGO Scientific Collaboration and Virgo Collaboration. *Abbott, R., Abbott, T. D.* (LIGO Scientific Collaboration and Virgo Collaboration). Observation of gravitational waves from a binary black hole merger. *Phys. Rev. Lett.,* **2016,** *116*(6), 061102.
[http://dx.doi.org/10.1103/PhysRevLett.116.061102] [PMID: 26918975]

[96]   Budd, G.E. The cambrian fossil record and the origin of the phyla. *Integr. Comp. Biol.,* **2003,** *43*(1), 157-165.
[http://dx.doi.org/10.1093/icb/43.1.157] [PMID: 21680420]

[97]   Gereke, M.; Luppold, F.W.; Piecha, M. The type locality of the Kellwasser-Horizons in the Upper Harz Mountains, Germany. *Z. Dtsch. Ges. Geowiss.,* **2014,** *165*(2), 145-162.
[http://dx.doi.org/10.1127/1860-1804/2014/0066]

[98]   Sallan, L.; Galimberti, A.K. Body-size reduction in vertebrates following the end-Devonian mass extinction. *Science,* **2015,** *350*(6262), 812-815.
[http://dx.doi.org/10.1126/science.aac7373] [PMID: 26564854]

[99]   Piran, T.; Jimenez, R. Possible role of gamma ray bursts on life extinction in the universe. *Phys. Rev. Lett.,* **2014,** *113*(23), 231102.
[http://dx.doi.org/10.1103/PhysRevLett.113.231102] [PMID: 25526110]

[100]   Mángano, M.G.; Buatois, L.A. Decoupling of body-plan diversification and ecological structuring during the Ediacaran-Cambrian transition: evolutionary and geobiological feedbacks. *P. Roy. Soc. Lond. B Biol.,* **2014,** *281*, 1780.

[101]   Shakun, J.D.; Clark, P.U.; He, F.; Marcott, S.A.; Mix, A.C.; Liu, Z.; Otto-Bliesner, B.; Schmittner, A.;

Bard, E. Global warming preceded by increasing carbon dioxide concentrations during the last deglaciation. *Nature,* **2012**, *484*(7392), 49-54.
[http://dx.doi.org/10.1038/nature10915] [PMID: 22481357]

[102] Martin, P.; Archer, D.; Lea, D.W. Role of deep sea temperature in the carbon cycle during the last glacial. *Paleoceanography,* **2005**, *20*, PA2015.
[http://dx.doi.org/10.1029/2003PA000914]

[103] Schmittner, A.; Galbraith, E.D. Glacial greenhouse-gas fluctuations controlled by ocean circulation changes. *Nature,* **2008**, *456*(7220), 373-376.
[http://dx.doi.org/10.1038/nature07531] [PMID: 19020618]

[104] Andreassen, K.; Hubbard, A.; Winsborrow, M.; Patton, H.; Vadakkepuliyambatta, S.; Plaza-Faverola, A.; Gudlaugsson, E.; Serov, P.; Deryabin, A.; Mattingsdal, R.; Mienert, J.; Bünz, S. Massive blow-out craters formed by hydrate-controlled methane expulsion from the Arctic seafloor. *Science,* **2017**, *356*(6341), 948-953.
[http://dx.doi.org/10.1126/science.aal4500] [PMID: 28572390]

[105] Haug, G.H.; Günther, D.; Peterson, L.C.; Sigman, D.M.; Hughen, K.A.; Aeschlimann, B. Climate and the collapse of Maya civilization. *Science,* **2003**, *299*(5613), 1731-1735.
[http://dx.doi.org/10.1126/science.1080444] [PMID: 12637744]

[106] Yancheva, G.; Nowaczyk, N.R.; Mingram, J.; Dulski, P.; Schettler, G.; Negendank, J.F.; Liu, J.; Sigman, D.M.; Peterson, L.C.; Haug, G.H. Influence of the intertropical convergence zone on the East Asian monsoon. *Nature,* **2007**, *445*(7123), 74-77.
[http://dx.doi.org/10.1038/nature05431] [PMID: 17203059]

[107] Becker, M.; Meyssignac, B.; Xavier, L. Past terrestrial water storage (1980–2008) in the Amazon Basin reconstructed from GRACE and *in situ* river gauging data. *Hydrol. Earth Syst. Sci.,* **2011**, *15*, 533-546.
[http://dx.doi.org/10.5194/hess-15-533-2011]

[108] Nie, N.; Zhang, W.; Zhang, Z. Reconstructed terrestrial water storage change (ΔTWS) from 1948 to 2012 over the Amazon Basin with the latest GRACE and GLDAS products. *Water Resour. Manage.,* **2016**, *30*(1), 279-294.
[http://dx.doi.org/10.1007/s11269-015-1161-1]

[109] Wright, J.D.; Schaller, M.F. Evidence for a rapid release of carbon at the Paleocene-Eocene thermal maximum. *Proc. Natl. Acad. Sci. USA,* **2013**, *110*(40), 15908-15913.
[http://dx.doi.org/10.1073/pnas.1309188110] [PMID: 24043840]

[110] Matter, J.M.; Stute, M.; Snæbjörnsdottir, S.Ó.; Oelkers, E.H.; Gislason, S.R.; Aradottir, E.S.; Sigfusson, B.; Gunnarsson, I.; Sigurdardottir, H.; Gunnlaugsson, E.; Axelsson, G.; Alfredsson, H.A.; Wolff-Boenisch, D.; Mesfin, K.; Fernandez de la Reguera Taya, D.; Hall, J.; Dideriksen, K.; Broecker, W.S. Rapid carbon mineralization for permanent disposal of anthropogenic carbon dioxide emissions. *Science,* **2016**, *352*(6291), 1312-1314.
[http://dx.doi.org/10.1126/science.aad8132] [PMID: 27284192]

[111] Bertau, M. *Energie und Rohstoffe - Gestaltung unserer nachhaltigen Zukunft*; Kausch, P.; Bertau, M.; Gutzmer, J., Eds.; Springer Spektrum Akademischer Verlag: Heidelberg, **2011**, pp. 135-149.

[112] Zhang, S.; Wang, X.; Wang, H.; Bjerrum, C.J.; Hammarlund, E.U.; Costa, M.M.; Connelly, J.N.; Zhang, B.; Su, J.; Canfield, D.E. Sufficient oxygen for animal respiration 1,400 million years ago. *Proc. Natl. Acad. Sci. USA,* **2016**, *113*(7), 1731-1736.
[http://dx.doi.org/10.1073/pnas.1523449113] [PMID: 26729865]

[113] Government, A.U. http://www.bom.gov.au/watl/about-weather-and-climate/australian-climate-influences.shtml? bookmark=enso **2008**.

CHAPTER 2

The Anthropocene

Abstract: The concept of the Anthropocene and differing views about its onset and discrimination from the Holocene are given. Its principal difference from strictly deterministic geological processes is explained. It is proposed here to subdivide it into: Earliest and Early Anthropocene: from 3.3 Ma - intentional creation of the first functional stone artefact - to ca. 1000 AD, when the erosion rate caused by anthropogenic land use began to exceed the natural one. Middle Anthropocene: from ca. 1855 AD (anthropogenic CO_2 emissions equal volcanogenic CO_2 emissions; human caused extinctions equal natural background extinction) to 1918 AD (invention of synthetic ammonia). Late Anthropocene (Great Acceleration): Since 1945: First above ground test of a nuclear fission bomb; global synchronous deposition of carbonaceous fly ash spherules in lacustrine deposits and creation of temporary niches in the lowermost exosphere.

Keywords: Anthropogenic transformations, Artificial niches, Atmospheric warming, Boundary layers, Emissions, Extinctions, Fossil fuels, Greenhouse gases, Holocene, Homogenisation, Human development, Hybridisation, Ocean dynamics, Stratigraphic subdivision, Technosphere, Tools, World population.

Not later than the onset of industrialisation, anthropogenic transformations of ecosystems took on speeds, qualities, and quantities that clearly differed from and exceeded the natural ones [1 - 4]. They also inverted the natural global oceanic cooling trend of preindustrial time that was caused by orbital and volcanic forcings [5]. According to prognoses based on intermediate-complexity modelcalculations, current CO_2-emissions from the combustion of fossil carbon will postpone the onset of the next glaciation for at least 200,000 years [6]. According to the evaluation of an absolutely dated and annually resolved marine delta^{18}O archive (953-2000 AD), derived from 1492 samples of the bivalve species *Arctica islandica*[1] that were recovered from the north-Icelandic shelf in 80 m water depth, reversal of the timing of ocean-atmosphere coupling has occurred since the onset of industrialisation: between 953-1800 AD, near-surface atmospheric temperature variability lagged changes in delta^{18}O shell sea water density anomalies by 40 ± 30 a, indicating a lead role for ocean variability in modulation of major climate variability, driven by solar and volcanic forcings.

Since 1800 AD, ocean dynamics have lagged increases in atmospheric surface air temperatures, caused by rising amounts of anthropogenic greenhouse gas emissions [7].

The Anthropocene is functionally, stratigraphically and mineralogically discernible from the Holocene, because its

- Deposits contain new materials like Silicon, Aluminium, minerals, mineral-like compounds, concrete, plastics, *etc.*;
- Naturally occurring minerals and gemstones were redistributed on large scale by mining and processing activities;
- Direct and indirect matter flux exceeds the natural one by double;
- Geochemical signatures - *e.g.* polychlorinated biphenyls, biocides, lead-tetraethylene, nitrates, ^{3}H-spike, heavy metal radionuclides, *etc.* - are specific and unique;
- Physical and chemical atmospheric developments are distinct from Quaternary patterns and are characterised by a significant rise in concentrations of trace gases like CO_2 and CH_4;
- Biotic development is characterised by the extinction rates and alterations far above background, due to habitat restrictions and the fatal Columbian Exchange [8, 9].

The biosphere developing in the Anthropocene is distinguished from the preceding one by humans, which homogenise biota; dominate marine and terrestrial primary production; have escaped photosynthetic energy constraints by utilising fossil fuels; control bioevolution *via* breeding, cultivation, and bioengineering; and by creating a hybrid bio-technosphere [10].

Proposals were made to define the time-boundary separating the Holocene from the Anthropocene[2], as well as to characterise the diachronous surface separating both kinds of deposits [11]. This boundary could be established at several points such as the year 1610, the minimum of post-glacial atmospheric CO_2, or the 1964 spike of maximum global artificial radiocarbon (^{14}C) and fallout of radioactive particulate matter from above ground nuclear bomb tests [12].

According to the Quaternary Subcommission of the International Commission of Stratigraphy, the Anthropocene should be formalised with an onset in the mid-20th century [13].

That boundary could be justifiably established at a few points elsewhere, whereby events are considered, which find their expressions in the stratigraphic record and also events not manifested in it:

Earliest and early Anthropocene:

- 3.3 Ma ago: The first functional stone artefacts: Hominins intentionally created sharp edged stone wedges [14];
- 12,000-10,000 a ago: Neolithic Revolution: Hunters and gatherers became farmers and breeders; start of anthropogenic transformation of ecosystems by creating agrarian land and domestication of animals; onset of interference in the carbon-, nitrogen-, and water-cycles; forest clearance since ca. 8000 a ago; rice irrigation since ca. 5000 a ago [15]; beginning of alteration of atmospheric CO_2, N_2O, and CH_4 concentrations;
- ca. 1000 a ago: average anthropogenic erosional rate overtook the average geological erosional rate of ca. 16 m/Ma [16, 17].

Middle Anthropocene:

- 1855: Annual anthropogenic CO_2 emissions, resulting from a world population totally ca. 1.3 billion, began to exceed the natural, volcanogenic annual amount of 260 MT (see chapter atmosphere).
- The time, when the intensity of Anthropocene Defaunation [18] began to exceed that of natural background extinction.
- 1860: Invention of artificial fertilisers: Onset of alteration of the nitrogen, phosphorus and potassium cycles; acceleration of world population growth; and on-average improvement of nutrition due to increase in crop yield.
- 1886[3]: Invention of the gas firing motor: First use of fossil hydrocarbons to facilitate transportation and to power machines of protoindustrial production plants; onset of accumulation of anthropogenic CO_2 in atmo- and hydrosphere due to combustion of fossil hydrocarbons in machines;
- 1918: Invention of synthetic ammonia: Industrial production of increasing amounts of artificial fertilisers, blasting agents, sulfonamids, and nitric acid began to modify the natural nitrogen cycle.

Late Anthropocene (Great Acceleration [19]):

- 07/16/1945, 11.45 GMT: First above ground test of a 21 kT nuclear fission bomb (^{235}U); beginning of global dispersion and deposition of artificial radionuclide fallout, defining an unambiguous marker [20].
- 1946: Annual anthropogenic CH_4 emissions, resulting from a world population totalling ca. 2.5 billion, began to exceed the natural CH_4-emission of ca. 170 MT/a [21].
- 1950: Global synchronous deposition of spheroidal carbonaceous fly ash particles (size: up to a few tens of μm) - recorded from 75 different lake deposits spread over the globe -, which originated from industrial fossil fuel combustion and which represent another anthropogenic marker [22].

- 1961: First human being orbiting in exosphere of earth; first creation of a niche in vacuum/outer space;
- 1969: First missile leaving earth's gravitational field and mankind's first step on the moon.

NOTES

[1] The continuous record of oxygen isotope signatures was determined by analysing the annually-formed growth increments in the aragonite shells of that bivalve.

[2] This book argues that it is illogical to attribute the Anthropocene to a geological time scale, because depositional modes and stratigraphic boundaries in the geological record originate due to deterministic natural forces, processes, and events, whereas anthropogenic impacts often occur indeterministically. Humans are able to create artefacts, but are unable to create geological products like a mountain chain or an ocean. On the other side, geological processes never resulted in artefacts like open cast pits, cities, or waste dumps. So it would be better to assign the Anthropocene to an archaeological time scale. Notabene: it is obvious that artefacts and geological products have begun to intermix by forming amalgamated, hybrid deposits.

[3] This datum has been also defined as the beginning of the Hyper-Anthropocene [23].

CONFLICT OF INTEREST

The author (editor) declares no conflict of interest, financial or otherwise.

ACKNOWLEDGEMENTS

Declare none.

REFERENCES

[1]   Hooke, R. LeB. On the history of humans as geomorphic agents. *Geology,* **2000**, *28*(9), 843-846.
       [http://dx.doi.org/10.1130/0091-7613(2000)28<843:OTHOHA>2.0.CO;2]

[2]   Crutzen, P.J. Geology of mankind. *Nature,* **2002**, *415*(6867), 23.
       [http://dx.doi.org/10.1038/415023a] [PMID: 11780095]

[3]   Steffen, W.; Crutzen, J.; McNeill, J.R. The Anthropocene: are humans now overwhelming the great forces of Nature? *Ambio,* **2007**, *36*(8), 614-621.
       [http://dx.doi.org/10.1579/0044-7447(2007)36[614:TAAHNO]2.0.CO;2] [PMID: 18240674]

[4]   Mann, M.E. *The Hockey Stick and the Climate Wars: Dispatches from the Front Lines*; Columbia University Press: New York, Chichester, **2013**.

[http://dx.doi.org/10.7312/columbia/9780231152556.001.0001]

[5]   McGregor, H.V.; Evans, M.N.; Goosse, H. Robust global ocean cooling trend for the pre-industrial Common Era. *Nat. Geosci.,* **2015**, *8*, 671-677.
[http://dx.doi.org/10.1038/ngeo2510]

[6]   Archer, D.; Ganopolski, A. A movable trigger: Fossil fuel CO_2 and the onset of the next glaciation. *G3 (Bethesda),* **2005**, *6*(5), Q05003.

[7]   Reynolds, D.J.; Scourse, J.D.; Halloran, P.R.; Nederbragt, A.J.; Wanamaker, A.D.; Butler, P.G.; Richardson, C.A.; Heinemeier, J.; Eiríksson, J.; Knudsen, K.L.; Hall, I.R. Annually resolved North Atlantic marine climate over the last millennium. *Nat. Commun.,* **2016**, *7*, 13502.
[http://dx.doi.org/10.1038/ncomms13502] [PMID: 27922004]

[8]   Waters, C.N.; Zalasiewicz, J.; Summerhayes, C.; Barnosky, A.D.; Poirier, C.; Gałuszka, A.; Cearreta, A.; Edgeworth, M.; Ellis, E.C.; Ellis, M.; Jeandel, C.; Leinfelder, R.; McNeill, J.R.; Richter, Dd.; Steffen, W.; Syvitski, J.; Vidas, D.; Wagreich, M.; Williams, M.; Zhisheng, A.; Grinevald, J.; Odada, E.; Oreskes, N.; Wolfe, A.P. The Anthropocene is functionally and stratigraphically distinct from the Holocene. *Science,* **2016**, *351*(6269), aad2622.
[http://dx.doi.org/10.1126/science.aad2622] [PMID: 26744408]

[9]   Hazen, R.M.; Grew, E.S.; Origlieri, M.J. On the mineralogy of the "Anthropocene Epoch". *Am. Mineral.,* in press

[10]   Williams, M.; Zalasiewicz, J.; Haff, P.K. The Anthropocene biosphere. *Anthropocene Review,* **2015**, *2*(3), 196-219.
[http://dx.doi.org/10.1177/2053019615591020]

[11]   Edgeworth, M. deB Richter, D.; Waters, C. Diachronous beginnings of the *Anthropocene: The lower bounding surface of anthropogenic deposits. Anthropocene Rev.,* **2015**, *2*(1), 33-58.
[http://dx.doi.org/10.1177/2053019614565394]

[12]   Lewis, S.L.; Maslin, M.A. Defining the anthropocene. *Nature,* **2015**, *519*(7542), 171-180.
[http://dx.doi.org/10.1038/nature14258] [PMID: 25762280]

[13]   ICS International Union of Geological Sciences, International Commission on Stratigraphy. *Annual Report,* **2014**.http://iugs.org/uploads/ICS%202014.pdf

[14]   Hovers, E. Archaeology: Tools go back in time. *Nature,* **2015**, *521*(7552), 294-295.
[http://dx.doi.org/10.1038/521294a] [PMID: 25993954]

[15]   Ruddiman, W.F. The anthropogenic Greenhouse Era began thousands of years ago. *Clim. Change,* **2003**, *61*(3), 261-293.
[http://dx.doi.org/10.1023/B:CLIM.0000004577.17928.fa]

[16]   Wilkinson, B.H. Humans as geologic agents: a deep-time perspective. *Geology,* **2005**, *33*(3), 161-164.
[http://dx.doi.org/10.1130/G21108.1]

[17]   Wilkinson, B.H.; McElroy, B.J. The impact of humans on continental erosion and sedimentation. *Geol. Soc. Am. Bull.,* **2006**, *119*(1-2), 140-156.
[http://dx.doi.org/10.1130/B25899.1]

[18]   Dirzo, R.; Young, H.S.; Galetti, M.; Ceballos, G.; Isaac, N.J.; Collen, B. Defaunation in the Anthropocene. *Science,* **2014**, *345*(6195), 401-406.
[http://dx.doi.org/10.1126/science.1251817] [PMID: 25061202]

[19]   McNeill, J.R.; Engelke, P. *The Great Acceleration: An Environmental History of the Anthropocene since 1945*; Belknap Press: Cambridge, MA, **2014**.

[20]   Zalasiewicz, J.; Williams, M.; Fortey, R.; Smith, A.; Barry, T.L.; Coe, A.L.; Bown, P.R.; Rawson, P.F.; Gale, A.; Gibbard, P.; Gregory, F.J.; Hounslow, M.W.; Kerr, A.C.; Pearson, P.; Knox, R.; Powell, J.; Waters, C.; Marshall, J.; Oates, M.; Stone, P. Stratigraphy of the Anthropocene. *Philos. Trans. A Math. Phys. Eng. Sci.,* **2011**, *369*(1938), 1036-1055.

[http://dx.doi.org/10.1098/rsta.2010.0315] [PMID: 21282159]

[21]   Asadoorian, M.O.; Sarofim, M.C.; Reilly, J.M. *Historical Anthropogenic Emissions Inventories for Greenhouse Gases and Major Criteria Pollutants.*, **2006**. http://globalchange.mit.edu/files/document/MITJPSPGC_TechNote8.pdf

[22]   Rose, N.L. Spheroidal carbonaceous fly ash particles provide a globally synchronous stratigraphic marker for the Anthropocene. *Environ. Sci. Technol.,* **2015**, *49*(7), 4155-4162.
[http://dx.doi.org/10.1021/acs.est.5b00543] [PMID: 25790111]

[23]   Hansen, J.; Sato, M.; Hearty, P. Ice melt, sea level rise and superstorms: evidence from paleoclimate data, climate modeling, and modern observations that 2C global warming could be dangerous. *Atmos. Chem. Phys.,* **2016**, *16*, 3761-3812.
[http://dx.doi.org/10.5194/acp-16-3761-2016]

CHAPTER 3

The Industrialisation[1]: Its Origination and Development

Abstract: Intellectual movements, which originated during the Age of Enlightenment (rationalism, scientific recognition, inductive method) created important preconditions for the industrial era. Calvinism, scientific progress, discoveries, and inventions led to the industrial revolution. It replaced human and animal work with machines and transformed agrarian societies into remote supplied societies. Fossil energy carriers became the core of industrial development. Improvements in food and medicine supply, as well as social reforms, set the beginning of demographic transformations. Economic developments, organisation, division of labour, innovations (*e.g.* electric current), and specialisations were accompanied by severe political and societal crises (*e.g.* rural exodus, machine runners, pauperism, and industrial slavery). Despite rising disparities caused by agglomeration, cartelism, and accumulation of assets and wealth, a slow improvement in the degree of education, qualification, literacy, average standard of living, and life expectancy was realised. Escalating world population growth since ca. 1960 implied increasing amounts of supply (mineral commodities, food, *etc.*). Accelerated techno-scientific progress, partially fostered after 1945 by space and arms races created fast developments in nuclear technology, electronics, digitalisation, automation, internet, genomics, bio- and nanotechnology, and resulted in globalised optimum physical and virtual mobility.

Keywords: Cultural development, Dependencies, Digitalisation, Fossil carbon, Globalisation, Growth, Industrial society, Innovations, Inventions, Matter flux, Mobility, Rationalism, Risk society, Science, Societal transformations, Technology, The Great Acceleration, World economy, World population.

The Age of Enlightenment, which existed from the end of 17th to end of 18th centuries, is characterised by reforms[2], rationalism[3,] and scientific recognition[4]. These lead to the strongest intellectual movement in Europe since The Reformation: the adoption of the scientific method resulted in technological power over nature and the utilisation of its inherent laws to serve human needs and wealth. This fundamental change in mentality and creativity generated important preconditions for the industrialisation [1]. This techno-economic process nucleated ca. 1790 in England[5] and began to spread in western Europe at the beginning of the 19th century. Realisation and application of epoch-making technical inventions[6], innovations, scientific discoveries and findings[7] caused fundamental and multiple structural changes *via* the introduction of industrial techniques to existing economic structures. The human and animal powered agrarian society was successively replaced by machines, first of all in textilemanufacturing. Cheap and abundant fossil energy carriers [2] facilitated the production of metals, cement, glass, ceramics, *etc.* Higher demands and refined pretensions accelerated innovation and created a favourable milieu for detections and inventions[8] in the 19th century. These innovations raised the quantity of products by implementing more and more machine-based technology and gradually brought an end to a long period of subsistence living, dearth, and general scarcity. This process of early industrialization lasted until 1870 [3] and proceeded under severe social crises [4 - 7] (pauperism, poverty mass-feedings, machine breakers, child labour), political crises [8], societal changes (transformation of self-supplying rural - autarchical - to remote-supplying - importing - societies, as well as of clerically and manorially regulated, estate-based society into class society [9]), liberation of trade, creation of homogeneous internal economic areas by creation of monetary units, creation of standard weights and measures, abolishment of inland duty, standardisation of time (universal time: Greenwich Mean Time [10]), extinction of antiquated and radiation of new professional fields, and accumulations of capital and trading assets (industrial capitalism). Mass slave work, lasting until 1888 in the colonies, contributed substantially to that development [11]. Economic and plantage slavery, also called industrial slavery, spread because of increased demand from industrial centres in the homelands. Mechanisation of production in Europe drove faster manufacturing of the raw materials from overseas [12].

Industrialisation rapidly accelerated and diversified between 1870 and 1914 - a second phase of industrialisation [3] -, influenced by the application of supremely important techno-scientific detections, inventions, and achievments[9]. This phase was characterised by progress in management [13], growth of factories, mechanisation of sequences of operations by set-up of prototype of assembly lines, intensified division of labour, specialisation, diffusion, selection of new production sites [14], expensive technical implementations, rationalisation in enterprises, agglomeration and cartelism, rising production, wholesale, export, and foundation of new economic centres in overseas. Global propagation of industrialisation was realised by the technologically, scientifically, organisationally, and economically superior Europeans [15]. Industrialisation spread asynchronously over the globe, beginning in the 19th century with Europe and North America, and the first half of the 20th century with Canada, South Africa, India, Australia, Japan, China, and the Soviet Union. These transitions were often accompanied by severe societal and political crises (Holodomor, years of famine) [16, 17]. By the second half of 20th century it had taken root in eastern

Asia, South America, and central Africa [3, 14, 18, 19]. The different diffusionspeeds and responses to innovation depended on the quality of existing means of communication, interconnected lines of transportation, individual economic histories, structural heterogeneities, as well as from the quantity, value, and quality of natural resources. On average, a slow increase in degree of education, literacy, and standard of living in developed countries was achieved. The matter and energy flux necessary for this development was made possible by intensified exploration and extraction of mineral raw materials [20]. An analysis of extraction trends of 18 metallic mineral raw materials since 1900 resulted in a consistent positive correlation between per-capita demand and time. This can be explained by increasing world population (ca. 430%), increasing per-capita consumption, and introduction of novel industrial and technical applications [20]. Global resource extraction (for supplying technical energy, food and feed, stock building materials, other dissipative uses) nearly elevenfolded from 7 GT in 1900 to 76 GT in 2010 [21].

Subsequent to the world wars, which caused a widespread stagnation in industrial output except for war material, the highly dynamic process of industrialisation spread and diversified very fast - the "Great Acceleration" [22, 23] - by applying inventions, innovations, and discoveries[10] to new branches like nuclear technology, production of plastics and synthetic materials, electronics, software development, automatisation, computers, information and communication technology, digitalisation, robotisation, internet, genomics, bio- and nanotechnology, *etc*. The arms race during the Cold War (1947-1991) and proxy-wars in the third world, and the aspiration to superiority in spaceflight, artificial intelligence, and information-processing boosted technological/scientific development and yielded many technical spin-off products for civil use, *e.g.* radar speed checker, captive balloon, hydrophone (sonar[11]), jet airplane, helicopter [24], handy, drones, cross-country cars, Teflon, *etc*. By mobilising gigantic fluxes of matter and the preparation of predominantly fossil energy carriers, as well as their economisation, consumption, and dissipation, this third phase of industrialisation became globalised due to cheap, seemingly boundless physical and virtual mobility and created, depending on established economic relations and application of subtle project management techniques [25], mass products for a rising number of individuals with rising pretensions and standard of living. Since the beginning of industrialisation in ca. 1760, world population has increased by ca. 6.4 billion to ca. 7.4 billion at the end of 2015 [26, 27]. 7.5 billion persons have made the turnover of huge amounts of material and energy carriers necessary; the figures for 2014 are addressed in the following chapter.

Deindustrialisation occurred due to outsourcing, focussing on mass services (*e.g.* tourism, logistics), downsizing or depletion of resources. New, often more "green" industrial branches developed in the sectors of demolition, recycling (gaining secondary resources, *e.g.* by fine-sorting of plastics by molecular identification), generation of renewable energy carriers and their infrastructures, as well as installation of waste repositories.

The fourth stage of industrialisation is still in an early state of development. It consists of design and creation of "intelligent" technical systems, in which all hierarchical levels are connected in real time for self-adaptation, self-optimisation, self-regulation, self-diagnosis, and self-cognition. This will result in so-called "smart[12] factories", organised for networking and controlling system complexity with robotics and automation [28].

The world has been globalised by industrial development promoting webs of trade and communication several times over: first by steam ships, followed by steam railways, the setup of transoceanic telegraph cables, motor vehicles, airplanes, and finally by wireless communication *via* satellite and internet.

NOTES

[1] Etymology: From industruus (adj.) [early Latin] = diligent. Latin word compounds: indu = in, within; struere = constructing, producing, building. Semantic change: 1530: Industria = diligence, zeal, energy, effort; 1560: trade, manufacture; 1610: systematic work. (http://www.etymonline.com).

[2] *e.g.* political economy, economic individualism and liberalism by Adam Smith 1776. Guild-coercion was soon replaced by free enterprise and trade, first in 1791 France.

[3] *e.g.* Benedictus Spinoza 1677: Ethics; Gottfried W. Leibniz 1695: Système nouveau de la nature et de la communication des substances.

[4] *e.g.* Francis Bacon 1620: Novum Organum Scientiarum; René Descartes 1637: Discours de la méthode pour bien conduire sa raison, et chercher la vérité dans les sciences; Isaac Newton 1687: Principles of deterministic mechanics, 1704: Optics; Pierre S. de Laplace 1799: Traité de mecanique celeste.

[5] The beginnings of industrialisation found advantageous growth-conditions in England, because Calvinism predominated there. According to this doctrine, the predestination of believers is recognisable by their worldly affluence. Economic gains, making fortune, diligence and frugality were in accordance with that

ideology (sociologist Max Weber defined that situation as protestant work ethics). Therefore aristocrats invested in new economic branches to step up productivity and the British Government set up protective duties. But as already mentioned in the text, also other factors like individualism, rationalism and literacy boosted the later growth-development of industrialism [29].

[6] *e.g.* Thomas Newcomen 1712: mine-drainage by steam engine; R. Arkwright 1769: spinning-machine; James Watt 1783: low-pressure steam engine; A. Darby 1790: coke blaste furnace; Philip Vaughan 1791: roller bearing.

[7] *e.g.* Charles A. Coulomb 1785: fundamentals of electrostatic laws; Antoine L. de Lavoisier 1789: Traité élémentaire de Chimie; Alois Senefelder 1797: lithography; J. N. Louis Robert 1799: paper-machine; Amedeo Avogadro 1811: molecular hypothesis; Augustin J. Fresnel 1815: polarisation of light; Joseph von Fraunhofer 1817: diffraction grating; Karl F. Draise 1817: bicycle; M. Faraday 1822: electro-magnetism; George Stephenson 1825: steam railway; Johann N. von Dreyse 1835: needle rifle; Christian J. Friedrich Wöhler 1828: anorganic urea, 1872: aluminium; Carl F. Gauss 1832: geo-magnetism; Samuel Colt 1836: revolver; Louis Daguerre 1837: photography; Thomas R. H. Thomson 1841: quinine; Christian J. Doppler 1842: dependency of wave frequency from relative motion between wave source and receiver; William Grove 1842: fuel cell; Ada Lovelace 1843: programming; Abraham Gesner 1846: distillation of kerosene from coal; Ascanio Sobrero 1847: nitroglycerine; Ignaz P. Semmelweis 1847: antiseptic prophylaxis of puerperal fever; Cyrus W. Field 1858: transatlantic telegraph cable; Robert Bunsen & Gustav R. Kirchhoff 1859: spectrochemical analysis; Julius Plücker 1859: cathode rays; Charles R. Darwin 1859: natural selection; James C. Maxwell 1859: kinetic theory of gases; Gregor J. Mendel 1865: basic genetics; Ernst H. P. A. Haeckel 1866: ecology; Jakob Monier 1867: ferro-concrete; Carl Zeiß, Ernst Abbe, F. Otto Schott 1867: microscope optics; J. Friedrich Miescher 1869: nucleic acid; Dimitri I. Mendelejew 1869: periodic system of chemical elements.

[8] Justus von Liebig 1832: organo-chemical radicals, 1860: artificial fertiliser; Georg C. K. Hunaeus and Edwin L. Drake 1858/59: discovery of oilfields; Alfred Nobel 1867: dynamite.

[9] Joseph Bazalgette 1858: communal sewer system; Ernst Mach 1866: supersonics; Robert Koch 1876: bacteriology; Carl P. G. von Linde 1876: gas liquefaction; Augustin Mouchot 1878: vapour absorption refrigerator; Thomas A.

Edison 1879: electric bulb; Werner Siemens 1879: electric railway; Lester A. Pelton 1880: hydroturbine; Louis Pasteur 1880: vaccination; Svante Arrhenius 1882: electrolytic dissociation; Henry Bessemer 1885: steel; Alfred Benz 1886: gas firing motor; Nikola Tesla 1887: biphase electric current; Friedrich Bayer 1892: synthetic plant-protective agent; F. Wilhelm Ostwald 1894: autocatalytics; Paul Ehrlich 1894: antidiphtheric serum, 1909: chemotherapy; Wilhelm C. Roentgen 1895: X-rays; Joshua A. Slocum 1895-1898: solo circumnavigation of the globe; Henry Bequerel 1896: radioactivity; Heinrich R. Hertz 1897: wireless telegraphy; Marie S. Curie 1898: Radium; David Hilbert 1899: fundamentals of geometry; Max Planck 1900: quantum theory; Ferdinand Zeppelin 1900: airship; Orville & Wilbur Wright: 1903: aviation; Michail S. Zwet 1903: chromatography; Albert Einstein 1905: photoelectric effect, 1907: $E = mc^2$; Christian Hülsmeyer 1904: radar.

[10] Casimir Funk 1912: Vitamine B1; William H. & William L. Bragg 1913: X-ray diffractometer; Henry Ford 1913: assembly line; Alfred Wegener 1915: continental drift; Fritz Haber 1918: synthesis of ammonia; Ernest Rutherford 1919: nuclear transmutation; Francis W. Aston 1919: mass spectrometer, isotopes; Hermann Staudinger 1920: polymer chemistry; Oleg V. Losev 1927: LED; Werner Heisenberg 1927: uncertainty relation; Charles A. Lindbergh 1927: nonstop flight New York - Paris; Alexander Fleming 1928: penicillin; Linus C. Pauling 1931: quantum chemistry; Ernst A. F. Ruska *et al.* 1931: electron microscope; James Chadwick 1932: neutron; Carl D. Anderson 1932: antimatter; Lise Meitner *et al.* 1939 nuclear fission; Hans von Ohain 1939: jet engine; Igor I. Sikorsky 1939: helicopter; Konrad E. O. Zuse 1941: computer; Enrico Fermi 1942: nuclear reactor; J. Robert Oppenheimer 1945: A-bomb; Harry H. Hess 1946: sea floor spreading; Michail T. Kalaschnikow 1947: AK-47; William B. Shockley 1947: transistor; Alan Turing 1950: artificial intelligence; John C. Eccles 1951: signal transfer in synaptic gap; Grace M. Hopper 1951: Compiler A-O; Edward Teller & Stanislaw Ulam 1952: H-bomb; Andrei D. Sacharow 1952: Tokamak; Stanley L. Miller 1952: abiogenesis: aminoacids from inorganic molecules; Rosalind E. Franklin, James D. Watson *et al.* 1953: DNA-structure; Melvin A. Cook 1956: slurry explosive; Rudolf L. Mößbauer 1957: nuclear gamma ray resonance; Artur Fischer 1958: dowel, Richard Feynman 1959: nanotechnology; Theodor H. Maiman 1960: ruby laser; Vostok Program 1961: human Earth orbit flight; Edward N. Lorenz 1963: deterministic chaos effect; Christiaan N. Bernard 1967: heart transplantation; Ilya Prigogine 1967: dissipative structure; Lynn Margulis 1967: endosymbiosis; Margaret Hamilton 1968: software engineering; NASA 1969: mankind's first step on moon; Martin Schadt & Wolfgang Helfrich 1970: LCD; Paul C. Lauterbur 1971: magnetic resonance tomography; Soviet Mars Program 1971: soft landing on Mars; Wu, R.

& Taylor, E. 1971: genetic engineering; Ray Tomlinson 1971: email; Jaques Vidal 1973: brain-computer-interface; Roger L. Easton 1977: GPS; Peter Vail 1977: sequence stratigraphy; Robert G. Edwards 1977: *in vitro* fertilisation; Chuck Hall 1983: 3D print; Dan Shechtman 1984: quasicrystals; Robert Curl *et al.* 1985: Buckminsterfullerene C_{60}; Johannes G. Bednorz 1986: high temperature supraconduction; Ulrich Beck 1986: risk society; Karlheinz Brandenburg 1989: MP3; Tim Berners-Lee 1989: WWW; Sumio Iijima *et al.* 1991: single layer carbon nanotube; Hamilton O. Smith 2003: synthetic genomics; Konstantin Novoselov & Andre Geim 2004: single-layer graphen; Emmanuelle Charpentier 2012: targeted genome editing with Crispr-Cas9; ESA 2012: touch down of lander Philae on comet 67P/Tchurjumow-Gerassimenko; NASA 2015: missile New Horizon flies by close past Pluto; LIGO 2015: gravitational waves; Bertrand Piccard & Andre Borschberg 2015-2016: round-the-world flight of Solar Impulse; DLR 29.09.2016: H_2 fuel cell aircraft; European XFEL 01.09.2017: research institution for ultrashort X-ray laser pulses for visualising atomic processes.

[11] Sonar: Sound navigation and ranging: depth sounding used for underwater orientation and localisation; hydrophones record reflected signals.

[12] Smart: self-monitoring, analysing and reporting technology.

CONFLICT OF INTEREST

The author (editor) declares no conflict of interest, financial or otherwise.

ACKNOWLEDGEMENTS

Declare none.

REFERENCES

[1]   Magee, B. *The Story of Philosophy,* 2nd ed; Dorling Kindersley Ltd.: London, UK, **2010**.

[2]   Clark, G.; Jacks, D. Coal and the Industrial Revolution, 1700–1869. *EREH,* **2007**, *11*(1), 39-72.

[3]   Tilly, R.H. *Industrialization as an Historical Process,* **2010**. http://www.ieg-ego.eu/tillyr-2010-en

[4]   Hauptmann, G. *Die Weber, S*; Fischer Verlag: Berlin, **1892**.

[5]   Zola, E. *Les Rougon-Macquart: Germinal, G. Charpentier et Cie, Paris* , **1895**.

[6]   Allen, R.C. Progress and poverty in early modern Europe. *Econ. Hist. Rev.,* **2003**, *56*(3), 403-443. [http://dx.doi.org/10.1111/j.1468-0289.2003.00257.x]

[7]   Knick Harley, C. *British and European Industrialization,* **2013**. http://www.economics.ox.ac.uk/ materials/papers/12644/harley111.pdf

[8]   Beckert, S. *Empire of Cotton: A Global History*; Alfred A. Knopf, Random House LLC: New York, **2014**.

[9] Ogilvie, S. Choices and Constraints in the Pre-industrial Countryside. In: *Plenary Lecture for Population, Economy and Welfare, c.1200-2000*; a Conference in Honour of Richard Smith, Faculty of Economics, CWPESH no.1: Cambridge: UK, **2011**.

[10] Conrad, S. *Geschichte der Welt 1750-1870: Wege zur modernen Welt 1750-1870*; Conrad, S.; Osterhammel, J., Eds.; C. H. Beck: Munich, **2016**, Vol. 4, pp. 512-559. [http://dx.doi.org/10.17104/9783406641145]

[11] Mbembe, A. *La critique de la raison nègre*; La Découverte: Paris, **2013**.

[12] Zeuske, M. Globale Sklavereien: Geschichte und Gegenwart. *APuZ,* **2015**, *50-51*, 7-14.

[13] Pollard, S. *The Genesis of Modern Management. A Study of the Industrial Revolution in Great Britain*; Harvard University Press: Cambridge, **1965**.

[14] Pierenkemper, T. *Die Industrialisierung europäischer Montanregionen im 19. Jahrhundert*; Pierenkemper, T., Ed.; Franz Steiner Verlag: Stuttgart, **2002**, Vol. 3, pp. 3-18.

[15] Reinhard, W. *Die Unterwerfung der Welt. Globalgeschichte der europäischen Expansion 1405-2015*; C. H. Beck Verlag: München, **2016**.

[16] Davies, R.W.; Wheatcroft, S.G. *The Industrialisation of Soviet Russia – The Years of Hunger – Soviet Agriculture, 1931-1933*; Palgrave Macmillan: London, **2004**.

[17] Chan, A.L. *Mao's Crusade. Politics and Policy Implementation in China's Great Leap Forward, Studies on Contemporary China*; Oxford University Press: Oxford, **2001**.

[18] Allen, R.C. *The British Industrial Revolution in Global Perspective (New Approaches to Economic and Social History)*; University Press: Cambridge, **2009**. [http://dx.doi.org/10.1017/CBO9780511816680]

[19] Voppel, G. *Die Industrialisierung der Erde*; B. G. Teubner: Stuttgart, **1990**.

[20] Patiño Douce, A. E. Metallic mineral resources in the twenty-first century. I. Historical extraction trends and expected demand. *Nat. Resour. Res.,* **2016**, *25*(1), 71-90. [http://dx.doi.org/10.1007/s11053-015-9266-z]

[21] Krausmann, F.; Wiedenhofer, D.; Lauk, C.; Haas, W.; Tanikawa, H.; Fishman, T.; Miatto, A.; Schandl, H.; Haberl, H. Global socioeconomic material stocks rise 23-fold over the 20th century and require half of annual resource use. *Proc. Natl. Acad. Sci. USA,* **2017**, *114*(8), 1880-1885. [http://dx.doi.org/10.1073/pnas.1613773114] [PMID: 28167761]

[22] McNeill, J.R.; Engelke, P. *The Great Acceleration: An Environmental History of the Anthropocene since 1945*; Belknap Press: Cambridge, Ma, **2014**.

[23] Steffen, W.; Broadgate, W.; Deutsch, L. The trajectory of the Anthropocene: The Great Acceleration. *Anthropocene Rev.,* **2015**, *2*(1), 81-98. [http://dx.doi.org/10.1177/2053019614564785]

[24] Boyne, W. *How the Helicopter Changed Modern Warfare*; Pelican Publishing Company: Gretna: Lousiana, USA, **2011**.

[25] Badiru, A.B. *Managing Industrial Development Projects: A Project Management Approach*; John Wiley & Sons: New York, **1993**.

[26] UN Department of Economic and Social Affairs. **2015**. http://esa.un.org/unpd/wpp/Download/ Probabilistic/Population/

[27] United States Census Bureau.   http://www.census.gov/population/international/data/worldpop/ table_history.php

[28] Lee, J.; Kao, H-A.; Yang, S. Service Innovation and Smart Analytics for Industry 4.0 and Big Data Environment. *Procedia CIRP,* **2014**, *16*, 3-8.
[http://dx.doi.org/10.1016/j.procir.2014.02.001]

[29] Becker, S.O.; Wößmann, L. *Was Weber Wrong? A Human Capital Theory of Protestant Economic History, Discussion paper no. 2007-7*; Department of Economics: Munich, **2007**.

CHAPTER 4

Its Impacts on and Transformations of Ecosystems and Life

Abstract: Rising affluence and world population demands increasing supply. Consequently, mass extraction, processing, and shipping of mineral raw materials and other goods has spread over the world. The huge anthropogenic turnover of materials caused by socio-economic metabolism and by all kinds of effects and interactions resulted in problems for the ecosphere. The transformations of natural environments turned out to be detrimental to biodiversity, to ecosystem services, and to life in general. Fauna and flora were exposed to greater adaptive pressure. Dissipative and dispersive use of natural products created huge amounts of waste and of toxic sites, resulting in health problems, fatalities, and enormous economic costs. Several planetary-scale tipping points, which occurred with further biodiversity loss, radiative CO_2 forcing, dieback of rainforests and loss of the West Antarctic ice sheet, were approached and partly transgressed, because regulatory natural forces cannot compensate for the magnitudes of human activities. Emerging system pressures and rising global connectedness have resulted in risk societies, which are characterised by latent conflicts with sustainability, dependencies, higher disparities, environmental degradations, and less precise system predictabilities, as well as the awareness of these problems, documented *e.g.* by the creation of Earth Overshoot Day.

Keywords: Adaptive pressure, Anthropogenic turnovers, Contamination, Dependencies, Ecologic counterparts, Environmental health, Risk, Socio-economic metabolism, Tipping points, Waste.

Industrialisation created not only scientific/technological progress and wealth, but also caused deep economic and societal transformations, and innumerable impacts on the environment like climate warming, rising sea level, biodiversity loss, *etc.* [1]. The effects and changes discussed in the following are of essential concern because of the monadic[1] coexistence of human beings and nature[2]. The following attempt at a holistic view considers intentional and unintentional, direct and indirect, predicted and unpredictable, proximity and long-range, spatial and temporal effects and interactions that originated during demand, supply, and socio-economic metabolisation of mineral commodities produced in increasing quantities by the extractive industry. The global impacts on ecosystems and life caused by 250 years of industrialisation, by the resulted extraction and utilisation

Hubert Engelbrecht

of mineral raw materials, as well as by the predominantly dissipative use of the products, are ambivalent, complex, and irreversible. They are problematic and detrimental because of having given rise to environmental degradation, dependencies, risks, and disparities, but they also improved wellbeing, nutrition, life expectancy, mobility, literacy, and cultural development. The following is a well-balanced overview regarding the "benefits" and "detriments" of these transformations and metamorphoses, which, in many cases, are causally related and complementary. The reasons for all the facts, states, and developments mentioned in the following lie in the broad and deep effects of 250 years of mass generation, application, and waste of products fabricated from mined and extracted mineral raw materials. But also the inefficiency of the current industrial global food production system - *e.g.* 44% of harvested crops dry matter lost prior to human consumption -, which resulted from consumer behaviour and production practices, is detrimental to ecosystems and biodiversity [2]. The damage to ecosphere, which is discussed first, occurred because of industrial activities and interferences that have substantially modified the physical, chemical, and biochemical states of natural systems [3]. The resulting adaptive stress on many species resulted in range shifts, creation of new niches, phenotypic changes resulting from genetic change or individual plasticity (adaptation by modifying physical and/or biochemical states), and changes in patterns of communication and behaviour (see chapter biosphere 4b) [4 - 8].

3241 most toxic sites were identified in 49 low- to middle-income countries, in which ca. 94 million people are exposed to contaminants like lead, radionuclides, mercury, hexavalent chromium, pesticides, and cadmium. Globally, ca. 300,000 toxic sites require remediation [9]. 8.4 million premature deaths were estimated in 2012 due to exposure to polluted air (indoor and outdoor), water and soil in these countries [10]. Globally, ca. 8.9 million premature deaths result from industrial environmental pollution, causing enormous economic costs [11].

Human activities have pushed the states of natural subsystems towards some of their planetary-scale tipping points. These include the impending loss of the West Antarctic Ice Sheet, melting of Arctic sea ice and permafrost areas, deceleration of Atlantic Meridional Overturning Circulation, dieback of Amazonian rainforest [12]. Beyond these critical and sensitive thresholds, these subsystems will have irreversibly switched - probably during the 21st century - into qualitatively different states and modes of operation, because the regulatory capacities of the earth system will not be enough to compensate for the magnitudes and/or intensities of impacts [13]. Of all CO_2 and CH_4 emitted by the 90 largest corporate investor-owned and state-owned producers of fossil fuels and cement released between 1854-2010, 50% (457 GT) was generated before 1986 and 50% since [14].

It has been calculated that melting parts of the East Antarctic coastal ice, equivalent to a further ca. 80 mm sea level rise, will cause the destabilisation of several inland glaciers, because they cover subglacial, inland sloping basins up to 1500 m below present sea level. The resulting self-sustained discharge of the low lying Wilkes Basin is estimated to contribute 3-4 m to future sea level rise [15]. A similar situation occurred in middle Pliocene, as estimated from geochemical characteristics of dated glaciomarine deposits 310 km off Adélie Land margin [16].

It has been asserted that several safe operating spaces for humanity, concerning *e.g.* the rate of biodiversity loss, rate of consumption of nitrogen and phosphorus, radiative CO_2 forcing, material flux, *etc.*, have been transgressed [17, 18]. It has also been pointed out that current biodiversity loss is globally of rising concern, because increasing the quality of biodiversity is directly connected with improving the quality of ecosystem functions and human health [19].

The recognition of risky human activities in and overuse of ecosystems gave rise to the creation of the Earth Overshoot Day. It is defined as the date of a given year when the maximum of global annual supply of natural resources and ecosystem services[3]-*i.e.* the global annual biocapacity - is spent because of anthropogenic economic demands. Environmental problems originate when overuse of ecosystems impairs and weakens their functions and effects, retarding their regenerations and replenishments, depleting their biocapacities, and the ecological balance is disarranged and ecological debts increase. Growing productivity of the world industrial economy on the one side, and rising dysfunctionality of ecosystems on the other, result in Overshoot Day occurring earlier in each year (1971: December-24; 1987: December-19; 2016: August-08). According to the average global ecological footprint[4], the natural resources of ca. 1.6 Earths are necessary to sustain the present global supply [20, 21].

The following concentrate, in which mentioned impacts on the spheres of earth and embedded ecosystems and life are detailed, evidences the huge impetus and momentum of 250 years of mass-extraction and socio-economic metabolisation on nature.

NOTES

[1] Monad: monas (Classical Greek): unit. Here it means that human being and nature form one indivisible unit. Both interfere in manifold ways, causing permanent reciprocal transformations. Concerning human beings, all is taken from and finally left back to nature. But compensation for and/or recovery from human-made changes occurs only to some extent.

[2] Ecologist Rachel Carson interpreted that situation as follows: "*... we have now acquired a fateful power to alter and destroy nature. But man is a part of nature, and his war against nature is inevitably a war against himself.*" [22].

[3] Ecosystem services: provide food, water, timber, and fibre; regulate climate, floods, disease, wastes, and water quality; grant recreational, aesthetic, and spiritual benefits; support by means of soil formation, photosynthesis, and nutrient cycling [23].

[4] Ecological footprint: It is a measure of global interdependence. Its indicators measure the ecological weight of space, mass and value of matter, land and embedded energy necessary to supply a consumer. It measures the absolute ecological impact of a product or an activity. This can be expressed, for instance, in land area or amount of energy carriers necessary to sustain one person at present level of lifestyle [24].

CONFLICT OF INTEREST

The author (editor) declares no conflict of interest, financial or otherwise.

ACKNOWLEDGEMENTS

Declare none.

REFERENCES

[1]   Mgbemene, C.A.; Nnaji, C.C.; Nwozor, C. Industrialization and its backlash: Focus on climate change and its consequences. *J. Environ. Sci. Technol.,* **2016,** *9*(4), 301-316.
[http://dx.doi.org/10.3923/jest.2016.301.316]

[2]   Alexander, P.; Brown, C.; Arneth, A.; Finnigan, J.; Moran, D.; Rounsevell, M.D. Losses, inefficiencies and waste in the global food system. *Agric. Syst.,* **2017,** *153*, 190-200.
[http://dx.doi.org/10.1016/j.agsy.2017.01.014] [PMID: 28579671]

[3]   Goudie, A.S. *The Human Impact on the Natural Environment. Past, Present and Future,* 7th ed; Wiley-Blackwell: Oxford, **2013.**

[4]   Walther, G.R.; Post, E.; Convey, P.; Menzel, A.; Parmesan, C.; Beebee, T.J.; Fromentin, J.M.; Hoegh-Guldberg, O.; Bairlein, F. Ecological responses to recent climate change. *Nature,* **2002,** *416*(6879), 389-395.
[http://dx.doi.org/10.1038/416389a] [PMID: 11919621]

[5]   Chen, I-C.; Hill, J.K.; Ohlemüller, R.; Roy, D.B.; Thomas, C.D. Rapid range shifts of species associated with high levels of climate warming. *Science,* **2011,** *333*(6045), 1024-1026.
[http://dx.doi.org/10.1126/science.1206432] [PMID: 21852500]

[6]   Merilä, J.; Hendry, A.P. Climate change, adaptation, and phenotypic plasticity: the problem and the evidence. *Evol. Appl.,* **2014,** *7*(1), 1-14.
[http://dx.doi.org/10.1111/eva.12137] [PMID: 24454544]

[7]   Morueta-Holme, N.; Engemann, K.; Sandoval-Acuña, P.; Jonas, J.D.; Segnitz, R.M.; Svenning, J.C.

Strong upslope shifts in Chimborazo's vegetation over two centuries since Humboldt. *Proc. Natl. Acad. Sci. USA,* **2015,** *112*(41), 12741-12745.
[http://dx.doi.org/10.1073/pnas.1509938112] [PMID: 26371298]

[8]   Lenoir, J.; Svenning, J.-C. Climate-related range shifts - a global multidimensional synthesis and new research directions. *Ecography,* **2015,** *38*(1), 15-28.
[http://dx.doi.org/10.1111/ecog.00967]

[9]   Bernhardt, A. *World's Worst Pollution Problems,* **2015.** http://www.worstpolluted.org/docs/ WWPP_2015_Final.pdf

[10]   Landrigan, P.J.; Fuller, R. Environmental pollution: An enormous and invisible burden on health systems in low- and middle-income counties. *World Hosp. Health Serv.,* **2014,** *50*(4), 35-40.
[PMID: 25985560]

[11]   Landrigan, P.J.; Fuller, R. Global health and environmental pollution. *Int. J. Public Health,* **2015,** *60*(7), 761-762.
[http://dx.doi.org/10.1007/s00038-015-0706-7] [PMID: 26135237]

[12]   Lenton, T.M.; Williams, H.T. On the origin of planetary-scale tipping points. *Trends Ecol. Evol. (Amst.),* **2013,** *28*(7), 380-382.
[http://dx.doi.org/10.1016/j.tree.2013.06.001] [PMID: 23777818]

[13]   Lenton, T.M.; Held, H.; Kriegler, E.; Hall, J.W.; Lucht, W.; Rahmstorf, S.; Schellnhuber, H.J. Tipping elements in the Earth's climate system. *Proc. Natl. Acad. Sci. USA,* **2008,** *105*(6), 1786-1793.
[http://dx.doi.org/10.1073/pnas.0705414105] [PMID: 18258748]

[14]   Heede, R. Tracing anthropogenic carbon dioxide and methane emissions to fossil fuel and cement producers, 1854-2010. *Clim. Change,* **2014,** *122,* 229-241.
[http://dx.doi.org/10.1007/s10584-013-0986-y]

[15]   Mengel, M.; Levermann, A. Ice plug prevents irreversible discharge from Eastern Antarctica. *Nat. Clim. Chang.,* **2014,** *4,* 451-455.
[http://dx.doi.org/10.1038/nclimate2226]

[16]   Cook, C.P.; van de Flierdt, T.; Williams, T. Dynamic behaviour of the East Antarctic ice sheet during Pliocene warmth. *Nat. Geosci.,* **2013,** *6,* 765-769.
[http://dx.doi.org/10.1038/ngeo1889]

[17]   Rockström, J.; Steffen, W.; Noone, K.; Persson, A.; Chapin, F.S., III; Lambin, E.F.; Lenton, T.M.; Scheffer, M.; Folke, C.; Schellnhuber, H.J.; Nykvist, B.; de Wit, C.A.; Hughes, T.; van der Leeuw, S.; Rodhe, H.; Sörlin, S.; Snyder, P.K.; Costanza, R.; Svedin, U.; Falkenmark, M.; Karlberg, L.; Corell, R.W.; Fabry, V.J.; Hansen, J.; Walker, B.; Liverman, D.; Richardson, K.; Crutzen, P.; Foley, J.A. A safe operating space for humanity. *Nature,* **2009,** *461*(7263), 472-475.
[http://dx.doi.org/10.1038/461472a] [PMID: 19779433]

[18]   Steffen, W.; Richardson, K.; Rockström, J.; Cornell, S.E.; Fetzer, I.; Bennett, E.M.; Biggs, R.; Carpenter, S.R.; de Vries, W.; de Wit, C.A.; Folke, C.; Gerten, D.; Heinke, J.; Mace, G.M.; Persson, L.M.; Ramanathan, V.; Reyers, B.; Sörlin, S. Sustainability. Planetary boundaries: guiding human development on a changing planet. *Science,* **2015,** *347*(6223), 1259855.
[http://dx.doi.org/10.1126/science.1259855] [PMID: 25592418]

[19]   Baillie, J.E.; Collen, B.; Amin, R. Toward monitoring global biodiversity. *Conserv. Lett.,* **2008,** *1,* 18-26.
[http://dx.doi.org/10.1111/j.1755-263X.2008.00009.x]

[20]   Simms, A.; Johnson, V.; Smith, J. *The Consumption Explosion, London,* **2009.** http://b.3cdn.net/nefoundation/41a473dfbe880a0742_ ucm6i4n29.pdf

[21]   Global Footprint Network - Advancing the Scinece of Sustainability. **2016.** http://www.footprint network.org/en/index.php/GFN/page/world_footprint/

[22]   Carson, R. Of man and the stream of time.*Scripps College Bull,* **1962,** *36*(4) 11 pages

[23]   UNEP. *Marine and coastal ecosystems and human well-being: A synthesis report based on the findings of the Millennium Ecosystem Assessment,* **2006**.http://www.unep.org/pdf/Completev6_LR.pdf

[24]   Wackernagel, M.; Rees, W.E. *Our Ecological Footprint: Reducing Human Impact on the Earth; New Catalyst Bioregional Series*; New Society Publishers: Gabriola Island, BC, Canada, **1998**.

CHAPTER 5

Extraction of Mineral Raw Materials and its Utilisations[1]

Abstract: Mineral commodities are the backbone of the global economy and the base of secondary and tertiary industries. Growing economic demand and wealth founded on the life cycles of mines (from exploration to closure) and the life cycles of extracted matter (processing, refining, trading, consumption, and disposal). Industrial mining started in the second half of the 18th century in England, predominantly with coal and iron. This development spread over the world and got diversified; quantities of some commodities have increased a thousandfold since then. The quantity of mineral raw materials, mined and put into use in the USA increased by thirty-fold in the 20th century. Technical development since the 1980s increased the demand for diversity of mining products considerably. Global mining is controlled by ca. 10 enterprises. Description of geological settings, development, major extraction and mining sites, recent production numbers, estimated resources and reserves of mineral commodities, as well as numerous applications follow. They include energy carriers (hard coal, lignite, natural gas, petroleum, uranium, geothermal heat), ferro-alloy metals (*e.g.* iron, chrome, nickel), and non ferrous metals (*e.g.* copper, aluminium). The problem of strategic metals is addressed. Finally, the cumulative annual mass of extraction necessary to supply more than 7 billion persons is given.

Keywords: Economic backbone, Energy carriers, Geological settings, Material diversity, Material flows, Metals, Mineral commodities, Mining companies, Mining sites, Production reserves, Strategic metals.

Mineral raw materials, which occur throughout the earth's crust and mantle, have ever been the backbone and lifeblood of economies. Mineral commodities are - beside agrarian products - the basics from which substantially all other processed products are derived. Primary industry (mining and extracting) fills the feedstocks for the various manufacturing processes in the secondary and tertiary industries, which beneficiate, refine, treat and finish products. Growing economic demands of global societies forced the exploitation of mines and the extracted material, which is processed and refined by industry and then traded, consumed, and disposed of by their users [1]. Mineral commodities form the basis of wealth [2]. The growth of economies has ever been connected with the discovery of mineral resources and the development of mineral production [3].

Hubert Engelbrecht

Industrial mining and utilization of the mineral raw materials started in the second half of the 18th century in England with coal and iron ore and spread from there over the globe [4]. In 1790, the cumulative output of hard coal in Great Britain amounted to 7 MT and 68,000 T of iron [5, 6]. Industrial coal mining was the nucleation point of all further industrial development [7]. Despite large disparities, mine output and industrial development increased since then a thousandfold and more and raised cultural and living standards [3]. For instance, growth of mineral raw materials (primary metals, industrial metals, construction materials [clay, sand, gravel, limestone, granite, basalt, *etc.*] and fossil organics (coal, tar, peat, crude oil, natural gas), which were industrially produced and put into use in the USA, increased thirtyfold during the last century. The reasons for this are manifold: population growth, technology development, and wealth development (as described in general in chapters 3 and 19), adequate resources of supply, competition, as well as reduced production costs and decline of long-term price of key U.S. mineral raw materials [8]. In 1900, the quantity of mineral raw materials from mining and extractive industry amounted to 7.80×10^7 T. Apart from the economic drawbacks during the Great Depression, World War II, oil crises, and subsequent depressions, the quantity of industrial raw material production has grown and in 2006, reached a maximum of 3.64×10^9 T, followed by a sharp downturn to 2.36×10^9 T, caused by the global financial crisis [9].

Concerning recent developments in primary industry (manufacturing components of *e.g.* semiconductors, photovoltaic moduls, flat screens, hybrid electromotors, i-phones), the demand for a diversity of mining products has increased. In the 1980s, the basic demand consisted of rock material containing 12 chemical elements; in the 2000s, this demand has increased fivefold and includes *e.g.* titanium, tantal, tungsten, platinum, gold, cadmium, molybdenum, gallium, antimony, indium, yttrium, and the lanthanides group [10].

The top ten companies controlling global mines and minerals are as follows: BHP Billiton (Au/UK), Rio Tinto (Au/UK), CVRD/Inco (Brazil, Canada), Anglo American (UK), Xstrata/Falconbridge (Swiss/UK/Canada), Alcoa (USA), Barrick/Placer (Canada), Newmont (USA), Norilsk (Russia), and Phelps Dodge (USA). Their cumulative market capitalization amounted in 2007 to ca. 460 billion USD [11].

Mineral commodities extracted in 2014 will be discussed here and include the following groups: fossil mineral fuels (*e.g.* coal, petroleum, uranium), iron and ferro-alloy metals (*e.g.* iron, chrome, nickel), and non-ferrous metals (*e.g.* copper, aluminium). Production numbers do not refer to the amount of crude ore or concentrate, but indicate the content of recovered valuable elements and compounds [12].

Extraction and production figures 2014 and most important sites:

Fossil mineral fuels (fossil energy carriers):

Hard Coal (coking and steam coal): 7.1 GT, of which more than 50% was produced in China. Some of the mining sites are located in the Bowen Basin (Queensland, Australia), Donetz-Basin (eastern Ukraine), Inner Mongolia (China), and in Cerejón (Colombia). The latter is regarded as the world's largest open cast mine, measuring ca. 700 km^2 and estimated resources were 5.2 GT. Coal is indispensable in smelting plants, heat, and electricity plants, as well as in the chemical and pharmaceutical industries. 41% of electricity, 90% of cement and 70% of steel were produced by using coal, either as an energy carrier or chemical component. It is also used as chemical reduction agent and in gasification and hydration processes [13]. In early industrialization, coal was also used to distil kerosene, used for lighting purposes.

Lignite: 0.8 GT, of which 21.7% was produced in Germany (first rank). Large mining sites are situated in Lusatia and Northrhine Westphalia (Germany), in North Dakota and Montana (USA), as well as in Victoria (Australia). Estimated economically recoverable resources in the latter amount to 430 GT [14]. Lignite is utilized in gasification, hydration, and as fuel in electricity and heat plants. Based on the published geochemical data, it is estimated that minimum amount of recoverable Germanium, a strategic metalloid hosted in coal, is up to 112 kT in proven global reserves, where Ge concentration is $\geq$ 200 ppm [15].

Natural Gas: It occurs in several kinds of deposits including porous reservoirs, into which gas has migrated from source rocks; in coal beds containing methane; in tight gas sands, carbonates, and oilshales (which may need to be extracted using hydraulic fracturing, described in the Petroleum section below); in methane-hydrates in continental slope sediments at 300-2000m water depth; in permafrost areas; and in aquifers: 3.5 10^{12} m^3 of natural gas is extracted annually, 20.4% was extracted in the USA (first rank). These energy carriers - predominantly methane - contributed 22% of global production of electricity [13]. It is used for heating purposes and as an agent to propel motors. In the chemical and metallurgical industry, it is applied as a reduction medium and for the production of hydrogen. The plays containing the largest reserves are situated in Siberia (Yamburg, Urengoy, Medvezh'ye), Iran (Pars, Kish, Kangan), Qatar (North Field), and the USA (Barnett shale play in Texas Montana, Marcellus shale play in the Appalachian Basin, *etc.*) [16]. Flare-down programmes have improved the eco-balances of petroleum industry.

Petroleum: Oil-plays are divided into conventional and unconventional varieties: Conventional reservoirs are characterized by porous lithologies, into which petroleum has migrated from source rocks, which consist predominantly of oil shales. Oil accumulates in porous lithologies, when they are sealed stratigraphically or tectonically by impermeable rock (oil trap). These resources can be localized by artificial seismic rays, tested by exploration drills and, eventually recovered. But, on average, nine drills out of 10 are dry. The second type of play consists mostly of oil shales, oil schists, and oil sands, characterized by low to very low porosity. Hydraulic fracturing techniques involving the injection of large amounts of frack fluids containing proppants and chemical additives (surfactants, friction reducers, biocides, solvents, scale and corrosion inhibitors, *etc.*) that increase permeability by stimulating artificial fractures and diminish viscosity of *in situ* petroleum (and natural gas), thus advancing profitable extraction. Petroleum contributed 16% of global electricity in 2014 [13]. Petroleum is used in the production of fuels for motors, machines and heating systems; in plastics, synthetic materials, varnishes, and medicaments and in the chemical and pharmaceutical industries. The first successful oil wells were made in 1858/59 and were 35m deep at Wietze (Lower Saxony, Germany, by Georg C. K. Hunaeus) and 21 m deep at Titusville (Pennsylvania, USA, by Edwin L. Drake). The 2014 industrial output amounted to ca. 4.32 GT [12]. The largest proven oil reserves are situated in the Maracaibo Basin and Orinoco Belt of Venezuela, in the Ghawar and Safaniya plays in Saudi Arabia, in the Athabasca oil sands and the offshore Jeanne d'Arc Basin of Canada, and in the Marun, Ahwaz and the Aghajari oil fields of Iran. Venezuela holds 297.5 billion barrels 1/5, of the worlds total proven oil reserves [17].

Uranium: The lithophile geochemical property of uranium affects its accumulation predominantly in acidic magmatic rocks and related differentiates. In its hexavalent state, it forms, many stable chemical complexes under oxidizing conditions in hydrous solutions. Therefore Uranium deposits originate in a great variety of geological settings, including

- Disseminated deposits in acidic to intermediary intrusives (alaskites, granites, monzonites);
- Pitchblende veins in granite and surrounding contact-metamorphosed sedimentary host rocks;
- Mineralization in polymetallic hematite-rich breccias within granitic breccias, which were formed by phreatomagmatic activity, like the Olympic Dam deposit (Australia);
- Accumulations in fault- and shear-zones formed in acidic and intermediary volcanics;
- Disseminated deposits within rocks transformed by Ca, Na, and K metasomatism;
- Accumulations within skarns and metamorphic, hydrothermally altered

sediments and volcanics;

- Structure-controlled, locally strata-bound deposits at Proterozoic unconformities;
- Mineralization in altered sandstones present in hydrothermally altered, collapsed pipes;
- Chemically precipitated and biogenic deposits in medium- to coarse-grained, fluvial and marine-littoral sandstones. The ore is hosted in basal palaeochannels and is arranged in tabular- or roll-front bodies, indicating fossil redox-boundaries and palaeo-groundwater flow directions. Uranium deposits in sandstones also occur close to permeable fault zones and mafic dykes;
- Detrital uraninite deposits of Archaean to early Palaeoproterozoic, fluvial quartz-pebble conglomerate, unconformably covering metamorphic basement. It is recovered as a byproduct of gold-mining;
- Tertiary to recent surficial uranium deposits, precipitated in palaeosols or cemented by calcretes and gypcretes above granite or sandstone formations, deeply weathered under semiaride palaeoclimate conditions;
- Accumulations in lignite and associated carbonaceous mud layers (so-called thucholithes);
- Deposits related to karst and tectonic features in limestones and dolostones;
- Low concentrations in apatite grains forming marine sedimentary phosphate deposits;
- Disseminated mineralization adsorbed on organic matter and clay, which form marine black shales [18 - 21].

Industrial uranium mining began in the 19th century - in *e.g.* Jachymov, Slovakia - in order to produce sodium diuranate, which was applied as dye and glaze in glass and porcelain factories [22]. Uranium mining commenced in Katanga Province (Congo) to extract radium in 1921. Later, uranium was mined to propel the nuclear arms race during the Cold War and as an energy carrier to fuel power stations. As a source of γ-radiation and neutrons, uranium was used in nuclear science to generate transuranides, in medicine to generate radionuclides for therapeutic or analytical purposes, and in chemistry and solid body physics for structural analytics [7]. It contributed 11% of 23,318 TWh to global electricity production in 2014 [13]. In 2014, 67,944 T of U_3O_8 were produced globally [12]. Uranium mining sites containing the largest reserves are located in Australia (Olympic Dam, Ranger, Four Mile), in Kazakhstan (Tortkuduk, Budenovskoye 2), and in Canada (McArthur River Mine, Athabasca Basin). The output of top producer Kazakhstan amounted to 26,920 T of U_3O_8 in 2014 [12, 20].

Geothermal Heat: Plays are characterized by elevated terrestrial heat flow, which occurs in foreland basins, rift zones, volcanic arcs and plate margins in general. Low enthalpy (20-150 °C) and high enthalpy (> 150 °C) systems are used to exploit heat from the earth's interior. Concerning the latter system, two main geothermal play types located at shallow depth (< 3 km) can be utilized. Geothermal fluids or steam above 150-200 °C that convects in a hydrothermal reservoir, can be exploited in a closed circle: reinjection of spent mineralized fluids prevents scaling of well casings by oxidation and the escape of toxic ingredients (B, F, As, Hg, H_2S) into the environment. If natural hot aquifers are absent, a second method may be employed by creating artificial geothermal fluids (the hot dry rock method). This consists of an injection well, through which cold water is pushed into the artificial reservoir, and an extraction borehole, used to recover hot fluids or steam having formed inside stimulated fractures. The entire system, complete with surface utilization plant, forms a closed loop. Industrial-scale use of geothermal heat started in 1892 in Boise (Idaho, USA), 1910 in Larderello (Tuscany, Italy), and 1928 in Iceland. Installed global geothermal capacity in January 2016 came up to 13.3 GW, less than 6.7% of the estimated global hydrothermal potential of > 200 GW available, according to the current knowledge and technology. Top global performers in 2015 were the USA (3.57 GW), Indonesia (3.31 GW) and Mexico (1.07 GW) [23, 24].

Iron and ferro-alloy metals:

Iron: Ore deposits formed during the Precambrian at first in relatively deep marine, sedimentary exhalative, anoxic milieu (Algoma type: 3.8-2.5 Ga ago) and later in anoxic shelf areas (Superior type: 2.5-1.8 Ga ago), both are called Banded Iron Formations, and yield the largest known amounts of ore deposits and contain high-grade ore > 60%. Liquid magmatic iron ores (*e.g.* vanadium containing titanomagnetite) originated in ultrabasic and basic layered intrusions like the anorthosites of the Bushveld Complex and in the carbonatites of the peninsula Kola (Russia). Magnetite and magnetite hematite apatite ores formed in contact-metasomatic and volcanogenic oxide deposits, like Kiruna-Gällivara (Sweden). Siderite ore deposits, like those in Erzberg (Austria), all formed by metasomatism in limestones affected by hydrothermal fluids. Sedimentary iron ores occur in lateritic deposits (ferricretes), in oolithic deposits (minette-type), and in recent littoral placers [25]. Sulphidic iron ores formed in *e.g.* volcanogenic massive sulphide deposits like the Iberian Pyrite Belt [26] and polymetallic hydrothermal veins or stockworks.

Iron has been utilized by humans since ca. 3400 years ago. Industrial production of iron started in 1790 in Great Britain and yielded ca. 0.07 MT [5]. The 2014 global output of industrial iron amounted to 1.55 GT [12]. 2014 top iron ore producers were China with 31.2% (banded iron formations in provinces Hebei and Shandong, skarns in Anhui province and oolithes in Hubei province), followed by Australia with 27.25% (Hamersley banded iron formations in the

Pilbara area) and Brazil with 12.9% (banded iron ores near Itabira and Carajas Mine Complex) [27, 28]. It is noteworthy that, in 2004, iron stocks in use from domestic iron mining in the USA amounted to 3.2 GT, which is ca. 50% more than the economically recoverable domestic iron reserve (2.1 GT). In use iron stocks per person in the USA increased fourfold from 2.7 T/capita in 1900 to 11 T/capita in 2004. Global anthropogenic iron stocks were estimated to come up to 1/3 of the world global reserves (79 GT) [29].

Chromium: Chromite, the only profitable mineral species containing the chemical element chromium, originated as cumulates by fractional crystallisation in ultrabasic magmatic melts and occurs in the following types of liquid magmatic ore deposits:

- Stratiform seams, contained in anorthosites and peridotites of layered magmatic intrusions of Proterozoic age, like those of Bushveld (South Africa), Stillwater (Montana) and the Great Dyke (Zimbabwe);
- Lens- and pocket-shaped (podiform), irregular occurrences in dunite-harzburgite sequences of Phanerozoic ophiolites, like those at Guleman (Turkey) and Kempirsai (Kazakhstan);
- Detrital chromite, exploitable in littoral or fluvial placer deposits at western Indian coastal areas [25, 30].

Chromite mining started 1811 in Baltimore (USA) [31]. In 2014, global production amounted up to 13.2 MT of Cr_2O_3 [12]. 2014 top chromite ore producers were South Africa (46.7%), Kazakhstan (16.2%) and Turkey (13.7%).

Chromium is applied in industry as an alloy material to produce ferrochrome, refractory and stainless steel, and armour plating of metals and cermets. The latter are heavy-duty materials made up of chrome alloyed with other metals and ceramics, are temperature- and pressure-resistant, and can be applied in jet-engines and as a coating material for turbine blades. Although toxic, chrome compounds have many applications: Chromium(III)-sulphate was used in leather tanning. Potassium dichromate has been applied as tanning mordant in dye-works. Chromated copper arsenate was used as a wood preservative in timber treatment. Chrome yellow was used for calico printing. Diverse chrome compounds have been used as pigments in dyes, glasses, glazes, and paints [31].

Nickel: Ni-limonite, garnierite, pentlandite, millerite, and awaruwite form the most profitable nickel ores, which occur in the following types of deposits:

- Residual deposits that formed by lateritisation of serpentinised ultramafic rocks, in which chemically precipitated, Ni-containing hydroxides and hydrous silicates (Ni-limonite and garnierite) are contained. This type of deposit accounts

for ca. 60% of the land-based global resources.
- Sulphide-rich deposits of the liquid magmatic suite, in which *e.g.* pentlandite and millerite precipitated *via* fractional crystallisation from sulphide-supersaturated mafic to ultramafic magmas (picritic-tholeiitic dykes and sills) of large igneous provinces, and which have assimilated external sulphur as vapour or fluid phase from intruded host rocks. Examples are Noril'sk-Talnakh (Russia), Voisey's Bay (Canada), and Katanga (Tanzania).
- Sulphide-poor, reef-type deposits of the liquid magmatic suite. These are layered intrusions in which nickel-sulphides associated with copper, gold and PGE are precipitated at the basal contact of the ultrabasites. Examples are the Platreef in South Africa and the Duluth intrusion in the USA.
- Liquid magmatic sulphidic Ni ores, locally associated with Cu - PGE mineralisations, precipitated in komatiitic lava flows and sills of Archean greenstone belts present at *e.g.* Norseman-Wiluna (western Australia), the Great Dyke (Zimbabwe), and Cape Smith Belt (Canada).
- Liquid magmatic Ni ores, which occur in mafic-ultramafic rocks of ophiolithes. The serpentinised peridotites contain either Ni-Cu-PGE sulphides or Awaruwite, a naturally occurring alloy (*e.g.* British Columbia).
- Liquid magmatic sulphidic Ni-ores, co-precipitated with Cu, Fe, Au, and PGE ores in the basal parts of a ultramafic to noritic igneous complex at Sudbury (Canada), which is interpreted as early Proterozoic astrobleme.
- Alluvial placer deposits containing awaruwite (*e.g.* New Zealand).
- Oceanic sea floor, on which Ni- and cobalt-bearing manganese crusts and nodules have formed [25, 32, 33].

Industrial nickel production started in the first half of the 19th century in the Ore Mountains (Germany), where it was mined as byproduct of cobalt ores. Then new sites of production emerged in Norway and New Caledonia [34].

Global nickel production amounted to 2,119,875 T in 2014 ; top producers were the Philippines (19.4%), Australia (11.5%), and Canada 11.0%) [12]. Global 2012 reserves of Nickel were estimated to be 75 MT [35]. Nickel is used in industrial production of austenitic stainless steel alloys (65%), which are characterised by hardness, ductility, and corrosion- and oxidation-resistance; other steels, and non-ferrous alloys (20%). Nickel was applied in electronics and in the Jungner Ni-C--accumulator, a further development of the Edison Fe-Ni accumulator (6%). Ni was used in the chemical industry as catalyst to transfer carbon monoxide in ethane-chemistry (carbonylation). Ni compounds are applied in galvanic nickel plating, a coating to resist corrosion (9%). Nickel has been utilised as an alloy component in coins since the 19th century [34, 36].

Non-ferrous metals:

Copper: Copper-bearing minerals of economic relevance include chalcopyrite, copper glance, digenite, covellite, cuprite, and malachite. They occur in the following types of deposits:

- Sulphide-rich deposits of the liquid magmatic ultramafic suite, as described above. Examples are Sudbury (Canada) and Noril'sk-Talnakh (Russia).
- Porphyry copper deposits (60% of global Cu resources) occurring in granite, granodiorite, and diorite intrusions at active margins. These deposits contain on average 0.6-0.7% Cu. The Cu-sulphide ores, often in paragenesis with Sn, Mo, Ag, Au, and U mineralisations, precipitated in fissures, veins, stockworks, and hydrothermally brecciated zones. Examples are Chuquicamata (Chile) and Bingham (Utah, USA).
- Ophiolites, in which copper mineralisations are bound to hydrothermally altered pillow basalts. The copper sulphides, which occur in association with pyrite and sphalerite, are interpreted as exhalites: fossil black smoker deposits. The average Cu grade varies between 0.3-0.45%. Examples are Troodos in Cyprus[2] and Ergani Maden in Turkey.
- Volcano-sedimentary active margin deposits containing stratiform ores made up of chalcopyrite, pyrite, cassiterite, galenite, sphalerite, trace metals (Bi, Co), *etc.* They are interlayered within black schists and are associated with andesitic tuff. Examples are Rio Tinto (Spain) and Neves Corvo (Portugal).
- Magmatic-hydrothermal Cu-Au-Fe oxide deposits, represented by the Olympic Dam deposit. They are made up of hematite-granite breccias containing Cu, U, Au, Ag, and REE mineralisations.
- Carbonatite pipe mantled with phoscorite, which consists of the mineral phases F-apatite, magnetite, calcite, phlogopite, serpentine, and copper ores (on average 0.48-0.57%), as well as important byproducts like Zr, Hf, U, Ag, Au, and Pt mineralisations. A unique example is Phalaborwa (South Africa).
- Cupriferous sedimentary-exhalative deposits, which originated at passive margins in post-rift euxinic basins. These are stratabound within black shales or carbonaceous silt- and shalestones, into which metalliferous fluids and hydrotherms are driven by convective cells from older formations or from deeper metamorphic or magmatic sources, and enter *via* conduits at extensional faults. The precipitated ore minerals are zoned associations of copper sulphides, arsenopyrite, pyrrhotite, pyrite, sphalerite, galenite, argentite, gold, *etc.* Examples are Rammelsberg (Harz Mountains, Germany) and Sullivan (British Columbia, Canada).
- Retrograde metamorphic copper ore deposits (grade 3-4%), situated at the tectonic boundary between metabasalts and silificied, hydrothermally altered, marly metadolostones. Chalcopyrite, pyrite, pyrrhotite, and traces of As, Zn, Pb,

and Co ores form in paragenesis in zones of lodes, massive lenses, and irregular replacement bodies. The ores formed by forced brine convection in the footwall of a major reverse fault. A unique example is Mount Isa (Queensland, Australia).

• Sediment-hosted, stratabound copper ores (ca. 30% of global reserves), which originated in early post-rift settings. The deposits form epigenetically by transportation of metals in formation brines or deeper magmatic fluids and subsequent precipitation of copper ores and associated Co, Ni, Pt, Au, Pb, and Zn mineralisations at redox-boundaries. Examples are White Pine (Michigan, USA) and the Copper Belt of Katanga (Kongo, Zambia) [25].

Native copper has been utilised since 10,000 a ago - longer than any other metal. Copper minerals were mined and smelted since 8000 a ago to form weapons and tools. At the beginning of industrialisation, the main producers of copper were in England and Wales (Cornwall, Anglesey, Staffordshire). In 1780, the total output quantity of copper there amounted to ca. 5000 T [37]. Global copper production amounted to 18,435,342 T in 2014; top producers were Chile (31.2%), China (8.9%) and Peru (7.5%) [12]. In 2013 an estimation of identified (2.1 GT) and undiscovered (3.5 GT) global resources came up to 5.6 GT [38]. In addition, deep-sea nodule fields contain unknown quantities of copper sulphide ores.

Elementary copper is used in many ways because of its tensile strength, ductility, corrosion resistance, and low thermal expansion as tubes and sheets in the building industry and machine factories and - due to its very low electric resistivity - as wire and cable conductors in electrical and electronic industries. Its importance will rise due to the shift to physical mobility driven by electrical motors. Because of its excellent heat conductivity, it is used for heating and cooling coils, as well as cooking and brewing devices. Copper is used to fabricate alloys with zinc (brass), tin (bronze), aluminium, and nickel (*e.g.* constantan, Monel), which are applied in machineries, mints, in foundries, made into bells and statues, and in copperplate printing. Copper and copper alloys are applied as anti-biofoulings and antimicrobials in aquacultures and healthcare facilities. Cupric sulphate is applied in galvanic copper plating, as a constituent in galvanic elements (accumulators), as a component in colorants, as a fungicide in agriculture, and as an algaecide. Copper chloride is applied in carbon monoxide gas analytics. Some other copper compounds are used in pyrotechnics, wood preservation, ceramic glazes, stain glass work, dyes, and colorants.

Nano-scale technological applications were realised in nano copper interconnections on silicon chips for computers to achieve speeds in the gigahertz range, in metal nanofluids to increase thermal conductivity, in highly porous Cs-selective sorbents - consisting of a nano-sized copper ferrocyanide complex on the surface of mesoporous silicate ceramics - suitable to treat radioactively contaminated matter, and in porous metal-organic nanostructures capable of adsorbing and separating larger quantities of gases [39].

Aluminium: Ore minerals of economic relevance are gibbsite, boehmite, diaspore, and alunite. These minerals are constituents in predominantly autochthonous residual deposits called silicate bauxites[3], which originated from aluminosilicate source rocks (*e.g.* basalt, phyllites) by allitic weathering and the formation of soils - enriched in Al (lateritisation) - in the humid, tropical climate zone. Aluminosilicates decay to kaolinite and subsequent hydrolysis generates colloidal Al hydroxides and colloidal silicic acid, which is drained off. Al hydrates, precipitated in the enrichment zone, form nodules and pisolites consisting predominantly of the ore minerals gibbsite and boehmite. Lateritic bauxites containing > 35% Al_2O_3 are recoverable. Important laterite-bauxite deposits are present in Australia (Weita), in Guinea (*e.g.* Boké), and Brasilia (Amazonas Basin). Carbonate (or karst) bauxites originate *via* fluvial or aeolian import of suspended particulate aluminosilicate matter onto emersion unconformities, major omissions, or erosive surfaces marked by karstified zones. Transport occurred from distal elevated source areas exposed to allitic weathering to karst areas subsided to sea level [25]. Karst bauxite deposits of major economic significance, yielding the ore minerals boehmite and diaspore, are present in Viet Nam (Tây Nguyên), Jamaica (*e.g.* St. Ann, Russell Place), and Ural (Severouralsk).

Industrial bauxite mining started in the second half of the 19th century, when the most complicated and very energy-intensive refining, fused salt electrolytic and purification techniques (Bayer and Hall-Héroult processes, *etc.*) were established. In 1890, global extraction of bauxite amounted to 22,000 T, equivalent to ca. 4400 T of aluminium. Production of that light metal soon rose steeply because of its suitable physical properties (ductility and strength, low specific weight) and chemical (corrosion resistance by protective passivation) properties. At least since the beginning of World War I, aluminium was considered as strategic metal, important to win military and/or economic superiority [40]. Estimated global in-use aluminium stocks amounted to 635 MT in 2009, equivalent to 85 kg/person, with a maximum in the USA at 540 kg/person [41]. In 2014, bauxite extraction yielded ca. 261.9 MT; and about 53.5 MT of aluminium was manufactured. Top global producer countries were Australia (30%), China (24.8%) and Brazil (13.5%) [12].

Aluminium and its alloys are widely used in the automobile, aviation, spacecraft, electronic, and building industries. It is used in a wide range of household items. Powdered aluminium is a component in paints, anticorrosive varnishes, letterpress printing, flash-powder, blasting agents, and pyrotechnics. Al granulate is applied in the thermite and thermate processes for iron and copper welding. Mirror finishes made of Al films are characterised by high reflectivity. Al coatings on iron utensils and instruments protect against corrosion. Super pure Al is used in electronic devices, CDs, and wires. It has been used in research on the fabrication of novel superconducting quantum bits.

To meet the energy needs of 7 billion people in 2014, 540 EJ was produced predominantly from fossil energy carriers: 4.3 GT crude oil (32.6%), 8.17 GT coal (30.0%), 3,483 Tm^3 natural gas (23.7%), 65.91 kT uranium (4.4%), but also from green energy carriers like hydropower (6.8%) and renewables (2.5%) [42]. The corresponding resource flux in 2014 was not available, but it very probably reached an amount slightly higher than in 2008, which was ca. 111 GT of matter extracted [43], of which ca. 70 GT was economically processed [44]. International physical trade of ca. 10 GT of resources occurred in 2008 [45]. Since 1950, ca. 500 MT of aluminium, 6 GT of plastic, and 450 GT of concrete have been produced globally. In 2014, ca. 1 trillion bricks were fabricated globally. The N and P contents of agricultural soils have, on average, doubled since 1950, due to application of fertilisers to raise crop yield [46]. Global consumption of aluminium, copper, lead, nickel, tin, and zinc grew from 30 MT in 1975 to 98 MT in 2014 [47]. The global average material consumption per capita in 2008 was 10.2 Tonnes [48]. According to a preliminary estimation, the total mass of active and residual parts of global physical technosphere, necessary to house and to sustain 7.4 billion persons, is ca. 30.11 teraT [49].

NOTES

[1] All resources, except solar irradiation, extracted to meet basic demands of life, but also to propel economic growth and affluence, originated in huge terrestrial and marine basins and in the earth's deeper interior (crust, mantle, lithosphere) in the form of natural deposits, concentrates, and differentiates formed during biogeological, geological, mineralogical, metamorphic, and geophysical processes. Prospection, localisation, characterisation, and exploitation of these fossil resources has occurred since at least since 250 years ago on behalf of geoscientists and economic geologists. Growing awareness about the problems of resource limitations and negative ecological correlates of resource exploitation and utilisation in industrial and economic processes impose rising responsibility on the global players in these activity-fields. This responsibility, which has already been assigned to the industry extracting fossil energy carriers [50], must be expanded to the other parts of primary industry.

[2] Type locality, where the term "copper ore" (Latin: *aes cyprium*) was coined.

[3] Termed after the location Les Baux-de-Provence (South France), where it was first discovered by the geologist Pierre Berthier in the year 1821.

CONFLICT OF INTEREST

The author (editor) declares no conflict of interest, financial or otherwise.

ACKNOWLEDGEMENTS

Declare none.

REFERENCES

[1]   Stephens, C.; Ahern, M. *Worker and Community Health Impacts Related to Mining Operations Internationally - A Rapid Review of the Literature, IIED and WBCSD,* **2002**. http://pubs. iied.org/pdfs/G01051.pdf

[2]   Wellmer, F-W. Rohstoffe, die Basis unsere Wohlstandes. *Gmit,* **2012**, *47*, 6-21.

[3]   Gregory, C.E. *A Concise History of Mining*; Pergamon Press: New York, Oxford, **1980**.

[4]   Voppel, G. *Die Industrialisierung der Erde*; B. G. Teubner: Stuttgart, **1990**.

[5]   Griffin, E. *A Short History of the British Industrial Revolution*; Palgrave: UK, **2010**. [http://dx.doi.org/10.1007/978-1-137-26727-6]

[6]   Rostow, W.W. *How it All Began - Origins of the Modern Economy*; Routledge Revivals: UK, **2014**.

[7]   Coulson, M. The history of mining - The events, technology and people involved in the industry that forged the modern world.*20ᵗʰ Century U.S. Mineral Prices Decline in Constant Dollars,* **2012**,

[8]   U.S. Geological Survey, Open File Report. U.S. Department of the Interior, **2000**, p. 00-389. http://pubs.usgs.gov/of/2000/of00-389/of00-389.pdf

[9]   Matos, G.R. *Use of raw materials in the United States from 1900 through 2010,* **2012**. http://pubs.usgs.gov/fs/2012/3140/pdf/fs2012-3140.pdf

[10]   Rolle, C. *Energie und Rohstoffe - Gestaltung unserer nachhaltigen Zukunft*; Kausch, P.; Bertau, M.; Gutzmer, J., Eds.; Springer Spektrum Akademischer Verlag: Heidelberg, **2011**, pp. 14-57.

[11]   Moody, R. *Rocks and Hard Places - The Globalization of Mining*; Zed Books Ltd.: London, **2007**.

[12]   Reichl, C.; Schatz, M.; Zsak, G. *Minerals Production,* **2016**. http://www.wmc.org.pl/ sites/default/files/WMD2016.pdf

[13]   World Coal Association. **2016**. http://www.worldcoal.org/

[14]   Mineral Councils of Australia. **2012**. http://www.communityovermining.org/MCA_VIC_-_ pre-budget2012-13-130312.pdf

[15]   Frenzel, M.; Ketris, M.P.; Gutzmer, J. On the geological availability of germanium. *Miner. Depos.,* **2014**, *49*(4), 471-486. [http://dx.doi.org/10.1007/s00126-013-0506-z]

[16]   *Hydrocarbon technology. The world's biggest natural gas reserves,* **2013**. http://www.hydrocarbons-technology.com/ features/feature-the-worlds-biggest-natural-gas-reserves/

[17]   *Hydrocarbon technology. Countries with the biggest oil reserves,* **2016**. http://www.hydrocarbons-technology.com/ features/feature-countries-with-the-biggest-oil-reserves/

[18]   Cuney, M. The extreme diversity of uranium deposits. *Miner. Depos.,* **2009**, *44*, 3. [http://dx.doi.org/10.1007/s00126-008-0223-1]

[19]   IAEA TECDOC-1629. *World Distribution of Uranium Deposits (UDEPO) with Uranium Deposit Classification, Vienna: Austria,* **2009**. http://www-pub.iaea.org/MTCD/publications/PDF/te_1629_web.pdf

[20]   World Nuclear Association. *Geology of Uranium Deposits,* **2015**. http://www.world-nuclear.org/information-library/nuclear-fuel-cycle/uranium-resources/geology-of-uranium-deposits.aspx

[21]   Bhattacharyya, A.; Campbell, K.M.; Kelly, S.D.; Roebbert, Y.; Weyer, S.; Bernier-Latmani, R.; Borch, T. Biogenic non-crystalline U$^{(IV)}$) revealed as major component in uranium ore deposits. *Nat. Commun.,* **2017**, *8*, 15538. [http://dx.doi.org/10.1038/ncomms15538] [PMID: 28569759]

[22]   Geipel, R.; von Philipsborn, H. *Natürliche Radionuklide in Gebrauchsgegenständen am Beispiel Urangläser und Uranglasuren*; Strahlenschutz-Praxis, **2001**, p. 1.

[23]   Dickson, M.H.; Fanelli, M. *What is Geothermal Energy?,* **2004**.https://www.geothermal-energy.org/

[24]   Matek, B. *Annual U.S. & Global Geothermal Power Production Report,* **2016**. http://geo-energy.org/reports/2016/2016%20Annual%20US%20Global%20Geothermal%20Power%20Production.pdf

[25]   Pohl, W.L. *W. u. W. E. Petrascheck's Lagerstättenlehre - Mineralische und Energie-Rohstoffe. Eine Einführung zur Entstehung und nachhaltigen Nutzung von Lagerstätten*; Schweizerbart'sche Verlagsbuchhandlung: Stuttgart, **2005**.

[26]   Tornos, F. Environment of formation and styles of volcanogenic massive sulfides: The Iberian Pyrite Belt. *Ore Geol. Rev.,* **2006**, *28*(3), 259-307. [http://dx.doi.org/10.1016/j.oregeorev.2004.12.005]

[27]   Xu, A. *Iron Mineral Deposits and Projects in People's Republic of China, srk consulting,* **2016**. http://www.srk.com/en/newsletter/focus-iron-ore/iron-mineral-deposits-and-projects-peoples-republic-china

[28]   Australian Gov. *Iron ore,* **2016**. http://www.ga.gov.au/scientific-topics/minerals/mineral-resources/iron-ore

[29]   Müller, D.B.; Wang, T.; Duval, B.; Graedel, T.E. Exploring the engine of anthropogenic iron cycles. *Proc. Natl. Acad. Sci. USA,* **2006**, *103*(44), 16111-16116. [http://dx.doi.org/10.1073/pnas.0603375103] [PMID: 17053079]

[30]   Gujar, A.R.; Ambre, N.V.; Mislanka, P.G. Ilmenite, Magnetite and Chromite Beach Placers from South Maharashtra, Central West Coast of India. *Resour. Geol.,* **2010**, *60*(1), 71-86. [http://dx.doi.org/10.1111/j.1751-3928.2010.00115.x]

[31]   International Chromium Development Association. *Chrome's colorful history,* **2011**. http://www.icdacr.com/

[32]   Schulz, K.J.; Chandler, V.W.; Nicholson, S.W. *Magmatic sulfide-rich nickel-copper deposits related to picrite and (or) tholeiitic basalt dike-sill complexes - A preliminary deposit model. U.S. Geological Survey Open-File Report 2010–1179,* **2010**.http://pubs.usgs.gov/of/2010/1179/

[33]   Naldrett, A. *Magmatic Sulfide Deposits. Geology, Geochemistry and Exploration*; Springer: New York, **2004**. [http://dx.doi.org/10.1007/978-3-662-08444-1]

[34]   McNeil, I. *An Encyclopedia of the History of Technology, Routledge Companion Encyclopedias*; McNeill, I., Ed.; Taylor & Francis: London, **1990**, pp. 96-100. [http://dx.doi.org/10.4324/9780203192115]

[35]   Kuck, P.H. *Mineral Commodity Summaries,* **2013**. http://minerals.usgs.gov/minerals/pubs/commodity/nickel/mcs-2013-nicke.pdf

[36]   Nickel Institute. *Nickel in British Columbia,* **2015**. http://www2.gov.bc.ca/ assets/gov/farming-natura--resources-and-industry/mineral-exploration-mining/documents/mineral-development-office/

nickel_september_2015.pdf

[37]   Symons, J.C. *The Mining and Smelting of Copper in England and Wales, 1760-1820. Masters thesis, Coventry University in collaboration with University College Worcester*, **2003**. http://eprints.worc.ac.uk/293/

[38]   Johnson, K.M.; Hammarstrom, J.M.; Zientek, M.L. *Estimate of undiscovered copper resources of the world, 2013.*, **2014**.https://pubs.usgs.gov/fs/2014/3004/pdf/fs2014-3004.pdf [http://dx.doi.org/10.3133/fs20143004]

[39]   Dresher, W.H. *Copper and Nanotechnology, Copper Applications in Innovative Technology Area*, **2006**. https://www.copper.org/publications/newsletters/innovations/2006/01/copper_nanotechnology.html#17

[40]   Gendron, R.S.; Ingulstad, M.; Storli, E. *Aluminium Ore. The Political Economy of the Global Bauxite Industry*, **2013**. http://www.ubcpress.ca/search/title_book.asp?BookID=299174015

[41]   Liu, G.; Bangs, C.E.; Müller, D.B. Stock dynamics and emission pathways of the global aluminium cycle. *Nat. Clim. Chang.*, **2013**, *3*, 338-342. [http://dx.doi.org/10.1038/nclimate1698]

[42]   BGR. *Energy Study*, **2015**. http://www.bgr.bund.de/EN/Themen/Energie/Downloads/energiestudie_2015_en.pdf?__blob=publicationFile&v=3

[43]   Schmidt-Bleek, F. *Grüne Lügen, 4. Aufl., Ludwig*; Random House: München, **2014**.

[44]   Lutter, S.; Giljum, S.; Lieber, M. *Global Material Flow database - Material extraction data, Vienna University of Economics and Business, Technical Report, Version 2014.1*, **2014**. http://www.materialflows.net/fileadmin/docs/materialflows.net/SERI_WU_MFA_technical_report_final_20140317.pdf

[45]   SERI Sustainable Europe Research Institut. *Distributing materials through global physical trade*, **2015**. http://www.materialflows.net/fileadmin/docs/materialflows.net/factsheets/matflow_FS7_2015.pdf

[46]   Zalasiewicz, J.; Waters, C. *The Anthropocene. Environmental Science, Oxford Research Encyclopedias*, **2015**.

[47]   *Worldbank. Commodity markets outlook. A World Bank Quarterly Report, Washington DC*, **2016**. http://pubdocs.worldbank.org/en/991211453766993714/CMO-Jan-2016-Full-Report.pdf

[48]   Dittrich, M.; Giljum, S.; Lutter, S. *Green economies around the world? Implications of resource use for development and the environment*, **2012**. http://seri.at/wp-content/uploads/2012/06/green_economies_around_the_world.pdf

[49]   Zalasiewicz, J.; Williams, M.; Waters, C.N. *Scale and diversity of the physical technosphere: A geological perspective*; The Anthropocene Rev, **2016**, pp. 1-14.

[50]   Frumhoff, P.C.; Heede, R.; Oreskes, N. The climate responsibilities of industrial carbon producers. *Clim. Change*, **2015**, *132*(2), 157-171. [http://dx.doi.org/10.1007/s10584-015-1472-5]

CHAPTER 6

Transformation of Ecosystems in the Spheres of Earth[1]

Abstract: Anthropogenic forcings caused complex ecosystem transformations in the pedo-, cryo-, hydro-, atmo-, and biospheres and global adaptive pressure on biota.

Keywords: Adaptation, Anthropogenic forcings, Ecosystem transformations.

INTRODUCTION

The enormous diversity of human-made ecosystem transformations is detailed in the following subchapters. These alterations, which have occurred in very short time, consist in overuse of ecosystems beyond their capacities to compensate for the changes of their physical and chemical states. Therefore many biota suffer from adaptive pressure, resulting from machine based global intrusion into the spheres, displacement of flora and fauna, and utilisation of resources. Soil and land surface were degraded due to the effects of mining and quarrying; immissions and fallout; waste and tailings disposal; pollution and contamination; compaction, drainage, and overfertilisation by agriculture; land take and urbanisation; desertification; erosion; and salinisation. Glaciers, ice sheets, and sea ice shrank at unprecedented rates. Thawing of permafrost areas resulted in more frequent rockfall and landslides, as well as emissions of CH_4 and CO_2. Groundwater resources diminished due to incongruous abstraction. Groundwater quality deteriorated due to leaking urban and industrial point sources as well as infiltration waters - charged with reactive nitrogen - from agrarian land. Rivers, lakes, and inland seas were exposed to atmospheric immissions, warming, contamination, deoxygenation, eutrophication, ingress of xenospecies, desiccation, and unsuitable water engineering. Marine coasts degraded due to elevated human pressure, inappropriate coastal engineering, subsidence and erosion. Littoral waters were loaded with effluents from desalination plants and aquafarms, spills, pollutants and toxins. Oceans and seas were exposed to warming, huge meltwater fluxes, eutrophication, deoxygenation, acidification, as well as spills and pollution.

The atmosphere was loaded with aerosols and greenhouse gases, resulting in warming and alteration of convection and circulation currents. The number of species and abundance of flora and fauna in the wild decreased because of harvesting and hunting, land take, habitat degradation, warming, pollution, contamination, infections, and intrusion of xenospecies. The creation of wildlife reserves and the Red Lists of the IUCN effected shelter of life from homogenisation, defaunation and extinction. Additional greening occurred because of deglaciation and fertilisation of the atmosphere with CO_2 and reactive N. Human beings profited from progress made in science, technology, medicine, and dietetics, resulting in increased life expectancy, food security, mobility, intellectual development, and wealth. The advers effects of that development include concentration due to population growth, accidents, civilised ailments, pollution, noise, water scarcity, dependencies, impending depletion of resources, contamination, overkill, industrial disasters, rising health costs, compensations for climate change-induced warming and sea-level rise, as well as rising risks.

NOTES

[1] It is obvious that the transformations of ecosystems and life are innumerable. Therefore, only a few examples will be mentioned in the following. Many others can be read in the scientific journals mentioned in the references.

CONFLICT OF INTEREST

The author (editor) declares no conflict of interest, financial or otherwise.

ACKNOWLEDGEMENTS

Declare none.

CHAPTER 7

Soil and Land Surface

Abstract: The impacts imply direct and indirect, physical, chemical and biochemical alterations of land surface, soil and rock. The impairments include: pollution by atmospheric immissions; degradation; compaction; tillage of monocultures; overuse by abandonment of the three field system; erosion; more frequent wildfires; unconfined landfills; biocide application and excess fertilisation in agricultural activity; degradation by deforestation and land use change; soil sealing by land take; drainage of peat land and mires; impairment of land surfaces by open cast and underground mining, by underground constructions, as well as extractive methods (*in situ* leaching); numerous nuclear underground tests; creation of pollution hot spots (brownfields) by industry, military, and in urban areas; contamination due to large scale disposal of waste from settlements, industry, and agriculture in landfills; toxic waste spread from dam failures of tailings and other deposit confinements; climate change induced heat and drought; and altered soil carbon uptake caused by the introduction of xenospecies plants. Soil impairment by industry disasters (*e.g.* Bhopal, Chernobyl) and by tailings and waste heaps created by the mining industry is emphasised. Contaminations persist over long periods, because of the stability of soil, its sorptive properties and predomiant immobility of its components.

Keywords: Agriculture, Brownfields, Compaction, Desertification, Drainage, Erosion, Fertilisation, Immissions, Land take, Mining, Pollution, Quarrying, Salinisation, Sealing, Soil degradation, Soil impairment, Waste disposal, Warming.

Soil has deteriorated due to exposure to acid rain, toxic dust, soot, and aerosols [1]. The European Convention on Long-range Transboundary Air Pollution [2] - signed in 1979 to reduce emissions by industry and consumers - resulted by 2011 in the reduction of SO_2 by 80%, of NH_3 by 30% and of NO_2 by 39%, as well as in the recovery of terrestrial and aquatic ecosystems and an increase in their neutralisation capacity. The following long-term impacts on soil chemistry of an Entic Podsol - covered by a primeval coniferous forest in western Ukraine, typical of European forest soils - were monitored over two years, analysed and modelled: acidic deposition - peaking in the 1980s - led to mobilisation of toxic Al^{3+}, leaching of $(NO_3)^-$, and depletion of base cations (Ca^{2+}, Mg^{2+}, K^+, Na^+) in soil water. Soil nitrogen accumulation decreased.

Soil chemistry is expected to slowly recover and re-equilibrate within the next 40 years [3]. 17 a of monitoring (1994-2010) of similar soil types in Belgium revealed that soil acidification has decreased, but that critical levels, *e.g.* very low relations between base cations and aluminium, are still present [4].

Deforestation has caused soil erosion [5], landslides [6], and general mobilisation of nutrients in soil, which is detrimental to surface and groundwater quality. Logging decreases nutrient uptake, fixation, and retention in temperate and tropical forests, because the process of natural cycling of reactive N in assimilating biomasses is interrupted, *e.g.* N fixation by diazotrophs on leaves and by symbiotic bacteria in rhizomes is disabled. Moreover, decreased evapotranspiration increases the amount of water through-flow. As a result, the load of reactive N in the surface waters and groundwater increases. However, conversion of tropical forests to pastures has caused a decrease of fluvial export of dissolved inorganic N, probably because of higher biological demand for that nutrient by grass species [7].

The consequences of both - deforestation and immissions - are explained in the following example: Excess logging at mining city, Sudbury (Canada), in concert with acid rain and immissions containing nickel, copper, aluminium *etc.*, caused degradation and local extinction of vegetation, giving rise to enhanced erosion of the soil, which consisted only of thin glacial till [8].

Agricultural soil has degraded due to compaction by grazers and harvesters, salinisation - caused by inappropriate irrigation -, tillage of monocultures, application of fertilisers, manure, biocides, and overuse (abandonment of the three field system) [9].

The global erosive loss of agricultural soil is estimated to ca. 33 GT/a, implying an average global dislocation of 860 MT/a of soil organic carbon from agrarian land [10]. In the EU, the mean annual rate of soil loss is estimated to 970 MT, which exceeds the average soil gain rate by factor 1.6 [11]. Sensor- and GPS-based smart- and precision farming for site-specific fertilisation, as well as soil protective measures have improved situations locally [1]. *e.g.* precision phosphorus management, which considers phosphorus input, crop yield, accumulation of soil phosphorus, and export of phosphorus into waterways, and the implementation of vegetation filter strips [12].

It was found out that 1/3 of the agrarian soil in Africa stands on the threshold to desertification and 52% of the global agrarian land is intermediately to heavily adversely affected. The economic damage due to soil loss in connection with its dysfunctional regulatory, filtering, and absorption capacities is estimated to annually cost ca. 7.5 billion euros [13]. Demographic growth, political conflicts and changing socio-economic practices - conversion of agricultural land to cashew orchards and later replacement of them by fields for cereals and peanuts - has caused considerable land cover change and landscape fragmentation in West Africa, as observed by remote sensing data, generated between 1990 and 2015 [14].

Since 1950 rising global population pressure has fostered the transition from individual small scale agriculture to more productive and intense industrial farming in order to keep pace with the growing demand for crops and meat. This spurred and application of higher quantities of more effective pesticides, veterinary drugs, and fertilisers, resulting in the alteration of soil quality, as explained in the following:

In that phase of agrohistory, pest management[1] consisted in the application of industrially produced novel synthetic organic pesticides[2] like dichlorodiphenyltrichloroethan (DDT), parathion, captan, and dieldrin. These were highly effective and seemed, at the time, to be safe to use. In the following years, due to their non-specified effects and the growing awareness of their bioaccumulation-based toxicity to humans and animals, intensified chemical research led to the development of more effective, more selective, and more easily handled pesticides characterised by improved environmental compatibility, improved biodegradability, and diminished toxicity of the products themselves and their by-products. Genetically engineered crops may be made resistant to pests or broad spectrum pesticides including the most popular glyphosate[3]. However, since 1962[4], environmentalists, chemists, biologists, nutritionists, and animal welfare proponents have pointed to the toxic and bioaccumulative effects of many of these chemicals, so that some have been officially restricted in use or banned [15]. 2.4 MT of pesticides were used globally in 2007 [16]. Ca. 4.6 MT of pesticides were applied globally in 2011. Pesticide consumption in China rose from 0.76 MT in 1991 to 1.46 MT in 2005 [17]. To keep the degree of food safety constant and to raise yields in agricultural production, applications of fertilisers/pesticides increased from 28 MT/0.8 MT in 1991 to 57 MT/1.8 MT in 2011 helping to boost grain yield by 130 MT to 510 MT [18].

Some side-effects of these chemicals on soil ecosystems are reduced N fixation in symbionts of glyphosate-resistant soybeans [19]; negative impact of pesticides on growth and reproduction of earthworms, which contribute to structuring and increasing nutrient content of soils [20]; and glyphosate was found to inhibit growth, respiration and nitrogen fixation of free-living heterotrophic bacteria [21]. Field experiments point to the fact that large amounts of biochar (> 30 T/ha) must be applied to reduce the ecotoxic effects of pesticides (*e.g.* 2,4-D, dicamba) on soil enzyme activity [22]. Key functions of agricultural soils were tested using microbial mRNA quantification. It found that some pesticides influence nitrification rate and that they transform microbial community structures and activities [23].

In Europe, ca. 10.64 MT of synthetic fertilisers and 11.3 MT of manure are annually applied in agriculture [24]. Global soil fertiliser production/consumption amounted to 170.6 MT/161.8 MT in 2008 [25]. In 2014, world fertiliser consumption came up to 187 MT, composed of 113 MT ammonia, 42.7 MT P_2O_5 and 31 MT potash. The averaged annual increase is 1.8% [26].

Average per capita world meat supply increased from 27kg/person/a in 1969 to 41.2 kg/person/a in 2005. The latter bases on a world supply of ca. 285 MT [27]. The number of food animals in 2009 was 2.0 billion cattle, buffalo, sheep and goat; 941 million pigs and 18.6 billion fowl [28], equalling the global biomass of humans. 90% of the global mammal biomass consists of humans and their food animals. Global consumption of the antibiotics used to maintain productivity and to prevent outbreaks of diseases in livestock was estimated at a minimum of 63,151 ± 1560 T in 2010, ca. twice the amount applied to humans [29].

In 2008, the cumulative size of arable land/floor with permanent crops was 1.4 billion ha/146 Million ha globally; 61 million ha need irrigation [25]. Global production of maize, wheat, and rice grew from 1.4 GT in 1990 to 2.15 GT in 2015. Edible oils grew from 55 MT to 170 MT in the same time interval [30].

In 2008, the cumulative area of pastures was estimated to be 3.4 billion ha globally, equivalent to ca. double the size of the Russian Federation [25].

According to remote sensing multiyear time series data (9-25 a), an overall pattern of increasing agricultural field sizes was observed. This indicates a decrease in landscape complexity and diversity through homogenisation caused by agroindustrial land use [31].

Chemical tests of 600,000 soil samples in 75,000 km² key regions of China's agricultural regions revealed eutrophication caused by the application of fertilisers. Compared to data from the 1980s, more potassium, phosphorus, and nitrogen were present; 29.3% of the soils were impacted by alkalinisation (pH increased by 0.64) and 21.6% degraded by acidification (pH reduced by 0.85) [32], thus increasing soil erodibility. The total area subjected to erosion by wind and water was 3.6 million km² [33].

Microelement fertilisation using lanthanides turned out to have problematic bioaccumulative effects and to be detrimental to soil microfauna [34].

In addition of recent climate warming, land use change and increasing livestock since 1951[5] as well as fencing and privatisation of pastures since 1980 entailed several detrimental consequences for large parts of the 1.5 million km^2 grassland of the Tibetan Plateau, which stores ca. 18.1 GT of C_{org}. Increased grazing pressure in fenced areas has led to soil erosion. In ungrazed pastures, the sedge *Kobresia* sp. was displaced by xenospecies that sequester less root biomass thus decreasing C-input in the upper 15 cm of soil [35, 36]. The response to a three year warming and snow adding experiment revealed that soil moisture constrained delay of reproductive phenology and decrease of number of inflorescences of *Kobresia pygmaea* C. B. Clarke. This species is a shallow rooted, early-flowering and dominant, but vulnerable sedge present in the semiarid ecosystem of Central Tibet [37].

30-year monitoring of soil organic carbon in alpine forests below 1150m in temperate humid climates revealed that increases in summer temperatures (warming by 0.5°C per decade) effected intensified soil organic carbon release by microbial metabolisation, so that these topsoils were depleted in soil organic carbon by 14% and switched from carbon sinks to carbon sources. Soil fertility and water storage capacities were adversely affected by this development [38]. It is obvious that, globally, other alpine regions characterised by similar settings and developments are also impaired by depletion of soil organic matter. According to experimental studies conducted in North America, Europe, and Asia concerning temperature sensitivity of soil carbon, it is estimated that the global carbon loss in upper soil horizons caused by an average of 1°C warming is 30 GT carbon, equivalent to ca. 110 GT of CO_2 released into the atmosphere [39].

Holocene to recent palaeoclimatic reconstructions of Mediterranean palaeoecosystems, based on abundant palynological data from sediment cores, warn that warming ≥ 2 °C above preindustrial levels (1780) will result in the desertification caused by recurring droughts of large parts of that cradle of civilisations [40].

Soil, affected and altered by more frequent wildfires, is more susceptible to rainfall-induced debris flow events [41].

The soil erosion rate caused by anthropogenic activities (agriculture, mining, construction, *etc.*) began to exceed the natural rate of 16 m/Ma (5 GT/a) ca. 1000 years ago and reached ca. 600 m/Ma (75 GT/a) in 2006, reaching a peak of 12,600 m/Ma in higher-order tributary channels and floodplains [6]. At present, soil loss by erosion is some orders of magnitude greater than soil generating processes [42]. Three examples concerning increased erosion: - Investigation of temporal and spatial development of debris flow fans in the Alpine region over the last 60 years through mapping and applying historic data from aerial photos and maps for reconstruction revealed its higher activity since 1980, as compared to its late glacial average, and its correlation with climate change-induced intensified rain storm events [43]. - In Mediterranean foothills, changes in land use and diminution of vegetation cover due to urbanisation and intensified industrial agriculture caused an increase in soil erosion by 3 T/ha/a to 34.7 T/ha/a, between 1999 and 2000 [44]. - Industrial agricultural activity, artificial drainage, and higher precipitation rate resulted in a tenfold increase of sedimentation rate in Lake Pepin, constituent of the Minnesota River Basin, over the past 170 a [45].

At present, fens, peatlands, mires, and moors cover ca. 400 million ha globally and contain approximately 455 GT of organic carbon, which slowly accumulated (1 mm/a, ca. 1 T/ha) since 11,000 years ago predominantly in the northern hemisphere. Since 1800, land use change (drainage) and industrial extraction of peat (for balneotherapeutic and horticulture purposes, as well as for fuel) effected that only about 30-45 million ha (10-12%) of global resources exist in an undisturbed state [46]. 14-20% of peatlands were utilised in agriculture (1), ca. 3% were transformed to forests (2), and ca. 5000 km^2 were extracted (3). The drainages necessary for all these utilisations and subsequent processes caused - beside subsidence of terrain -

1. A decrease of CO_2 sink, an increase of CO_2 emissions due to turf oxidation, CH_4 emissions in the ditches and enhances N_2O emissions because of N mineralisation and subsequent fertiliser application;
2. An increase of CO_2 uptake because of tree growth, aerobic decay of peat, and increase of CO_2 emissions, CH_4 emissions in the ditches and higher N_2O emissions;
3. Degradation of a CO_2 sink and increase of CO_2 production because of peat oxidation, combustion, and decomposition, CH_4 emissions in the ditches [46].

For instance, peat was exploited in the Netherlands from the 17th century until 1976 for heating and commercial purposes, resulting in a completely transformed Dutch landscape[6] - almost entirely deprived of its original biodiversity. The original peatlands were replaced by polderlands in the west and industrial agricultural land in the east; only a few nature reserves remain [47].

Anthropogenic climate change has induced more frequent droughts in peatlands and caused, in the originally water saturated zone of peat horizons, oxidation of phenolic compounds, which inhibit microbial activity. Subsequent enhanced bioproductivity effected decomposition of organic matter and the emission of additional CO_2 [48]. Dried peatlands, as a consequence of more frequent droughts or of drainage measures, are more vulnerable to fire and smouldering. The latter process oxidizes deeply buried fossil carbon, which contributes to atmospheric CO_2 [49]. It is estimated that CO_2 emissions from dried peatlands amount to 3 GT annually [50]. Extensive anthropogenic degradation (deforestation, drainage, fire) of tropical forests like those in Indonesia and Brazil, occurring since ca. 1960 because of industrial economic purposes (timber, palm oil, cotton, tobacco, rice, tea, fruits, *etc.*), led to a ca. 50% increase of the total fluvial organic carbon flux and to the mobilisation of dissolved fossil organic carbon deeply buried within the peat column [51, 52]. Because current agricultural practices are unsustainable and devastating, a more responsible tropical peatland agricultural technique has been recommended by ca. 120 scientists in order to avoid further peat carbon loss [53].

An estimated 21 million ha of peatland in European Russia and Fennoscandia have been mined and drained, making them more susceptible to wildfire and deep smouldering. C loss of 35 kg/m^2 through combustion is equivalent to ca. 1000 years of C sequestration; draining of peatland switches their function as C sink to a C source of about 2 T of C/ha/a [54].

Because of their social and environmental functions, the importance of intelligently and sustainably managing the functionalities of peatlands and mires is highly recommended. These ecosystems serve as locations for recreation and inspiration; as faunal and floral biodiversity pools; as special habitats; as regulators of heat, moisture, and local climates; as carbon sinks and oxygen sources; as stores and filters of water; and compensators of floods [55]. The same applies to forests (see chapter 15).

Annual land take in Europe - to create artificial surfaces for residential areas, recreation sites, industrial, and commercial centres, transport networks, dump sites, mines and quarries *etc.* - amounted to 111,788 ha/year between 2000 and 2006 [56]. Cities have grown and spread at the expense of valuable agrarian land [1]. In addition, land used for waste disposal degraded: in the industrialised countries of the EU, the wastes produced in 1995 totalled 2 GT, necessitating many facilities and sites, where the matter is deposited for recycling, combustion, or temporary or permanent storage: This included 400 MT from mining and power plants, 160 MT from industry, 90 MT from households, 1 GT from agriculture, 1.9 MT waste oil, 160 MT of rubble and 230 T of sewage sludge [57] (Details concerning waste see below).

On 12/22/2008, the Kingston Fossil Plant Harriman (Tennessee, USA) released 5.4 million m^3 of coal fly ash slurry over 4.1 km runout and contaminated ambient soil [58].

Soil and groundwater below brownfields (abandoned industrial sites) are often contaminated by hydrocarbons, polychlorinated biphenyls (PCBs), heavy metals *etc.* [59]. For example, an assessment concerning the extent of chemical contamination 25 years after the Bhopal industrial disaster revealed that analysed soil samples contained significantly elevated values of di-, tri-, and hexachlorobenzene; hexachlorocyclohexane; carbamate; arsenium; mercury; lead; and chromium [60].

Extraction of rock by humans was estimated at ca. 41.6 teraT [61]. Assuming an extraction rate of 110 GT/a [62], the total amount of rock extracted until 2016 was 43.3 teraT.

The origination of many thousand of quarries in the 19th century, to supply industry with building stones, sand, and gravel, caused substantial landscape transformations in Germany and caused conflict of interest between protection of nature and historical heritage on the one side and economy and industrial open cast mining on the other [63]. In 2010, ca. 549 MT of building stones, shale, sand, gravel *etc.* were mined in open cast pits in Germany and 7.8 million m^3 of turf was used [64]. The effected landscape degradation due to global surface mining are severe and have created increasing legacies and conflicts of interest.

Mining and extractive activities have caused ground sagging and collapse sink, resulting in loss of human life, enormous economic damage (infrastructure, buildings), and landscape degradation. According to the topographical and geological situation, to the depth of mining activity and groundwater level, to the architecture and dimensions of the workings, created by removal of solids, or to the amount of extracted fluid or gaseous matter, sudden rock failure or slow subsidence - that can occur contemporaneously or long after mine closure - of the roof strata into the generated voids occur [65]. A few examples are given:

Ca. 30 long wall underground coal mines (1-2 km long and 200-300 m wide) in Queensland and New South Wales (Australia) have been constructed since 1960, replacing the "bord and pillar method" used since ca. 1890. On average, 2-4.5 m thick seams were mined at depths of 200-600 m. This activity caused surface deformations including subsidence of the topography at values between 1-5 m. The area affected aboveground is typically larger than the mined out region. Mining of thick seams at shallow depths caused temporary surface cracks of 5-20 cm, sometimes up to 60 cm width [66].

During industrial mining activity in the Ruhr district (Germany), ca. 9.56 GT first grade coal was produced between 1800 and 1990. Long wall mining occurred beneath an urbanised area of 5.3 million people, spread across 4400 km^2. A cumulative subsidence volume of ca. 8 km^3 was created in workings, which had a maximum depth of 920 m. From ca. 1850 onwards, effects included ground sagging and uplift, collapse structures, displaced infrastructure, reversals of natural fluvial systems and of sewer drainage, and origination of subsidence lakes and polderlands. Maximum of observed subsidence was 24 m [67] while average subsidence concerning an area of 2700 km^2 was ca. 1.6 m [68]. The urbanised polderlands were prevented from flooding by permanent pumping [69][7]. Between 1980-2006, mined out coal workings in the Ruhr area were partly filled with ca. 1.6 MT of industrial waste, consisting of *e.g.* cinders and filtered dust from energy plants and residuals from communal waste incinerators. Additionally, ca. 10,000 T of hydraulic oils containing PCB were applied during operations and only 5% of this amount was disposed of properly and 10% was recycled; the remainder was dispersed underground *in situ* due to leaking or because of safety measures [70, 71][8].

During industrial carnallite and sylvite mining in the Verkhnekamskoye District (Urals) from 1933 onwards, ca. 550 MT of sylvite were extracted. 84 million m^3 of excavations at 200-400 m depth were created until 2008. More than 285 rockbursts have been registered since 1968. Major mine collapses occurred. The first one was in 1986, when an impermeable stratum was broken and uncontrolled flooding began. The second water inflow formed in 2007 when the city of Berezniki, situated above the workings, was exposed to induced seismic events (magnitudes < 4.7), sinkhole formation, ground sagging and uplift, flooding, severe infrastructure damage, and the accumulation of considerable amounts of CH$_4$ in several voids in the mine [72].

115 years of industrial open cast lignite mining in Lusatia (Germany) resulted in the creation of 104 open cast pits - distributed over an area of 2500 km^2 - and the unintentional co-extraction of 100 MT of sulphur, of which ca. 66% is sulphide-sulphur of microscopic ore grains. Oxygenation and hydrolysis, enhanced by exposure of intensely fragmented rock to the atmosphere, caused the origination of ca. 10 MT of sulphate-sulphur, contributing, *via* discharge, to substantial acidification of groundwater, rivers, and of the largest open pit post-mining lake district of Europe [73, 74].

43 GT of crushed and/or milled barren rock from open cast and underground mining were produced and accumulated globally in 2008 [62]. Most of it was dumped on numerous spoil areas. Subaerially exposed, fragmented rock material treated in this way is susceptible to chemical alteration and sulphide oxidation. It generates acid seepage waters containing mobilised toxic metals, which infiltrate and contaminate soil and groundwater.

Industrial production of Zechstein-aged potassium salts in the Werra valley (Germany) commenced in 1901 for industrial application in fertilisers. Waste matter was also produced as brines (which were either injected into subsurface dolostone layers or fed *via* sewers into the fluvial system) and as solid matter, deposited as spoil heap, which at present is 200 m high, occupies an area of ca. 20 ha, and consists of 188 MT of overburden salts (22,000 T are accreted per day) [75, 76].

A global inventory of 18,401 mining sites revealed that between 1910 and 2009 tailings[9] dam failures - in sum 218 events - occurred at a rate of 1.2%, which is ca. two magnitudes higher than the failure rate of conventional water retention dams. The failure events were caused by unusually high meteoric precipitations, inappropriate management (poor dam construction, improper drainage and long-term surveillance), foundation subsidence, slope instability, overtopping, seepage, structural defects, *etc.* The post-2000 increase of tailings dam failures caused by excess rain is partly attributable to climate change and are concerned with constructions close to marine shores and in equatorial regions. The impact of failures, during which usually 1/5 of the contained tailing mass (up to several million m^3) is released, is long-term environmental damage (contamination of soil, surface and ground water, destruction of fauna and flora, *etc.*), infrastructure damage, deterioration of public health, and loss of life [77]. A few examples of such hazards include: January 1992 Padcal, Luzon, Philippines: collapse of dam wall, 80 MT of copper tailings; 01/30/2000 Baja Mare: Gold tailings dam crest failure: 100000 m^3 of cyanide-containing tailings; 11/05/2015 Germano Mine, Minas Gerais, Brazil: 32 million m^3 of iron tailings [78].

The more frequent occurrences of serious and very serious tailings dam failures (>100,000 m^3 release) since 1950 is seen as consequence of mining metrics, mining lower-grade ore, and falling prices of many metals, resulting in larger and higher tailings storage facilities and increasing risk potential by a factor of 20 every 35 years [58]. New guidelines containing stricter recommendations and improvement proposals have been laid down to diminish risks of tailings dam failures [79].

From 1942 to 2004, 2.2 MT of uranium (115,790 m^3) has been produced by industry globally. It is used for nuclear power production, for institutional research activities, and for construction of warheads, of which 70,000 have been fabricated in the time considered[10]. Corresponding waste originates in form of mining residues (classified as naturally occurring radioactive material), mill tailings (containing the chemical elements of the decay chains of uranium and thorium, metallogenetically associated heavy metals like copper, arsenic, molybdenum, vanadium, *etc.*, and hazardous extractive chemicals like sulphuric acid or cyanide), low and intermediate levels waste, spent fuel (from reactors, reprocessing facilities, decontamination, and decommissioning measures), and high level waste from reprocessing. Cumulative global radioactive waste that originated from these sources, is estimated at 7.3 million m^3 of low and intermediate level waste, 180,000 T of spent fuel (heavy metals of both uranium decay chains and unstable fission products), 830,000 m^3 of high level waste, 1.8 billion m^3 of mine waste and mill tailings; its cumulative activity is estimated at 2.81×10^{22} Bq. The waste has been dumped near surface facilities or stored in geological repositories [91].

The 2014 world supply of the ca. 400 nuclear reactors[11] produced 66,117 T of yellowcake (U_3O_8 concentrate). Beside reprocessing and recycling of used fuel, as well as the dismantling and diluting of military uranium since 1995, large amounts of uranium bearing mineral raw material was extracted by underground and open cast mining, as well as by *in situ* leaching. The latter method - which has been applied since ca. 1960 and is the most cost effective extraction method - recovers uranium by injecting hydrogen peroxide and solvents (an alkaline solution or sulphuric acid) into *e.g.* a permeable, ore bearing, sandstone aquifer below the wellfield, oxidising and mobilising uranium *via* chemical complexation and recovery in extraction wells in a closed circulation loop. Turnover per 1 kg recovered uranium is up to 40 kg acid and up to 33 kWh. However, after completion of extraction, the restoration of original aquifer conditions of the exploited formation was neither technically nor economically feasible [80, 81].

Production of 1 T of yellowcake implies - when applying the first two methods - mining of an average of 913 T of uranium bearing rock by blasting, mechanical crushing, milling to powder, leaching with sulphuric acid or an alkaline solution, and dumping of 912 T of tailings [82]. Annual fuel supply of a 1 GW nuclear power plant - yielding 8.76 TWh - requires 27 T of UO_2 enriched in ^{235}U to 4.5%, extracted from 230 T of yellowcake. Annual tailings production at the uranium milling site is ca. 210,000 T, equivalent to 105,000 m^3 of slurry containing above-mentioned chemicals. Annual waste production at the site of the 1 GW reactor amounts to 27 T, containing 23 T uranium (0.8% ^{235}U), 240 kg transuranics (mainly plutonium) and 1.1 T of nuclear fission and activation products [82].

The global amount of artificial plutonium rose from a few milligrams in 1941 (due to the discovery by G. Seaborg shortly prior to the onset of the Nuclear Age[12]) to ca. 1350 T in 1996 and has increased annually by 75 kg [83]. 2009 global production of weapon-grade Plutonium amounted to ca. 300 T [91].

At present, exploration has not yet resulted in the identification of optimal geological sites as repositories for final disposal of used nuclear fuel and waste from reprocessing[13] [82], although storage and retention of uranium and fission products over geological times is evident from Oklo natural nuclear reactors [84].

Between 1908-1964, exploitation of a salt diapir in shaft Asse II (Germany) yielded 4.9 million m³ of halite and carnallite at ca. 480-730 m depth. Parts of the workings at Asse II were reused after plant decommissioning as disposal sites of low- and intermediate-level radioacitve waste: storage of 125,878 drums (volumes: 100-400 L) of the first type occurred between 1967-1978 at a depth of 750 m; emplacement of 16,100 drums (volume: 200 L) of the second type was carried out between 1972-1977 at a depth of 511 m. Radioactivity was estimated at the time of storage to ca. 4,6 × 10³ TBq. In the following decades, problems arose concerning brines containing ^{137}Cs, leaking of some drums, emanation of gases (H₂, Rn) and incipient induced seismicity. Detailed rock mechanic analysis of the stability of the workings revealed that mining activity has resulted in deconsolidation (brittle and ductile deformation) of the pillars and failure of several impermeable salt barrier layers, the latter caused by the 1988 percolation of up to 12 m³/d halite brine. The identified instability of the workings and the unintentional fatal hydraulic connections to aquifers present in the overburden compels the retrieval of the disposed radioactive waste prior to the mechanical collapse of the workings, forecasted in above-mentioned analysis. Partial mechanical stabilisation of the workings was achieved by import of 2.15 MT of salt fillrock. Estimated costs amount to several billion euros [85, 86].

The above mentioned extraction and disposal methods cause degradation of ecosystems, because natural land, surface-, and groundwater was adversely affected and poisonous chemicals and concentrates are dissipated above and underground [87]. Dry stackings and highly alkaline and sodic tailings (toxic red mud) from aluminium refining amounted globally in 2007 to ca. 2.7 GT with an increase of about 120 MT/a [88]. Since 1945, 37 million m³ of radioactive waste (activity concentration: 273 billion Bq/m³), stored at Hanford Nuclear Reservation, Lawrence Livermore National Laboratory, Los Alamos National Laboratory, *etc.*, and 8.52 million m³ of radioactive mill tailings have been stored at interim dump sites set up at nuclear power and reprocessing plants in the USA. Contamination of the environment due to leaking into aquifers, the atmosphere and surface waters and consequent bioaccumulation in the food web have been observed [89]. Similar environmental deterioration and problematic brownfields occur - due to equivalent industrial operations - in the USA (*e.g.* Oak Ridge, Tennessee), Canada (*e.g.* Sudbury), Australia (*e.g.* Hamersley, Philbara), South Africa (*e.g.* Witwatersrand, Sishen), Kazakhstan, *etc.*

Ca. 1400 subterranean, submarine and cratering nuclear tests were carried out at several locations[14]. Poorly contained underground explosions resulted in radioactive contamination of rock, soil, and groundwater [90]. The above ground tests resulted in soil contamination due to fallout: Global deposition of radionuclides from above ground nuclear tests (1945-1963), conducted at Nevada Test site, Marshall Islands, Semipalatinsk, Novaya Zemlya, Maralinga, French

Polynesia, Lop Nor, *etc.*, is estimated at ca. 17.5×10^5 TBq, considering the radionuclides ^{90}Sr, ^{137}Cs, ^{14}C, and ^{239}Pu. Nuclear accidents caused widespread contamination with radionuclides, including the dispersal of 7.3×10^4 TBq at Kryshtyn (Chelyabinsk) in 1957, 1.1×10^7 TBq at Chernobyl in 1986 [91] and of 1×10^8 TBq at Fukushima-Daiichi in 2011 [92].

The nuclear accident of Chernobyl reactor nr. 4 on 04/20/1986 consisted in several explosions, conflagration of its graphite moderator, and release of ca. 4% $(7.6 \text{ T})^{15}$ of nuclear fuel (190.2 T) into the environment. Fallout contaminated wide areas surrounding Chernobyl, 146,110 km^2 in which the activity of ^{137}Cs in soil exceeds 37 kBq/m^2, including 3110 km^2 where contamination surpassed 1480 kBq/m^2 [93].

Dissipation and dispersion of other toxic metal species due to anthropogenic activities is given her concerning mercury: The preindustrial global reservoir of Hg in terrestrial ecosytems, including the uppermost soil, is calculated to be 28,800 T. The present day amount is 61,000 T, larger by a factor 2.12 [94]. According to the results of a global land-ocean-atmosphere model (including the transported species Hg^0, Hg^{2+}, and particulate Hg), anthropogenic Hg sources have risen from negligible preindustrial values to 3400 T/a in 2007. An example is given, how that contaminant has been dissipated by human activities: elevated concentrations of Hg and its biogeochemically transformed derivative methyl Hg were found *e.g.* in soil and soil gas within a radius of 24 km of the dump sites^{16} of abandoned Hg mines that were active between 1888 and 1973 in Big Bend National Park in southwest Texas, USA [95].

Sediment core geochemical data revealed that global deposition of Hg has, on average, risen by factor of three above the preindustrial value. The global annual inventory of Hg concentration in soil has increased from 43 ng/g preindustrially to 50 ng/g. The global Hg enrichment factor has at least doubled compared to preindustrial times. Enrichment has increased fivefold near industrial centres in the USA, Europe, South Africa, and India and at artisanal mining centres in central Africa. In the industrial centres of China, enrichment has increased tenfold [96]. Long-term application in technical instruments (barometers, mirrors, clinical thermometers, relais, blood pressure apparatuses, mercury vapour discharge lamps, accumulators, calomel electrodes, *etc.*), dentistry (amalgam fillings), hat fabrication, the chemical industry (*e.g.* chloralkali electrolysis, production of acetaldehyde), extractive industry (amalgamation with gold grains), analytical chemistry (Nessler's reagents), china painting, and as a blasting agent, disinfectant, and pesticide contributed to mentioned increased presence of that contaminant in the pedosphere [97].

Induced (or stimulated) seismicity and surface deformations were concomitant symptoms, which occurred

• After implementation of reservoirs (water, tailings);
• During mining, hydraulic fracturing, and extraction of geothermal heat and fluids, as well as during injection of sequestration of fluids or gases;
• During and after flooding of open cast mines and underground workings [98 - 102].

Monitoring of intensity and frequencies of earthquakes before and after impoundment of the Three Gorges Reservoir revealed that, compared to natural seismicity, the reservoir-induced ones have shallow sources, high intensities, quick attenuation and smaller affected areas; a remarkable increase in seismicity (magnitude < 2) was observed after impoundment [103].

Surface and subsurface deformation can also occur due to inappropriate water usage. Anthropogenically fostered decline of water volume in the Dead Sea, caused by water abstraction from its affluents, effected lowering of the groundwater level and subsidence of terrain, causing the intrusion of fresh groundwater - undersaturated with respect to halite - *via* buried tectonic faults, which cut impermeable clay and silt strata interlayered with rock salt. Voids, generated by subsequent subrosion (chemical dissolution of rock salt by undersaturated waters), caused the collapse of the overburden and formation of sinkholes at the land surface. This effect, which first appeared in the early 1980s, has intensified since the 2000s, so that hundreds of collapse sinkholes exist [104].

Other anthropogenic subterranean (and subsea) modifications - anthroturbations - consist in the creation of boreholes for extractive, injective, or scientific purposes. The cumulative length of global oil drillholes is estimated at ca. 50 million km [105]. The overall length of submarine telecommunication cables measured ca. 406,000 km in 1903 [106]. The cumulative lengths of subterranean conduits (for water, seepage, gas, steam, *etc.*), copper and glass fibre cables (for electric current and signals), tunnels (for auto- and locomotives), pipelines, and mine workings created globally since the onset of industrialisation, as well as the cumulative void generated in soil and solid rock (*e.g.* basements; bunkers built for protecting life, data, objects of value) have not been estimated up to now.

In industrialised countries between the early 20th century until ca. 1970, municipal and industrial waste[17] has been sent to abandoned quarries and landfills. These open and unlined dumps were not supplied with cover, drainage for seeping water, sequestration of harmful gases (CO_2, CH_4, H_2S, N_2O, CFC, *etc.*), or basal and lateral confinements, which now consist of technical or geological barriers. Therefore adjacent soil and sediment has not been protected from infiltration with leached contaminants [107]. In the meantime, many of these poorly organised historic repositories had to be worked, excavated, and remediated in Europe and the USA because of stricter environmental legislation. In 2016, post-consumer municipal solid waste amounted approximately to 1.3 GT, of which at least 510 MT were dumped or sent to landfills [108]. It depends on the quality of existing environmental regulations to what degree these dump sites act as point sources of contamination of soil, water, and air. According to the D-Waste Group, which cooperates with the University of Leeds and the International Solid Waste Association, the overall annual amount of municipal solid waste has been estimated at ca. 1.9 GT. 30% of that remains uncollected, because 3.5 billion individuals do not have access to most elementary waste management systems. 40% of the global waste is deposited even now at open dumpsites close to urban areas. The 50 biggest active open dumpsites - nearly all of them situated in poor countries in Africa, Asia, Latin America, and the Caribbean - make up an actually estimated total volume of 0.6-0.8 km^3, a total weight of 258-368 MT and cover a cumulative area of 2175 ha. The overall annual waste disposed of at these dumpsites is estimated at ca. 21.5 MT [109].

The cumulative mass of electric and electronic equipment put on the global markets in 2012 came up to 56.56 MT. The mass of e-waste discarded in the world's largest dump sites in Southeast Asia rose from ca. 7.5 MT in 2009 to 13.1 MT in 2016. This steep increase was caused by a rising number of products and consumers, decreasing duration of use per product, and rising quantities of imported new and second-hand products. Informal recycling imposes rising environmental and health problems in these countries [110].

The impairment of soil by military activities can be summarised as follows: Physical alterations of the pedosphere consist in sealing; excavation; tunnelling; compaction by traffic; and cratering by exploding bombs, shells and land mines; as well as crashing bombers. Chemical impacts consist in the scattering of liquid hydrocarbons (lubricating oil, diesel, kerosine, *etc.*); metal fragments made up of lead and depleted uranium; dynamite; TNT; napalm; absorption of diverse war gases like chlorine[18], mustard gas, organophosphorous nerve agents, *etc.*; dioxins from herbicides, *etc.* Biological deteriorations of soil occur due to the above-mentioned physical and chemical impacts, and because of the dispersal of toxic microorganisms like botulin, anthrax, *etc.* The legacies of these impacts endure in areas where past hostile confrontations and military trainings have taken place: *e.g.*: In 1979, treatment of radioactively contaminated material on Runit Island of the Enewetak Atoll resulted in removing and dumping of topsoil, vegetation, concrete, and metal pieces (with ca. 545 GBq radioactivity) into an unlined crater, produced by a 18 kT surface test, and sealing that waste[19] with a concrete dome measuring 107 m in diameter [111].

Often long-term contaminations and threats (blind shells, land mines) resulted due to the stability of soil and immobility of its components [112]. Ca. 72 million litres of dioxin[20]-containing herbicides were released in Vietnam between 1961-1971 as part of the strategic destruction of forest cover and food crops. According to conservative calculations, at least 221 kg of dioxin was dissipated [113]. A survey conducted between 1994-1996 revealed that, beside dioxin-contaminated hotspots at several military aircraft bases, soils of sprayed areas were found to have dioxin levels above background values [114]. At Bien Hoa Airbase and surrounding areas, analysed soil samples revealed levels 46 times above the standard background value for soil in 2003 [115].

It is obvious that industrial mass production and application of war materials (ammunition, repetition guns, tanks, bombers, war ships, submarines, fuel, explosives, *etc.*) greatly increased detrimental impacts on the pedosphere. 3168 bilateral military conflicts occurred between 1870 and 2001 [116]. The existence of ca. 110 million land mines, buried below the killing fields in 64 countries, was estimated in 1996 [117]. A further problem is blind bombs, when struck during construction or road works.

NOTES

[1] It includes the application of insecticides, herbicides, fungicides, and bactericides to protect the health of plants, animals, and human beings against pests. Until 1870, protection consisted in using natural chemicals (*e.g.* sulphur, pyrethrum); between 1870 and 1945, natural organic substances (*e.g.* naphthalene, petroleum oil) and inorganic synthetic pesticides (*e.g.* sodium chlorate, sulphuric acid, ammonium sulphate, sodium arsenate) were applied.

[2] Main constituents of synthetic organic pesticides are chemical derivatives of petroleum.

[3] Formulation of *N*-(phosphonomethyl)glycine originated in 1970; it is a component of the trade product "Roundup". Estimated use of it on US-agricultural land rose from 9000 T in 1992 to 136,000 T in 2013 [119].

[4] Biologist and environmentalist Rachel Carson's book "Silent Spring (1962)", in which she described the detrimental effects of *e.g.* DDT on plants, animals, and human beings can be seen as start of the global environmental movement.

[5] Onset of industrialisation and economic utilisation of natural resources in Tibet.

[6] Some of the original landscapes got preserved in public memory by several art painters of the Baroque period ca. 1600-1710), *e.g.* E. van de Velde, J. van Goyen, S., and J. van Ruysdael, H. Seghers.

[7] During and subsequent to the mine closure phase ending in 2018, expensive follow-up-costs were calculated. Shutdown: 1 billion euros, acid mine drainage measures: 5 billion euros, mining damage repairs: 3.5 billion euros, shaft stabilisations: 600 million euros, groundwater purification: 500 million euros. Eternal burdens: acid mine drainage (100 million m³/a) and draining polderlands by means of ca. 1100 pumpstations: minimum 100 million euros/a [120].

[8] Post-closure reuse options for these former mining sites and their workings are as follows: Utilisation of coal bed methane for generation of electricity and heat; use of medium depth geothermal heat from acid mine drainage; utilisation of spoil heaps and brownfields for wind turbine sites; utilisation of mine shafts for storage from hydro power stations; use of large relict industrial buildings as bases for photovoltaic units; utilisation of remediated brownfields for short rotation forestry, consisting of cultivation of fast-growing tree species like poplar, willow, *etc.* [121].

[9] Tailings facilities, probably the largest man-made physical structures on earth, consist of fine grained residues (slurry) from crushed rock, which contain finely dispersed ore. Ore separation from the crushed parent rock occurs by flotation, washing, and dissolution by cyanide or sulphuric acid. Tailings slurries can be discharged under water or may be discarded subaerially into impoundments. The embankment of tailings ponds, which can measure several km in length, consist either of a dam constructed across a valley containing the tailings upstream or of dams encircling the tailings [87].

[10] 2016 the total number of strategic nuclear warheads amounted to 15,395 and expensive nuclear modernisation programmes are planned. 4120 of these warheads were deployed for action: placed on missiles or located on bases with operational forces [122].

[11] Technological development and rising demand on energy carriers since 1954 caused the atomic industry to construct 667 nuclear power plants globally. In 2015, 400 reactors were operating in 31 nations and 66 were under construction. The first reactor, Obninsk, in the former Soviet Union was equipped in 1954 with a capacity of 5 MW. Since then, the total capacity rose and peaked in 2005 at 378

GW. A steep capacity increase occurred between 1970 and 1990, when most of the reactors were built. In 2015, world capacity amounted to 354 GW and contributed ca. 11% (ca. 2600 TWh) to global electricity demand. Except in Southeast Asia, the age of the bulk of these plants ranges between 20 and 45 years [123]. According to a database, 754 nuclear reactors have been constructed since 1951 in 41 countries. As of 01/01/2017, 55 nuclear reactors were under construction in 13 countries [124].

[12] Natural occurrence of actinide ^{239} plutonium: in pitchblende: Pu:U = 10^{-11}:1 and in Carnotite: Pu:U = 10^{-13}:1. It originates because of neutron absorption by ^{238}U, which transmutes to ^{239}Pu.

[13] Suited subterranean repositories must be characterised by long-term physical robustness; imperviousness for at least 1 million years; tectonic and seismic stability; presence of impervious and confining host rocks like shalestone, rock salt, or granite; stable geological and technical barriers preventing ingression of groundwaters, formation waters, or other fluids; safe management of waste heat and emanating gases (*e.g.* H_2, Rn); retrievability of nuclear waste; and stable above-ground plants, monitoring sites and service sites against natural and environmental disasters. Another concern of deep bedrock radioactive waste disposal is corrosion by microbes coming from the deep biosphere [125]. Most countries have adopted dry cask spent-fuel storage as an interim strategy since no repository has yet been licensed [126]. The aspects of these measures are problematic in that they imply ultra-long-term expenditures, duties, responsibilities and risks, imposed on numerous future generations far beyond the age of cheap electricity generated in atomic power plants.

[14] Most important sites where nuclear tests were conducted prior to 1963: Nevada Test Site, Novaya Zemlja, Semipalatinsk, Johnston Atoll, Marshall Islands, and Christmas Islands.

[15] Gaseous fissure products (Kr, Xe) evaded completely and highly volatile products like J and Cs evaded almost entirely. The matters released in the following days and their activities (in PBq) were as follows: ^{85}Kr (33), ^{133}Xe(6500), ^{132}Tl (1000), ^{131}J (1800), ^{134}Cs (47), ^{137}Cs (85), ^{90}Sr (10), and $^{239-240}$Pu (0.031). The volume of contaminated rubble and added material to absorb heat, to prevent further critical states of nuclear fuel, and to shield radiation, is estimated to be several 100,000 m^3. To minimise further release of radioactivity, the defective reactor was sealed with a provisional sarcophage within six months. 30

years later, the shelter implementation plan scheduled the construction of a second sarcophage to achieve a new safe confinement; estimated costs amounted to several billion USD [93]. A new safe confinement - 110 m high, 165 m long, and spanning 260 m - was put in place in November 2016 [118].

[16] Elevated concentrations of Hg were also detected there in the hydrological system and ambient air.

[17] Waste begins to accumulate when biocapacities have been overstepped and the ecoservices of nature begin to fail to absorb and assimilate post-consumer products.

[18] The first application of industrially produced chlorine as tactical warfare was realised on 04/22/1915 at Ypern, Belgium. Chemist Otto Haber coordinated industrial production of 150 T of chlorine, strategic setup of this gas weapon, and instruction of Reichsgerman military forces specialised in gas-battle.

[19] Fatal hydraulic communication between radioactive waste and groundwater, lagoon water, as well as sea water has occurred because of the low topographic level of the dump site and rising sea level.

[20] Dioxin - polychlorinated dibenzodioxins - are a group of polyhalogenated, oxygen-containing organic compounds. Bioaccumulation in the food chain occurs because they are lipophilic and hardly biodegradable. Dioxins act acutely toxic and genetically toxic; they are designated as persistent organic pollutants. Exposure to dioxins has occurred since the beginning of industrial chlorine production in the second part of the 18th century. In 2008, remediation and securing costs of 24 contaminated sites was estimated to minimum of 2.5 billion dollars [127].

CONFLICT OF INTEREST

The author (editor) declares no conflict of interest, financial or otherwise.

ACKNOWLEDGEMENTS

Declare none.

REFERENCES

[1]   Goudie, A.S. *The Human Impact on the Natural Environment. Past, Present and Future,* 7th ed; Wiley-

Blackwell: Oxford, **2013**.

[2]   Bull, K.R.; Achermann, B.; Bashkin, V. Coordinated Effects Monitoring and Modelling for Developing and Supporting International Air Pollution Control Agreements. *Water Air Soil Pollut.,* **2001**, *130*(1), 119-130.
[http://dx.doi.org/10.1023/A:1012255604267]

[3]   Hruška, J.; Oulehle, F.; Šamonil, P. Long-term forest soil acidification, nutrient leaching and vegetation development: Linking modelling and surveys of a primeval spruce forest in the Ukrainian Transcarpathian Mts. *Ecol. Modell.,* **2012**, *244*, 28-37.
[http://dx.doi.org/10.1016/j.ecolmodel.2012.06.025]

[4]   Verstraeten, A.; Neirynck, J.; Genouw, G. Impact of declining atmospheric deposition on forest soil solution chemistry in Flanders, Belgium. *Atmos. Environ.,* **2012**, *62*, 50-63.
[http://dx.doi.org/10.1016/j.atmosenv.2012.08.017]

[5]   Walsh, R.P.; Clarke, M.A.; Bidin, K. *Soil Erosion and Sediment Redistribution in River Catchments*; CAB International: Wallingford, **2006**.

[6]   Wilkinson, B.H.; McElroy, B.J. The impact of humans on continental erosion and sedimentation. *Geol. Soc. Am. Bull.,* **2007**, *19*(1-2), 140-156.
[http://dx.doi.org/10.1130/B25899.1]

[7]   Valiela, I.; Barth-Jensen, C.; Stone, T. Deforestation of watersheds of Panama: nutrient retention and export to streams. *Biogeochemistry,* **2013**, *115*(1), 299-315.
[http://dx.doi.org/10.1007/s10533-013-9836-2]

[8]   Courtin, G.M. The last 150 years: a history of environmental degradation in Sudbury. *Sci. Total Environ.,* **1994**, *148*(2-3), 99-102.
[http://dx.doi.org/10.1016/0048-9697(94)90388-3]

[9]   Smil, V. *Harvesting the Biosphere: How Much We Have Taken From Nature*; The MIT Press: Cambridge, MA, **2013**.

[10]   Van Oost, K.; Quine, T.A.; Govers, G.; De Gryze, S.; Six, J.; Harden, J.W.; Ritchie, J.C.; McCarty, G.W.; Heckrath, G.; Kosmas, C.; Giraldez, J.V.; da Silva, J.R.; Merckx, R. The impact of agricultural soil erosion on the global carbon cycle. *Science,* **2007**, *318*(5850), 626-629.
[http://dx.doi.org/10.1126/science.1145724] [PMID: 17962559]

[11]   Panagos, P.; Borrelli, P.; Poesen, J. The new assessment of soil loss by water erosion in Europe. *Environ. Sci. Policy,* **2015**, *54*, 438-447.
[http://dx.doi.org/10.1016/j.envsci.2015.08.012]

[12]   Iho, A.; Laukkanen, M. Precision phosphorus management and agricultural phosphorus loading. *Ecol. Econ.,* **2012**, *77*, 91-102.
[http://dx.doi.org/10.1016/j.ecolecon.2012.02.010]

[13]   Initiative of Land Degradation. *The value of land. Prosperous lands and positive rewards through sustainable land management,* **2015**.   http://eld-initiative.org/fileadmin/pdf/ELD-main-report_ 05_web_72dpi.pdf

[14]   Cabral, A.I.; Lagos Costa, F. Land cover changes and landscape pattern dynamics in Senegal and Guinea Bissau borderland. *Appl. Geogr.,* **2017**, *82*, 115-128.
[http://dx.doi.org/10.1016/j.apgeog.2017.03.010]

[15]   Unsworth, J. *History of Pesticide Use,* **2010**. http://agrochemicals.iupac.org/index.php?option= com_sobi2&sobi2Task=sobi2Details&catid=3&sobi2Id=31

[16]   Grube, A.; Donaldson, D.; Kiely, T. *Pesticides Industry Sales and Usage - 2006 and 2007 Market Estimates: U.S. Environmental Protection Agency. Washington DC,* **2011**. https://www.epa.gov/ sites/production/files/2015-10/documents/market_estimates2007.pdf

[17]   Zhang, W.J.; Jiang, F.B.; Ou, J.F. Global pesticide consumption and pollution: with China as a focus.

Proc. Int. Acad. Ecol. Environ. Sci., **2011**, *1*(2), 125-144.

[18] Lu, Y.; Song, S.; Wang, R.; Liu, Z.; Meng, J.; Sweetman, A.J.; Jenkins, A.; Ferrier, R.C.; Li, H.; Luo, W.; Wang, T. Impacts of soil and water pollution on food safety and health risks in China. *Environ. Int.,* **2015**, *77*, 5-15.
[http://dx.doi.org/10.1016/j.envint.2014.12.010] [PMID: 25603422]

[19] Zablotowicz, R.M.; Reddy, K.N. Impact of glyphosate on the Bradyrhizobium japonicum symbiosis with glyphosate-resistant transgenic soybean: a minireview. *J. Environ. Qual.,* **2004**, *33*(3), 825-831.
[http://dx.doi.org/10.2134/jeq2004.0825] [PMID: 15224916]

[20] Yasmin, S.; D'Souza, D. Effects of Pesticides on the growth and reproduction of earthworm: A review. *Appl. Environ. Soil Sci.,* **2010**, 678360.

[21] Santos, A.; Flores, M. Effects of glyphosate on nitrogen fixation of free-living heterotrophic bacteria. *Appl. Microbiol.,* **2008**, *20*(6), 349-352.
[http://dx.doi.org/10.1111/j.1472-765X.1995.tb01318.x]

[22] Oleszczuk, P.; Jośko, I.; Futa, B. Effect of pesticides on microorganisms, enzymatic activity and plant in biochar-amended soil. *Geoderma,* **2014**, *214-215*, 10-18.
[http://dx.doi.org/10.1016/j.geoderma.2013.10.010]

[23] Jacobsen, C.S.; Hjelmsø, M.H. Agricultural soils, pesticides and microbial diversity. *Curr. Opin. Biotechnol.,* **2014**, *27*, 15-20.
[http://dx.doi.org/10.1016/j.copbio.2013.09.003] [PMID: 24863892]

[24] *EU Science for Environment Policy In-depth report - Nitrogen Pollution and the European Environment - Implications for Air Quality Policy,* **2013**. http://ec.europa.eu/environment/integration/research/newsalert/pdf/IR6_en.pdf

[25] Statistics, F.A. *Resources, Land use,* **2010**. http://www.fao.org/fileadmin/templates/ess/ess_test_folder/Publications/yearbook_2010/a04.xls

[26] FAO, Food and Agriculture Organization of the United Nations, World fertilizer trends and outlook to 2018, Rome. **2015**. http://www.fao.org/3/a-i4324e.pdf

[27] FAO. *The state of food and agriculture, Livestock in the balance,* **2009**. http://www.fao.org/docrep/012/i0680e/i0680e.pdf

[28] Statistics, F.A. *Resources, Number of animals,* **2010**. http://www.fao.org/fileadmin/templates/ess/ess_test_folder/Publications/yearbook_2010/a09.xls

[29] Van Boeckel, T.P.; Brower, C.; Gilbert, M.; Grenfell, B.T.; Levin, S.A.; Robinson, T.P.; Teillant, A.; Laxminarayan, R. Global trends in antimicrobial use in food animals. *Proc. Natl. Acad. Sci. USA,* **2015**, *112*(18), 5649-5654.
[http://dx.doi.org/10.1073/pnas.1503141112] [PMID: 25792457]

[30] Worldbank. Commodity markets outlook. A World Bank Quarterly Report, Washington DC. **2016**. http://pubdocs.worldbank.org/en/991211453766993714/CMO-Jan-2016-Full-Report.pdf

[31] White, E.V.; Roy, D.P. A contemporary decennial examination of changing agricultural field sizes using Landsat time series data. *Geo,* **2015**, *2*(1), 33-54.
[http://dx.doi.org/10.1002/geo2.4] [PMID: 27669424]

[32] China Geological Survey, Geochemical Survey Report on Chinese Farmland 2015, Ministry of Land and Resources. **2015**. http://en.cgs.gov.cn/UploadFiles/2015_06/30/r-SuntGeochemical%20Survey%20on%20Chinese%20Farmland.pdf

[33] China Report 2009, Ministry of Environmental Protection of the People's Republic of China, Report on the State of the Environment of China, Water Environment. **2009**. http://english.mep.gov.cn/down_load/Documents/201104/P020110411532104009882.pdf

[34] Charalampides, G.; Vatalis, K.I. Global production estimation of rare earth elements and their environmental impacts on soils. *J. Geosci. Environ. Protect.,* **2015**, *3*, 66-73.

[http://dx.doi.org/10.4236/gep.2015.38007]

[35]   Qiu, J. Trouble in Tibet. *Nature,* **2016**, *529*(7585), 142-145.
[http://dx.doi.org/10.1038/529142a] [PMID: 26762440]

[36]   Hafner, S.; Unteregelsbacher, S.; Seeber, E. Effect of grazing on carbon stocks and assimilate partitioning in a Tibetan montane pasture revealed by $^{13}CO_2$ pulse labeling. *Glob. Change Biol.,* **2012**, *18*, 528-538.
[http://dx.doi.org/10.1111/j.1365-2486.2011.02557.x]

[37]   Dorji, T.; Totland, O.; Moe, S.R.; Hopping, K.A.; Pan, J.; Klein, J.A. Plant functional traits mediate reproductive phenology and success in response to experimental warming and snow addition in Tibet. *Glob. Change Biol.,* **2013**, *19*(2), 459-472.
[http://dx.doi.org/10.1111/gcb.12059] [PMID: 23504784]

[38]   Prietzel, J.; Zimmermann, L.; Schubert, A. Organic matter losses in German Alps forest soils since the 1970s most likely caused by warming. *Nat. Geosci.,* **2016**, *9*, 543-548.
[http://dx.doi.org/10.1038/ngeo2732]

[39]   Crowther, T.W.; Todd-Brown, K.E.; Rowe, C.W.; Wieder, W.R.; Carey, J.C.; Machmuller, M.B.; Snoek, B.L.; Fang, S.; Zhou, G.; Allison, S.D.; Blair, J.M.; Bridgham, S.D.; Burton, A.J.; Carrillo, Y.; Reich, P.B.; Clark, J.S.; Classen, A.T.; Dijkstra, F.A.; Elberling, B.; Emmett, B.A.; Estiarte, M.; Frey, S.D.; Guo, J.; Harte, J.; Jiang, L.; Johnson, B.R.; Kröel-Dulay, G.; Larsen, K.S.; Laudon, H.; Lavallee, J.M.; Luo, Y.; Lupascu, M.; Ma, L.N.; Marhan, S.; Michelsen, A.; Mohan, J.; Niu, S.; Pendall, E.; Peñuelas, J.; Pfeifer-Meister, L.; Poll, C.; Reinsch, S.; Reynolds, L.L.; Schmidt, I.K.; Sistla, S.; Sokol, N.W.; Templer, P.H.; Treseder, K.K.; Welker, J.M.; Bradford, M.A. Quantifying global soil carbon losses in response to warming. *Nature,* **2016**, *540*(7631), 104-108.
[http://dx.doi.org/10.1038/nature20150] [PMID: 27905442]

[40]   Guiot, J.; Cramer, W. Climate change: The 2015 Paris Agreement thresholds and Mediterranean basin ecosystems. *Science,* **2016**, *354*(6311), 465-468.
[http://dx.doi.org/10.1126/science.aah5015] [PMID: 27789841]

[41]   De Graff, J.V. Improvement in quantifying debris flow risk for post-wildfire emergency response. *Geoenviron. Disasters,* **2014**, *1*, 5.
[http://dx.doi.org/10.1186/s40677-014-0005-2]

[42]   Amundson, R.; Berhe, A.A.; Hopmans, J.W.; Olson, C.; Sztein, A.E.; Sparks, D.L. Soil science. Soil and human security in the 21st century. *Science,* **2015**, *348*(6235), 1261071.
[http://dx.doi.org/10.1126/science.1261071] [PMID: 25954014]

[43]   Dietrich, A.; Krautblatter, M. Evidence for enhanced debris-flow activity in the Northern Calcareous Alps since the 1980s (Plansee, Austria). *Geomorphology,* in press

[44]   Ben Hamman Lech-hab. K.; Issa, L.K.; Raissouni, A. Effects of Vegetation Cover and Land Use Changes on Soil Erosion in Kalaya Watershed (North Western Morocco). *Int. J. Geosci.,* **2015**, *6*, 1353-1366.
[http://dx.doi.org/10.4236/ijg.2015.612107]

[45]   Belmont, P.; Foufoula-Georgiou, E. Solving water quality problems in agricultural landscapes: New approaches for these nonlinear, multiprocess, multiscale systems. *Water Resour. Res.,* in press
[http://dx.doi.org/10.1002/2017WR020839]

[46]   Strack, M. *Peatland and Climate Change.,* **2008**. http://www.peatsociety.org/sites/default/files/files/PeatlandsandClimateChangeBookIPS2008.pdf

[47]   Karel, E. *Circulating natures. Water - Food - Energy. Abstract book: 90-91, concurrent session 5: Deformation or transformation? Quarrying, mining, land, and water in the Netherlands. 7th conference of the European Society for Environmental History, Munich,* **2013**. http://www.abstractstosubmit.com/eseh2013/abstracts/

[48]   Fenner, N.; Freeman, C. Drought-induced carbon loss in peatlands. *Nat. Geosci.,* **2011**, *4*, 895-900.

[http://dx.doi.org/10.1038/ngeo1323]

[49]   Turetsky, M.R.; Benscoter, B.; Page, S. Global vulnerability of peatlands to fire and carbon loss. *Nat. Geosci.,* **2015**, *8,* 11-14.
[http://dx.doi.org/10.1038/ngeo2325]

[50]   Succow, M. *Moor- und Klimaschutz. Michael Succow Stiftung zum Schutz der Natur, Greifswald, Germany,* http://www.succow-stiftung.de/tl_files/pdfs_downloads/Buecher%20und%20Broschueren/ Moorschutz.pdf

[51]   Moore, S.; Evans, C.D.; Page, S.E.; Garnett, M.H.; Jones, T.G.; Freeman, C.; Hooijer, A.; Wiltshire, A.J.; Limin, S.H.; Gauci, V. Deep instability of deforested tropical peatlands revealed by fluvial organic carbon fluxes. *Nature,* **2013**, *493*(7434), 660-663.
[http://dx.doi.org/10.1038/nature11818] [PMID: 23364745]

[52]   Hoscilo, A.; Page, S.E.; Tansey, K.J. Effect of repeated fires on land-cover change on peatland in southern Central Kalimantan, Indonesia, from 1973 to 2005. *Int. J. Wildland Fire,* **2011**, *20*(4), 578-588.
[http://dx.doi.org/10.1071/WF10029]

[53]   Wijedasa, L.S.; Jauhiainen, J.; Könönen, M.; Lampela, M.; Vasander, H.; Leblanc, M.C.; Evers, S.; Smith, T.E.; Yule, C.M.; Varkkey, H.; Lupascu, M.; Parish, F.; Singleton, I.; Clements, G.R.; Aziz, S.A.; Harrison, M.E.; Cheyne, S.; Anshari, G.Z.; Meijaard, E.; Goldstein, J.E.; Waldron, S.; Hergoualc'h, K.; Dommain, R.; Frolking, S.; Evans, C.D.; Posa, M.R.; Glaser, P.H.; Suryadiputra, N.; Lubis, R.; Santika, T.; Padfield, R.; Kurnianto, S.; Hadisiswoyo, P.; Lim, T.W.; Page, S.E.; Gauci, V.; Van Der Meer, P.J.; Buckland, H.; Garnier, F.; Samuel, M.K.; Choo, L.N.; O'Reilly, P.; Warren, M.; Suksuwan, S.; Sumarga, E.; Jain, A.; Laurance, W.F.; Couwenberg, J.; Joosten, H.; Vernimmen, R.; Hooijer, A.; Malins, C.; Cochrane, M.A.; Perumal, B.; Siegert, F.; Peh, K.S.; Comeau, L.P.; Verchot, L.; Harvey, C.F.; Cobb, A.; Jaafar, Z.; Wösten, H.; Manuri, S.; Müller, M.; Giesen, W.; Phelps, J.; Yong, D.L.; Silvius, M.; Wedeux, B.M.; Hoyt, A.; Osaki, M.; Hirano, T.; Takahashi, H.; Kohyama, T.S.; Haraguchi, A.; Nugroho, N.P.; Coomes, D.A.; Quoi, L.P.; Dohong, A.; Gunawan, H.; Gaveau, D.L.; Langner, A.; Lim, F.K.; Edwards, D.P.; Giam, X.; Van Der Werf, G.; Carmenta, R.; Verwer, C.C.; Gibson, L.; Gandois, L.; Graham, L.L.; Regalino, J.; Wich, S.A.; Rieley, J.; Kettridge, N.; Brown, C.; Pirard, R.; Moore, S.; Capilla, B.R.; Ballhorn, U.; Ho, H.C.; Hoscilo, A.; Lohberger, S.; Evans, T.A.; Yulianti, N.; Blackham, G.; Onrizal, S.H.; Husson, S.; Murdiyarso, D.; Pangala, S.; Cole, L.E.; Tacconi, L.; Segah, H.; Tonoto, P.; Lee, J.S.; Schmilewski, G.; Wulffraat, S.; Putra, E.I.; Cattau, M.E.; Clymo, R.S.; Morrison, R.; Mujahid, A.; Miettinen, J.; Liew, S.C.; Valpola, S.; Wilson, D.; D'Arcy, L.; Gerding, M.; Sundari, S.; Thornton, S.A.; Kalisz, B.; Chapman, S.J.; Su, A.S.; Basuki, I.; Itoh, M.; Traeholt, C.; Sloan, S.; Sayok, A.K.; Andersen, R. Denial of long-term issues with agriculture on tropical peatlands will have devastating consequences. *Glob. Change Biol.,* **2017**, *23*(3), 977-982.
[http://dx.doi.org/10.1111/gcb.13516] [PMID: 27670948]

[54]   Granath, G.; Moore, P.A.; Lukenbach, M.C.; Waddington, J.M. Mitigating wildfire carbon loss in managed northern peatlands through restoration. *Sci. Rep.,* **2016**, *6,* 28498.
[http://dx.doi.org/10.1038/srep28498] [PMID: 27346604]

[55]   Lishtvan, I.I. *Role of peat deposits and peat in human life,* **2012.** http://www.peatsociety. org/document/role-peat-deposits-and-peat-nature-and-human-life

[56]   European Environment Agency. *Land take,* **2015.** http://www.eea.europa.eu/data-and-maps/ indicators/land-take-2/assessment

[57]   WHO. **2001.** https://www.yumpu.com/en/document/view/10554959/chapter-1-origin-of-nitra-e-in-drinking-water-world-health-

[58]   Newland Bowker, L.; Chambers, D.M. *The risk, public liability, and economics of tailings storage facilities failures,* **2015.** https://www.earthworksaction.org/files/pubs-others/BowkerChamber--RiskPublicLiability_EconomicsOfTailingsStorageFacility%20Failures-23Jul15.pdf

[59]   Beaulieu, M. *Post-mining landscapes. Reclamation, ecology, nature preservation and socio-economy*

in practice, Peckiana; Xylander, W.E.R., Ed., **2004**.

[60]   Johnson, S.; Sahu, R.; Jadon, N. Contamination of soil and water in and around the union carbide site at Bhopal. *Chim. Ind.*, **2010**, *8*, 122-129.

[61]   Hooke, R. LeB. On the history of humans as geomorphic agents. *Geology*, **2000**, *28*(9), 843-846.
       [http://dx.doi.org/10.1130/0091-7613(2000)28<843:OTHOHA>2.0.CO;2]

[62]   Schmidt-Bleek, F. *Grüne Lügen, 4. Aufl., Ludwig*; Random House: München, **2014**.

[63]   Kaiser, M. Mining in European history and its impacts on environment and human societies. *Proc. for the 2nd Mining in European History Conference of the FZ HiMAT, Session III, Societal Interaction and Ecology*, Anreiter, P.; Brandstätter, K.; Goldenberg, G. Eds. **2013**.

[64]   Börner, A.; Bornhöft, E.; Häfner, F., Eds. *Steine- und Erden-Rohstoffe in der Bundesrepublik Deutschland*; Sonderhefte Reihe D - Geol. Jahrb, Eds., **2012**, p. 10.

[65]   Bell, F.G.; Stacey, T.R. Mining subsidence and its effect on the environment: some differing examples. *Environ. Geol.*, **2000**, *40*(1), 135-152.
       [http://dx.doi.org/10.1007/s002540000140]

[66]   Jacobs, S.K. *Subsidence from coal mining activities, background review*, **2014**.
       http://dx.doi.org/www.iesc.environment.gov.au

[67]   Drecker, P.; Genske, D.D.; Heinrich, K. Subsidence and wetland development in the Ruhr district of Germany. *Proc. of the Fifth Int. Symp. on Land Subsidence, The Hague, October; IAHS Publ.*, **1995**, *234*, pp. 413-421.

[68]   Harnischmacher, S. Quantification of mining subsidence in the Ruhr District (Germany). *Géomorphologie*, **2010**, *2010*(3), 261-274.
       [http://dx.doi.org/10.4000/geomorphologie.7965]

[69]   Peters, R. *100 Jahre Wasserwirtschaft im Revier: die Emschergenossenschaft 1899-1999*; Verlag Peter Pomp: Bottrop, Essen, **1999**.

[70]   Duin, G. *Bericht der Landesregierung an die Präsidentin des Landtags*, **2015**.
       https://www.landtag.nrw.de/portal/WWW/dokumentenarchiv/Dokument/MMV16-2631.pdf

[71]   *ahu AG Aachen. Prüfung möglicher Umweltauswirkungen von Abfall- und Reststoffen zur Bruch-Hohlraumverfüllung in Steinkohlenbergwerken in Nordrhein-Westfalen*, **2016**.
       http://www.umweltauswirkungen-utv.de/

[72]   Malovichko, A.A.; Malovichko, D.A.; Dyagilev, R.A. *Proc. of the 8th International Symposium rockbursts and seismicity in mines*, **2013**, pp. 463-473.

[73]   Graupner, B.; Koch, C.; Werner, F. *Can EU-WFD be applied to former Lignite Mining Districts? A case study for the Lausitz Mining District (Germany), Dresdner Grundwasserforschungszentrum e.V., Dresden, Germany*, **2006**. https://www.imwa.info/docs/imwa_2008/IMWA2008_040_Graupner.pdf

[74]   Busch, S.; Grosser, R.; Schroeckh, B., Eds. *Energie aus heimischen Brennstoffen: Der Braunkohlentagebau Cottbus-Nord und die Lausitzer Landschaft nach der Braunkohle. EDGG*; **2015**, Vol. 254.

[75]   Hübner, G. Ökologisch-faunistische Fließgewässerbewertung am Beispiel der salzbelasteten unteren Werra und ausgewählter Zuflüsse *Ökologie und Umweltsicherung, vol. 27, dissertation, University of Kassel, Germany*, **2007**.

[76]   Werra Kalibergbau-Museum. **2016**. http://www.kalimuseum.de/

[77]   Azam, S.; Li, Q. Tailings dam failures: A review of the last one hundred years. *Geotech. News*, **2010**, *28*(4), 50-53.

[78]   World Information Service on Energy. *Uranium Project. Chronology of major tailings dam failures*, **2016**. http://www.wise-uranium.org/mdaf.html

[79]   Golder Associated. *Review of tailings management guidelines and recommendations for improvement.*, **2016**. http://www.icmm.com/publications/pdfs/2016/161205_golder-associates_review-of-tailings-management-guidelines.pdf

[80]   Catchpole, G.; Kirchner, G. *Proc. of the Int. Conference and Workshop, Sven von Loga Verlag,* Köln **1995**, pp. 81-89.

[81]   World Nuclear Association. *In situ* Leach (ISL) Mining of Uranium, **2014**. http://www.world-nuclear.org/information-library/nuclear-fuel-cycle/mining-of-uranium/in-situ-leach--ining-of-uranium.aspx

[82]   World Nuclear Association. *The nuclear fuel cycle,* **2015**. http://www.world-nuclear.org/ information-library/nuclear-fuel-cycle/introduction/nuclear-fuel-cycle-overview.aspx

[83]   Ewing, R.C. Nuclear waste forms for actinides. *Proc. Natl. Acad. Sci. USA,* **1999**, *96*(7), 3432-3439. [http://dx.doi.org/10.1073/pnas.96.7.3432] [PMID: 10097054]

[84]   Davis, E.D.; Gould, C.R.; Sharapov, E.I. Oklo reactors and implications for nuclear science. *Int. J. Mod. Phys. E,* **2014**, *23*(4), 1430007. [http://dx.doi.org/10.1142/S0218301314300070]

[85]   Bundesamt für Strahlenforschung. *Asse II, Radioactive waste in Asse mine,* **2016**. http://www.asse.bund.de/Asse/EN/topics/what-is/radioactive-waste/radioactive-waste_node.html

[86]   Minkley, W.; Kamlot, P. *Gebirgsmechanische Zustandsanalyse des Tragsystems der Schachtanlage Asse II, Kurzbericht,* **2007**. http://www.helmholtz-muenchen.de/fileadmin/ASSE/PDF/News/Kurzbericht-Zustandsanalyse-V-4.pdf

[87]   Bell, F.G.; Donelly, L.J. *Mining and Its Impact on the Environment*; Taylor & Francis: New York, **2006**.

[88]   Power, G.; Gräfe, M.; Klauber, C. Bauxite Residue Issues: I. Current Management, Disposal and Storage Practices. *Hydrometallurgy,* **2011**, *108*(1-2), 33-45. [http://dx.doi.org/10.1016/j.hydromet.2011.02.006]

[89]   Wilshire, H.; Nielson, J.E.; Hazlett, R.W. The American west at risk. In: *Science, Myths and Politics of Land Abuse and Recovery*; Oxford University Press: New York, **2008**.

[90]   Centers for Disease Control and Prevention, Feasibility Study of Weapons Test Fallout, Technical Report. *Fallout from nuclear weapons*; **2014**, Vol. 1, p. 2.

[91]   International Atomic Energy Agency. *Estimation of global inventories of radioactive waste and other radioactive materials, Waste and Environment Safety Section of IAEA,* **2009**. http://www-pub.iaea.org/MTCD/publications/PDF/te_1591_web.pdf

[92]   Koo, Y.-H.; Yang, Y-S.; Song, K-W. Radioactivity release from the Fukushima accident and its consequences: A review. *Prog. Nucl. Energy,* **2014**, *74*, 61-70. [http://dx.doi.org/10.1016/j.pnucene.2014.02.013]

[93]   Deutsches Atomforum, V. *Der Reaktorunfall in Tschernobyl - Unfallursachen, Unfallfolgen und deren Bewältigung; Sicherung und Entsorgung des Kernkraftwerks Tschernobyl, Berlin,* http://www.kernenergie.de/kernenergie-wAssets/docs/service/025reaktorunfall_tschernobyl2011.pdf**2013**.

[94]   Amos, H.M.; Jacob, D.J.; Streets, D.G. Legacy impacts of all-time anthropogenic emissions on the glovbal mercury cycle. *Global Biochem. Cycles,* **2015**3, *27*, 410-421. [http://dx.doi.org/http://dx.doi.org/10.1002/gbc.20040]

[95]   Gray, J.E.; Theodorakos, P.M.; Fey, D.L.; Krabbenhoft, D.P. Mercury concentrations and distribution in soil, water, mine waste leachates, and air in and around mercury mines in the Big Bend region, Texas, USA. *Environ. Geochem. Health,* **2015**, *37*(1), 35-48. [http://dx.doi.org/10.1007/s10653-014-9628-1] [PMID: 24974151]

[96]   Selin, N.E.; Jacob, D.J.; Yantosca, R.M. Global 3-D land-ocean-atmosphere model for mercury:

Present-day *versus* preindustrial cycles and anthropogenic enrichment factors for deposition. *Global Biogeochem. Cycles,* **2008**, *22*, GB2011.

[97]   Jakubke, H-D.; Karcher, R. *Lexikon der Chemie*; Spektrum Akademischer Verlag GmbH: Heidelberg, **1998**.

[98]   Guha, S.K. *Induced earthquakes*; Springer: Berlin, Heidelberg, New York, **2000**.
[http://dx.doi.org/10.1007/978-94-015-9452-3]

[99]   Trifu, C.I., Ed. *The Mechanism of Induced Seismicity*; Springer: Berlin, Heidelberg, New York, **2002**.
[http://dx.doi.org/10.1007/978-3-0348-8179-1]

[100]   Mulargia, F.; Bizzarri, A. Anthropogenic triggering of large earthquakes. *Sci. Rep.,* **2014**, *4*, 6100.
[http://dx.doi.org/10.1038/srep06100] [PMID: 25156190]

[101]   Marcak, H.; Mutke, G. Seismic activation of tectonic stresses by mining. *J. Seismol.,* **2013**, *17*(4), 1139-1148.
[http://dx.doi.org/10.1007/s10950-013-9382-3]

[102]   White, J.A.; Chiaramonte, L.; Ezzedine, S.; Foxall, W.; Hao, Y.; Ramirez, A.; McNab, W. Geomechanical behavior of the reservoir and caprock system at the In Salah CO_2 storage project. *Proc. Natl. Acad. Sci. USA,* **2014**, *111*(24), 8747-8752.
[http://dx.doi.org/10.1073/pnas.1316465111] [PMID: 24912156]

[103]   Yunsheng, Y.; Qiuliang, W.; Jinggang, L. Seismic hazard assessment of the Three Gorges Project. *Geod. Geodyn.,* **2013**, *4*(2), 53-60.
[http://dx.doi.org/10.3724/SP.J.1246.2013.02053]

[104]   Abelson, M.; Yechieli, Y.; Crouvi, O. Evolution of the Dead Sea sinkholes. *Geol. Soc. Am., Spec. Paper,* **2006**, 401.

[105]   Zalasiewicz, J.; Waters, C. The Anthropocene. *Environmental Science, Oxford Research Encyclopedias,* **2015**.

[106]   Müller, S.M. *Wiring the World. The Social and Cultural Creation of Global Telegraph Networks*; Columbia University Press, **2016**.

[107]   Spencer, K.; O'Shea, F.T. *The Hidden Threat of Historical Landfills on Eroding and Low-lying Coasts*; ECSA Bull, **2014**, pp. 16-17.

[108]   World Bank. *urban development series - knowledge papers, Waste disposal,* **2016**. http://siteresources.worldbank.org/INTURBANDEVELOPMENT/Resources/336387-1334852610766/Chap6.pdf

[109]   D-Waste. *Waste Atlas, Global Waste Generation Clock,* http://www.atlas.d-waste.com/**2016**.

[110]   Honda, S.; Khetriwal, D.S.; Kuehr, R. *Regional E-waste Monitor: East and Southeast Asia,* **2016**. http://ewastemonitor.info/pdf/Regional-E-Waste-Monitor.pdf

[111]   Davisson, L.; Hamilton, T.; Tompson, A.F. Radioactive waste buried beneath Runit Dome on Enewetak Atoll, Marshall Islands. *Int. J. Environ. Pollut.,* **2012**, *49*(3-4), 161-178.
[http://dx.doi.org/10.1504/IJEP.2012.050897]

[112]   Certini, J.; Scalenghe, R.; Woods, W. The impact of warfare on the soil environment. *Earth Sci. Rev.,* **2013**, *127*, 1-15.
[http://dx.doi.org/10.1016/j.earscirev.2013.08.009]

[113]   Stellman, J.M.; Stellman, S.D.; Christian, R.; Weber, T.; Tomasallo, C. The extent and patterns of usage of Agent Orange and other herbicides in Vietnam. *Nature,* **2003**, *422*(6933), 681-687.
[http://dx.doi.org/10.1038/nature01537] [PMID: 12700752]

[114]   Dwernychuk, W. *Agent Orange and Dioxin Hot Spots in Vietnam,* http://www-esd.worldbank.org/popstoolkit/POPsToolkit/POPSTOOLKIT_COM/ABOUT/ARTICLES/AODIOXINHOTSPOTSVIETNAM.HTM

[115]   Mai, T.A.; Doan, T.V.; Tarradellas, J.; de Alencastro, L.F.; Grandjean, D. Dioxin contamination in

soils of Southern Vietnam. *Chemosphere,* **2007**, *67*(9), 1802-1807.
[http://dx.doi.org/10.1016/j.chemosphere.2006.05.086] [PMID: 17222446]

[116] Harrison, M.; Wolf, N. *The frequency of wars,* **2011**. http://www2.warwick.ac.uk/fac/soc/ economics/staff/mharrison/public/ehr2011postprint.pdf

[117] *Unicef. Land mines: Hidden killers,* **1996**. http://www.unicef.org/sowc96pk/hidekill.htm

[118] WNA World Nuclear Association. *Chernobyl Accident 1986,* **2016**. http://www.world-nuclear.org/ information-library/safety-and-security/safety-of-plants/chernobyl-accident.aspx

[119] USGS. *Pesticide National Synthesis Project,* **2016**. http://water.usgs.gov/nawqa/pnsp/usage/ maps/show_map.php?year=2013&map=GLYPHOSATE&hilo=L&disp=Glyphosate

[120] Voß, H. W. *The Current Situation of German Coal Mining. AMS online,* **2015**.

[121] Lintker, S. *Energetic reuse of former mine sites in NRW,* **2014**. http://www.enviacon.com/ website/fileadmin/enviacon/Praesentationen/141205_Vortrag_IV_Chile_4_EnergieAgentur.NRW_Ste phanus_Lintker.pdf

[122] SIPRI Stockholm International Peace Research Institute. *Global nuclear weapons: downsizing but modernizing,* **2016**. https://www.sipri.org/media/press-release/2016/global-nuclear-weapons-downsizing-modernizing

[123] Carbon Brief. *Mapped: The world's nuclear power plants,* http://www.carbonbrief. org/mapped-th--worlds-nuclear-power-plants?utm_source=Daily+Carbon+Briefing&utm_campaign=e8467c-ae3-cb_daily&utm_medium=email&utm_term=0_876aab4fd7-e8467cfae3-303446941

[124] *Bulletin of the Atomic Scientists. Global nuclear power database,* **2017**. http://thebulletin.org/ global-nuclear-power-database

[125] Huttunen-Saarivirta, E.; Rajala, P.; Bomberg, M. Corrosion of copper in oxygen-deficient groundwater with and without deep bedrock micro-organisms: characterisation of microbial communities and surface processes. *Appl. Surf. Sci.,* **2017**, *396*, 1044-1057.
[http://dx.doi.org/10.1016/j.apsusc.2016.11.086]

[126] Feiveson, H.; Mian, Z.; Ramana, M.V. *Spent fuel from nuclear power reactors,* **2011**. http://fissilematerials.org/library/ipfm-spent-fuel-overview-june-2011.pdf

[127] Weber, R.; Tysklind, M.; Gaus, C. Dioxin-contemporary and future challenges of historical legacies. Dedicated to Prof. Dr. Otto Hutzinger, the founder of the DIOXIN Conference Series. *Environ. Sci. Pollut. Res. Int.,* **2008**, *15*(2), 96-100.
[http://dx.doi.org/10.1065/espr2008.01.473] [PMID: 18380226]

CHAPTER 8

Glaciers, Ice Sheets, Sea Ice, and Permafrost[1] Areas

Abstract: The global ice volume, thickness, and coverage has been shrinking since ca. 1900. Since 1950, this has been caused by the anthropogenic part of atmospheric warming. The following effects of ice mass wasting and deglaciation are discussed: diminution of ice albedo; input of huge meltwater masses into the oceans and its influence on oceanic circulation and contribution to sea level rise; more frequent landslides, rockfalls, and glacial lake outburst floods; switching of cryospheric sinks into sources; secondary transportation of released pollutants by meltwater and deterioration of drinking water quality; transition of alpine runoff regimes from icemelt to snowmelt domains; decrease of Arctic land ice and multiyear sea ice; diminution of terrestrial permafrost zones; thawing induced destabilisation of Arctic technical infrastructure; emission of CO_2 and CH_4 from permafrost areas because of enhanced microbial activity; labilisation of shelf permafrost zones due to boreal sea water warming; and increasing primary production adjacent to decreasing shelf ice areas and to drifting icebergs. Warming and deposition of water or wind driven particulate contaminants (*e.g.* microplastic, aerosol) in high latitudes are fostered by the Arctic Amplification effect and by the global distillation phenomenon. Shrinkage of high-latitude ice occurred due to heat transfer from the atmosphere and sea water (retreat of grounding line). Pollution of Arctic regions has occurred due to poor decommissioning of military facilities and to oil spills.

Keywords: Aerosol, Albedo, Arctic Amplification, Carbon dioxide, Cryosphere, Deglaciation, Global distillation, Greening, Grounding line, Meltwater, Methane, Microplastic, Permafrost, Pollution, Rock avalanches, Subsea permafrost, Thawing, Warming.

These ecologically sensitive snow and ice regions, which presently cover ca. 10% of the land surface and less than 1% of the sea surface, are recognised as high confidence climate indicators. Variations in sea ice coverage, in volume and areal extent of ice sheets, and glacier fronts have been monitored since 1894 and are measured by remote sensing, mapping, and straining *in situ* data acquisition. Documentation and interpretation of the data is conducted *e.g.* by the Global Cryosphere Watch Organisation, which set up a world glacier inventory in 1970. It states that globally, since the Little Ice Age (17th - 19th centuries), a general shrinkage of ice volume and extent has occurred, modified by short phases of local ice surges around 1920 and 1970, as well as phases of accelerated ice retreat ca. 1940 and since 1985 [1]. The global cumulative specific ice mass balance between 1945 and 2005 is -20 m (ca. -20%), equivalent to -35cm/a. The average rate of global ice loss - resulting from accumulation minus melting and iceberg calving - has increased, because the calculated rate between 1979 and 2009 was ca. 226 GT/a, increasing from 1993 and 2009 to ca. 275 GT/a [2]. According to more and improved data sources, the meltwater fluxes into the oceans in 2011 were estimated to 720 GT (Antarctic) and 360 GT (Arctic). Forcings due to the accelerated melting of polar ice masses causes surface cooling of the Southern Ocean and the Northern Atlantic, and increases ocean stratification, atmospheric temperature gradients, eddy kinetic energy, and baroclinicity, resulting in more powerful storms[2] [3].

There is a consensus that this ice shrinkage on a century time scale has a non-periodic, *i.e.* non-orbitally, forced, signature and that the accelerated ice mass wasting since 1950 is caused by the anthropogenic part of the greenhouse effect and will probably result in the deglaciation of many mountain ranges and global sea level rise [4, 5].

At present, ca. 2 km^3 of the remaining 70 km^3 of Alpine ice is lost annually due to melting[3]. This development will have detrimental effects on landscape evolution; security of fresh water supply; fluvial and limnic flora and fauna exposed to more glacial turbid meltwater loaded with rock milk; strength of frozen fractured rock masses in high-alpine regions, elevating risks of landslides, rock avalanches, and outburst floods from expanding glacial lakes [6, 7].

The effects on water reservoirs with glacierised catchments are given by means of one example (Mauvoisin Region, Switzerland): Ice volume wasting between 1900 and 2009 is estimated at 43% (6.49 km^3 to 3.69 km^3), which corresponds to the reconstruction of an irregular runoff increase from 216 million m^3/a (1900) to 265 million m^3/a (2009): the runoff regime feeding the reservoir began to switch from an icemelt to a snowmelt domain [8]. Lack of downstream snow meltwater in springtime and early summer for irrigating adjacent agricultural areas is of major concern [9].

Ice loss has also been observed in low latitude mountain ranges: Kilimanjaro lost 85% of its ice cover between 1912 and 2000, seen as a unique development over the last 4000 years [10].

Between 1875-2008, the average Arctic warming rate of the terrestrial near surface air temperature derived from land stations amounted to 1.36 °C/century, compared to 0.79 °C/century in the northern hemisphere. Between 1967-2008, land ice coverage decreased in May by 14% and in June by 46% [11].

Arctic amplification[4] caused September Arctic sea ice extent diminution from ca. 8.5 million km^2 in 1953 to 4.41 million km^2 in 2015 [12, 13]. A detailed, long-term reconstruction of the development of the Arctic sea ice extent since 1850, expressing reliably its monthly state, has been made available by compiling numerous historic datasets (*e.g.* from whale ship records, aeroplane surveys, meteorological offices and other archive sources). This synthesis reveals an overall, almost steady, but accelerating, decline of September sea ice extent from ca. 9 million km^2 (1850) by 55% to ca. 5 million km^2 (2013). March sea ice extent diminished during the same time span by 3.2% to ca. 15.5 million km^2. In addition, the data lay bare that in the past 150 years no precedents exist concerning recent September sea ice extent and present rate (13.4% between 1979-2015) of September sea ice retreat. The reasons for this development go back to climate change-induced rising air temperatures, arctic amplification, sea water warming, and altered wind patterns [14]. A linear relationship was determined between monthly mean September sea ice extent and cumulative anthropogenic CO_2 emissions, consisting in an average loss of 3 ± 0.3 m^2 of summer sea-ice per 1 T of emitted CO_2 [15]. Based on long-term data from drift buoys, weather stations, and satellites equipped with microwave instruments, Arctic multiyear sea ice (≥ 5 a) has shrunk in its areal extent from 1.86 million km^2 (20% of all sea ice) in Sept. 1984 to 0.11 million km^2 (3% of all sea ice) in Sept. 2016 [16].

Monitoring of the glaciers of the Greenland ice shield revealed acceleration of surface velocity, shrinkage of ice thickness, and retreat of grounding line, and lateral extent due to increasing temperatures of the sea water and atmosphere [17].

Long-term observation of the ice mass balance of Canadian glaciers by means of satellite based velocity and ice thickness time series (1991-2015) laid bare that since 2005 the average annual ice mass loss from surface meltwater runoff (29.6 ± 6 GT/a) dominated by far the average annual ice mass loss from ice discharge (3.5 ± 0.2 GT/a) into the sea [18].

According to remote sensing data, summertime surface albedo of the Greenland ice shield decreased between 1996-2012 at a statistically significant rate of 0.02/decade. Model calculations indicate that this fact results from several effects resulting from increased near-surface temperatures: 1) more frequent thawing/freezing cycles cause growth of snow grain size, 2) ablation of snow layers results in exposure of larger surfaces consisting of bare ice, 3) surficial accumulation of trapped dust particles. These effects contribute to diminution of Greenland ice shield albedo [19]. In addition, spring and summer blooms of cosmopolitan red snow algae, which colonise glaciated ecosystems, increase duration and areal extent of exposed bare ice surfaces, contribute to decrease of snow albedo during summer by ca. 13%, and boost ice melt rates [20].

Quantification of supraglacial Arctic meltwater extent and volume by remote sensing multispectral imagers revealed that, during the melt season (06/01 to 09/01), the water volume present in these glacial lakes and ponds, which have maximum depths of up to 7.7 m and cover an area of 25,246 km^2 at 68 °N and maximum elevation of 1200 m asl, stored on 07/15/2015 ca. 811 million m^3 of meltwater. Depending on the connectivity of channel and fracture systems, the heat content can be transferred *via* hydrofracturing to the base of the inland ice, thus increasing ice flow velocity [21].

Comparison between historic Antarctic sea ice edge positions, according to 11 ships' expedition logbook data (1897-1917), and daily mean satellite-derived sea ice edge concentrations (1989-2014), revealed that the pan-Antarctic ice edge latitude has shifted 0.41° southward on average, equivalent to sea ice loss of 14.2% [22].

However, calculations concerning the Earth system's sensitivity, responding to the processes resulting from decaying ice sheets, are in an early stage and the resulting uncertainties prevent accurate prognoses of events [23]. Because Antarctic ice masses have been implicated in the geological past as primary contributors to sea level rise during the last interglacial (0.130-0.115 Ma ago) - and 3 Ma ago -, it is supposed that this ice will respond more sensitively to future greenhouse conditions. The effects of the surface warming of shelf ice and the retreat of grounding lines - both contributors to the collapse and wasting of ice - were underestimated[5] [24].

Climate change-induced retreat, thinning and decay of western Antarctic (Amundsen-Bellinghausen) ice shelves, as well as forced surface melting and ponding, will double meltwater quantity - at present ca. 100 GT/a - until 2050. This incipient shelf ice labilisation entails accelerated flow velocity of inland ice glaciers and an increasing quantity of ice transported seawards [25].

The reasons for glacial retreat on the western side of the Antarctic Peninsula was found not only in the rise of air temperatures by 3 °C over the past 50 years, but also in the water masses of the Antarctic Circumpolar Current, which have warmed in recent decades. A positive correlation was ascertained between the degree of glacial retreat and the temperatures of the mid depth Circumpolar Deep Waters near their fronts [26]. Thinning of the western Antarctic Pine Island Glacier, observed since 1975, has been caused by sub-ice melting and landward shift over tens of km of the grounding line on bedrock with a steep retrograde slope, so that the glacier has become susceptible to the marine ice sheet instability mechanism: reduced buttressing exerted by the ice shelf causing acceleration of this ice stream flow. According to simulations, the corresponding output volume of meltwater was calculated for the time between 1992-2011 at an average of 20 GT/a [27]. According to computer simulations, sub-ice warming of marine-terminating glaciers can result also in massive iceberg discharges, also called Heinrich events [27, 28].

Borehole logging and ground penetrating radar data from the Antarctic ice shelf reveal that repeated refreezing of climate warming-induced melt ponds - a response to föhn winds blowing for a few days during Antarctic summer - caused the formation of massive, substantially warmer and denser subsurface ice layers within the compressed firn, which caused hydrofracturing of the ice shelf and modified its stability and flow [29].

Sea ice loss on the western Antarctic (Amundsen-Bellinghausen) shelf resulted in a negative feedback to anthropogenic climate warming through intensified phytoplankton blooms that caused accelerated growth of bryozoan colonies on the shelf bottom. In this way, an estimated 200,000 T more carbon/a were sequestered since 1980 (total amount ca. 2.9 MT carbon/a) [30]. A similar effect accounts for the rising number of melting giant Antarctic icebergs (> 1 km length); their input of terrigenous nutrients and Fe minerals increased marine primary productivity - monitored between 2003 and 2013 - and, in consequence, the capacity of the Southern Ocean to bury carbon *via* phytoplankton blooms [31].

Anthropogenically-induced thawing of permafrost soil has initiated the emanation of additional CO_2, N_2O, and CH_4 [32]. Particulate organic matter stored in permafrost soil is mobilised by thawing processes, transported by running water, and deposited and buried in boreal marine delta deposits for geologic time periods [33].

From boreal permafrost areas (tundra), an estimated 17-42 MT CH_4/a - ¼ of all global methane emissions - escape into the atmosphere [34]. Climate change-induced warming of boreal sea waters causes decomposition of vulnerable subsea permafrost areas[6] of the East Siberian Shelf. A pilot study proved that the response of microbial communities in thawing permafrost soil consists in the creation of biochemical pathways leading to metabolic processes that stimulated decomposition and respiration of deposited carbon, resulting in acetogenesis, methanogenesis, and the production of CO_2 [35]. Release of CH_4 *via* ebullition and diffusion of dissolved CH_4 into the atmosphere in the mentioned regions amounts to annual quantities of ca. 8×10^6 T, an equal magnitude of CH_4 flux from the world oceans into the atmosphere [36]. Ca. 1470 GT of carbon - ca. 70% more than that in the present atmosphere - is contained within the permanently frozen soils of the Arctic [37]. Sea level rise, as well as the amplified response of the Arctic to climate warming [38, 39], will expand the area of vulnerable subsea permafrost and the amount of emitted CH_4. Climate change-forced degradation, thawing, and thermal erosion of the terrestrial permafrost, including the variety rich in organic carbon (yedoma), have been observed at many locations in Siberia, Alaska, *etc.* and it is estimated that the areal loss of Arctic permafrost between 1980 and 2010 amounted to ca. $1,3 \times 10^6$ km² [40 - 42]. A substantial amount of organic matter - estimated to ca. 160 GT - was deposited during late Holocene at the bottom of anoxic thermokarst lakes and was transformed into perennially frozen lacustrine sediments as a consequence of draining of those lakes. These deposits are potential sources of CH_4 [43]. Photochemical reactions and microbial metabolic activities will effect the release of additional CO_2 and CH_4 from organic carbon accumulated in permafrost thaw ponds [44].

Monitoring of high-altitude permafrost ground temperature at 6 m depth between 1995 and 2006 revealed warming by +0.43 °C on average at the Qinghai Tibet Engineering Corridor. Infrastructure (railway and road connections, pipelines, electricity transmission lines, edifices, *etc.*) will supposedly be at elevated risk of thermal surface settlement hazards due to permafrost thawing [45]. Painting of the runway (surface size including approach roads = 0.37 million m²) of Thule Air Base (Greenland) white reduced Summer thawing depth by ca. 60 cm. This measure, carried out several times between 1953 and 2012, increased albedo, so that the formation of surface settlements above thawing massive ice wedges in permafrost soil was diminished. At present, insulation tests with rigid, 10.14 cm thick polystyrene sheets, buried horizontally in the embankment between the asphalt surface and the permafrost soil, are being carried out. This thaw mitigation measure effectively lessens heat transport to depth and is expected to perform under future climate conditions (+2.5 °C) [46].

Intentional pollution of the Arctic cryosphere has occurred by minimal decommissioning and dispersal of considerable amounts of diverse sorts of waste (*e.g.* rubble, diesel, PCBs, grey water, sewage, and coolants for one portable nuclear generator) in unlined sumps at Thule Airbase, Camp Century and dozens of radar bases along the Distant Early Warning Line, established 1955-1957 at the Arctic Circle during the Cold War [47]. Remediation of petroleum pollution (spilled or leaked in open cast mines, along roads, at drill sites, in abandoned dump sites, along pipelines, close to research and military facilities, *etc.*) has turned out to be a difficult task, if it occurred on permafrost or seasonally frozen ground. Low temperatures (implying altered physical properties of oil and reduced microbial activity in contaminated soil), remoteness, scant local resources, *etc.* complicate the application of proved physical, chemical, and biological treatments if applied to oil-polluted frozen soils, and make the measures more expensive [48].

Unintentional pollution of the polar areas, as well as other glaciated mountainous regions, occurred *via* the global distillation phenomenon. Episodic long-range aeolian transport of aerosol and low volatility compounds from warm, equatorial, subtropical, and temperate areas, *via* tropospheric circulation systems, to the cold regions mentioned above, where cold condensation of persistent organic pollutants and precipitation or surface deposition of other particulate matter containing sulphate, nitrate, and heavy metal ions occur, thus becoming trapped in the cold environments [49, 50].

Arctic ice cores covering the time between 1720 and 1993 revealed that 70% of atmospheric deposition of Hg can be attributed to anthropogenic sources and that a twentyfold increase of Hg input occurred between pre- and early industrial times (before 1840) and the mid-1980s [51]. Deposition of Hg in high-latitude regions is accomplished by the global distillation process and is enhanced there near coastal areas by Arctic mercury depletion events. These occur concurrently with tropospheric O_3 depletion. Due to the presence of atomic bromine (marine source), photochemical reactions, starting with polar sunrise, effected decomposition of O_3 and oxidation of gaseous Hg^0 to reactive Hg^{2+}. The shorter atmospheric lifetime of the latter promotes its deposition on snow surfaces [52]. Net atmospheric deposition of Hg in permanent snow regions of the globe is estimated at 140 T/a [53].

Physical analysis of four cores - 1-3.5 m length - recovered close from Arctic Sea ice surface revealed high amounts of microplastic pollutants. Fourier transform infrared spectroscopy resulted in the identification of polystyrene (2%), acrylic (2%), polyethylene (2%), polypropylene (3%), polyester (21%), polyamide (16%), and rayon (54%). Fragment concentrations, which ranged between 38 and 234 pieces/m^3 ice, were 1-2 magnitudes larger than average concentrations in Atlantic and Pacific sea water and point to the fact that sea ice acts as major global sink for microplastic, because growing ice crystals scavenge, concentrate, and trap fragments floating in the sea water[7]. The expected amount of melting Arctic sea ice during the next 10 years will probably release at least one trillion microplastic fragments into the sea water [54].

In general: Cryospheric sinks changed to sources. Anthropogenic atmospheric particulate pollutants, emitted during times prior to the implementation of filters, were deposited and fixed in the ice layers, and got remobilised *via* meltwaters, decreasing the quality of surface waters and drinking water resources [55].

At the escarpments in Greenland, climate change-induced deglaciation has exposed sensitive terrestrial Arctic ecosystems like tundra, permafrost soils, fens, moraines, periglacial lakes, glacial outwashes plains, as well as bare rock formations, in which mineral concentrations of economic grade were prospected (*e.g.* Precambrian magmatic rocks containing platinum group elements; nickel; lead; zinc; gold; uranium; rare earth elements like zirconium, niobium, and tantal; Palaeozoic sediments containing copper-, zinc- and molybdenum; and offshore oil and gas resources). Economic and ecological interests are in conflict [56].

Considerable reduction in summer Arctic sea ice caused interest in utilising the Arctic Sea as shortcut between North Atlantic and North Pacific harbours, because this option saves time, reduces fuel consumption and decreases emissions of exhaust gases and aerosols from shipping traffic [57].

NOTES

[1] Translated from the original Russian concept "Vechnaia Merzlota" = permafrost area, defined by Mikhail Ivanovich Sumgin. The concept of permafrost passed through semantic changes from the 19th to the 21st centuries: first it was seen in the context of natural history and geography, then it concerned civil engineering and geology, and currently it is an important point in climate change sciences [58]. M. I. Sumgin (1873-1942) coined the concept, who developed the science of frozen soils and fossil ice, and who mapped the areal extent of cryogenic soils: [Sumgin, M. I.: *Permafrost soils in the USSR* (in Russian), Vladivostok, 1927. In addition, he discussed geotechnical aspects and established engineering practises considering foundations and constructions on frozen soil [59]. Areal extent of permafrost comprises ca. 20% of the earth's soil ecosystems. Soil, continuously frozen over two years, is defined as permafrost soil. Permafrost can reach up to 1500 m down into the subsurface [60].

[2] Sedimentary deposits formed under similar palaeoclimatic conditions were described for the late part of the Eemian, the latest interglacial epoch between Saalian and Weichselian glaciations. Rapid global sea level rise caused Central Atlantic tropical high stand carbonate platform deposits to be erosively buried below high-energetic runup sediments (tempestites), containing megaboulders, eroded and redeposited by waves from older carbonate deposits present at the platform margins [3].

[3] A side-effect of this process is the release and uncovering of ice-buried archaeological objects. An important research object became an glacier-ic--mummified human corpse, later called "Ötzi" or "Frozen Fritz", a ca. 5300 a old (Copper Age) dead body, discovered on 09/19/1991 near the surface of the ice at

Tisenjoch (Eastern Alps). The mummy and artefacts found there were saved and got preserved in the South Tyrol Museum of Archaeology [61].

[4] Arctic amplification means that a change in the global radiation balance causes the near surface air temperature change in regions above 60 °N to be larger than the global average. This phenomenon, first described by S. Arrhenius in 1895, is an inherent mechanism of the global climate system, resulting from several intertwined forcings acting on diverse temporal and spatial scales. Arctic amplification is caused by poleward heat transport by air and water masses from the temperate and tropical zones and by sea ice coverage loss, which diminishes the insulation effect of relatively warm Arctic sea water from the cold atmosphere above, thus more heat will be transported vertically from the sea surface upward into the atmosphere. During polar summer, exposed Arctic sea water absorbs more solar irradiation, heating the mixed layer and causing further melt of sea ice, resulting in lowered production of new sea ice in the following polar winter. Decrease of ice albedo fosters the reduction of land and sea ice extent and adds heat to the exposed land and water masses. An increase in water-containing clouds and water vapour in the Arctic atmosphere fosters absorption of reflected long wave irradiation to the surface, thus adding more heat than under clear skies. Decreasing amount of sulphate aerosol and increasing amount of black carbon aerosol contributed to warming of the Arctic atmosphere over the past three decades [11].

[5] The latent instability of the West Antarctic Ice Sheet, first mentioned by John H. Mercer, lies in locally retrograde slopes of its bedrock bottom surface at the passage to the shelf and the fact that the grounding line is situated well below present sea level. He postulated that anthropogenic warming will result in deglaciation of large parts of the western Antarctic, as has occurred during the last interglacial [62].

[6] The water temperature immediately above the subsea permafrost area varies between -1.8 °C and +1.0 °C annually and is on average 12 °C to 17 °C warmer than the near-surface air temperature of the adjacent terrestrial permafrost zones [36]. The area examined - ca. 2.1×10^6 km^2 - measures according to the author's own estimations, less than 15% of the present arctic subsea permafrost area.

[7] In addition to the marine flux of microplastic into the Arctic zone discussed in the text, transport of microplastic fragments very probably occurs with the global distillation process (aeolian transport) described above.

CONFLICT OF INTEREST

The author (editor) declares no conflict of interest, financial or otherwise.

ACKNOWLEDGEMENTS

Declare none.

REFERENCES

[1]   Goudie, A.S. *The Human Impact on the Natural Environment. Past, Present and Future,* 7th ed; Wiley-Blackwell: Oxford, **2013**.

[2]   IPCC. *Climate change 2013: The physical science basis.,* Stocker, **2013**, http://www.ipcc.ch/pdf/assessment-report/ ar5/wg1/WG1AR5_SPM_FINAL.pdf

[3]   Hansen, J.; Sato, M.; Hearty, P. Ice melt, sea level rise and superstorms: evidence from paleoclimate data, climate modeling, and modern observations that 2C global warming could be dangerous. *Atmos. Chem. Phys.,* **2016**, *16*, 3761-3812.
[http://dx.doi.org/10.5194/acp-16-3761-2016]

[4]   UNEP & World Glacier Monitoring Service. *Global glacies changes: facts and figures,* **2008**. http://www.grid.unep.ch/glaciers/pdfs/glaciers.pdf

[5]   IPCC Working Group I. *Fourth Assessment Report. Climate Change 2007: The Physical Science Basis, Chapter 4.1,* **2007**. https://www.ipcc.ch/publications_and_data/ar4/wg1/en/ch4s4-es.html

[6]   Haeberli, W.; Gruber, S. *Permafrost soils, Soil Biology*; Margesin, R., Ed.; Springer: Berlin, Heidelberg, **2009**, Vol. 16, pp. 205-218.
[http://dx.doi.org/10.1007/978-3-540-69371-0_14]

[7]   Deline, P.M.; Gruber, S.; Delaloye, R. *Snow and Ice-related Hazards, Risks, and Disasters*; Haeberli, W.; Whiteman, C.; Shroder, J.F., Jr, Eds.; Elsevier: Amsterdam, **2015**, pp. 521-561.
[http://dx.doi.org/10.1016/B978-0-12-394849-6.00015-9]

[8]   Gabbi, J.; Farinotti, D.; Bauder, A. Ice volume distribution and implications on runoff projections in a glacierized catchment. *Hydrol. Earth Syst. Sci.,* **2012**, *16*, 4543-4556.
[http://dx.doi.org/10.5194/hess-16-4543-2012]

[9]   Sturm, M.; Goldstein, M.A.; Parr, C. Water and life from snow: A trillion dollar science question. *Water Resour. Res.,* in press
[http://dx.doi.org/10.1002/2017WR020840]

[10]   Gabrielli, P.; Hardy, D.R.; Kehrwald, N. Deglaciated areas of Kilimanjaro as a source of volcanic trace elements deposited on the ice cap during the late Holocene. *Quat. Sci. Rev.,* **2014**, *93*, 1-10.
[http://dx.doi.org/10.1016/j.quascirev.2014.03.007]

[11]   Serreze, M.C.; Barry, R.G. Processes and impacts of Arctic amplification: A research synthesis. *Global Planet. Change,* **2011**, *77*(1-2), 85-96.
[http://dx.doi.org/10.1016/j.gloplacha.2011.03.004]

[12]   Stroeve, J.; Holland, M.M.; Meier, W. Arctic sea ice decline: Faster than forecast. *Geophys. Res. Lett.,* **2007**, *34*, L09501.
[http://dx.doi.org/10.1029/2007GL029703]

[13]   Cole, S.; Gran, R. *Arctic Sea Ice Summertime Minimum Is Fourth Lowest on Record,* **2015**. http://www.nasa.gov/press-release/nasa-holds-media-briefing-on-carbon-s-role-in-earth-s-future-climate

[14]   Walsh, J. E.; Fetterer, F.; Stewart, J. S. *A database for depicting Arctic sea ice variations back to 1850. Geogr. Rev., Special Issue: Arctic 2017,* **2016**.

[15]   Notz, D.; Stroeve, J. Observed Arctic sea-ice loss directly follows anthropogenic CO_2 emission. *Science,* **2016**, *354*(6313), 747-750.
[http://dx.doi.org/10.1126/science.aag2345] [PMID: 27811286]

[16]   Meier, W. *Arctic sea ice is losing its bulwark,* **2016**. http://earthobservatory.nasa.gov/ IOTD/view.php?id=89038&src=iotdrss

[17]   Mouginot, J.; Rignot, E.; Scheuchl, B.; Fenty, I.; Khazendar, A.; Morlighem, M.; Buzzi, A.; Paden, J. Fast retreat of Zachariæ Isstrøm, northeast Greenland. *Science,* **2015**, *350*(6266), 1357-1361.
[http://dx.doi.org/10.1126/science.aac7111] [PMID: 26563135]

[18]   Millan, R.; Mouginot, M.; Rignot, E. Mass budget of the glaciers and ice caps of the Queen Elizabeth Islands, Canada, from 1991 to 2015. *Environ. Res. Lett.,* **2017**, *12*, 024016.
[http://dx.doi.org/10.1088/1748-9326/aa5b04]

[19]   Tedesco, M.; Doherty, S.; Fettweis, X. The darkening of the Greenland ice sheet: trends, drivers, and projections (1981–2100). *Cryosphere,* **2016**, *10*, 477-496.
[http://dx.doi.org/10.5194/tc-10-477-2016]

[20]   Lutz, S.; Anesio, A.M.; Raiswell, R.; Edwards, A.; Newton, R.J.; Gill, F.; Benning, L.G. The biogeography of red snow microbiomes and their role in melting arctic glaciers. *Nat. Commun.,* **2016**, *7*, 11968.
[http://dx.doi.org/10.1038/ncomms11968] [PMID: 27329445]

[21]   Pope, A. Reproducibly estimating and evaluating supraglacial lake depth with Landsat 8 and other multispectral sensors. *Earth Space Sci.,* **2016**, *3*(4), 176-188.
[http://dx.doi.org/10.1002/2015EA000125]

[22]   Edinburgh, T.; Day, J.J. Estimating the extent of Antarctic summer sea ice during the Heroic Age of Antarctic Exploration. *Cryosphere,* **2016**, *10*, 2721-2730.
[http://dx.doi.org/10.5194/tc-10-2721-2016]

[23]   Previdi, M.; Liepert, B.G.; Peteet, D. Climate sensitivity in the Anthropocene. *Q. J. R. Meteorolog. Soc. Part A,* **2013**, *139*(674), 1121-1131.

[24]   DeConto, R.M.; Pollard, D. Contribution of Antarctica to past and future sea-level rise. *Nature,* **2016**, *531*(7596), 591-597.
[http://dx.doi.org/10.1038/nature17145] [PMID: 27029274]

[25]   Trusel, L.D.; Frey, K.E.; Das, S.B. Divergent trajectories of Antarctic surface melt under two twenty-first-century climate scenarios. *Nat. Geosci.,* **2015**, *8*, 927-932.
[http://dx.doi.org/10.1038/ngeo2563]

[26]   Cook, A.J.; Holland, P.R.; Meredith, M.P.; Murray, T.; Luckman, A.; Vaughan, D.G. Ocean forcing of glacier retreat in the western Antarctic Peninsula. *Science,* **2016**, *353*(6296), 283-286.
[http://dx.doi.org/10.1126/science.aae0017] [PMID: 27418507]

[27]   Favier, L.; Durand, G.; Cornford, S.L. Retreat of Pine Island Glacier controlled by marine ice-sheet instability. *Nat. Clim. Chang.,* **2014**, *4*, 117-121.
[http://dx.doi.org/10.1038/nclimate2094]

[28]   Bassis, J.N.; Petersen, S.V.; Cathles, L.M. Heinrich events triggered by ocean forcing and modulated by isostatic adjustment. *Nature,* **2017**, *542*(7641), 332-334.
[http://dx.doi.org/http://dx.doi.org/10.1038/nature21069]

[29]   Hubbard, B.; Luckman, A.; Ashmore, D.W.; Bevan, S.; Kulessa, B.; Kuipers Munneke, P.; Philippe, M.; Jansen, D.; Booth, A.; Sevestre, H.; Tison, J.L.; O'Leary, M.; Rutt, I. Massive subsurface ice formed by refreezing of ice-shelf melt ponds. *Nat. Commun.,* **2016**, *7*, 11897.
[http://dx.doi.org/10.1038/ncomms11897] [PMID: 27283778]

[30]   Barnes, D.K. Antarctic sea ice losses drive gains in benthic carbon drawdown. *Curr. Biol.,* **2015**, *25*(18), R789-R790.

[http://dx.doi.org/10.1016/j.cub.2015.07.042] [PMID: 26394097]

[31]   Duprat, L.P.; Bigg, G.R.; Wilton, D.J. Enhanced Southern Ocean marine productivity due to fertilization by giant icebergs. *Nat. Geosci.,* **2016**, *9*, 219-221.
[http://dx.doi.org/10.1038/ngeo2633]

[32]   Schuur, E.A.; McGuire, A.D.; Schädel, C.; Grosse, G.; Harden, J.W.; Hayes, D.J.; Hugelius, G.; Koven, C.D.; Kuhry, P.; Lawrence, D.M.; Natali, S.M.; Olefeldt, D.; Romanovsky, V.E.; Schaefer, K.; Turetsky, M.R.; Treat, C.C.; Vonk, J.E. Climate change and the permafrost carbon feedback. *Nature,* **2015**, *520*(7546), 171-179.
[http://dx.doi.org/10.1038/nature14338] [PMID: 25855454]

[33]   Hilton, R.G.; Galy, V.; Gaillardet, J.; Dellinger, M.; Bryant, C.; O'Regan, M.; Gröcke, D.R.; Coxall, H.; Bouchez, J.; Calmels, D. Erosion of organic carbon in the Arctic as a geological carbon dioxide sink. *Nature,* **2015**, *524*(7563), 84-87.
[http://dx.doi.org/10.1038/nature14653] [PMID: 26245581]

[34]   Wagner, D.; Liebner, S. *Permafrost soils, Soil Biology*; Margesin, R.; Liebner, S. Springer: Berlin Heidelberg, **2009**, Vol. 16, pp. 219-236.

[35]   Coolen, M.J.; Orsi, W.D. The transcriptional response of microbial communities in thawing Alaskan permafrost soils. *Front. Microbiol.,* **2015**, *6*, 197.
[http://dx.doi.org/10.3389/fmicb.2015.00197] [PMID: 25852660]

[36]   Shakhova, N.; Semiletov, I.; Salyuk, A.; Yusupov, V.; Kosmach, D.; Gustafsson, O. Extensive methane venting to the atmosphere from sediments of the East Siberian Arctic Shelf. *Science,* **2010**, *327*(5970), 1246-1250.
[http://dx.doi.org/10.1126/science.1182221] [PMID: 20203047]

[37]   Tarnocai, C.; Canadell, J.G.; Schuur, E.A. Soil organic carbon pools in the northern circumpolar permafrost region. *Global Biogeochem. Cycles,* **2009**, *23*, GB2023.
[http://dx.doi.org/10.1029/2008GB003327]

[38]   Przybylak, R. Recent air-temperature changes in the Arctic. *Ann. Glaciol.,* **2007**, *46*(1), 316-324.
[http://dx.doi.org/10.3189/172756407782871666]

[39]   Pithan, F.; Mauritsen, T. Arctic amplification dominated by temperature feedbacks in contemporary climate models. *Nat. Geosci.,* **2014**, *7*, 181-184.
[http://dx.doi.org/10.1038/ngeo2071]

[40]   Kwong, Y.T.; Gan, T.Y. Northward migration of permafrost along the Mackenzie highway and climatic warming. *Clim. Change,* **1994**, *26*(4), 399-419.
[http://dx.doi.org/10.1007/BF01094404]

[41]   Osterkamp, T.E.; Romanovsky, V.E. Evidence for warming and thawing of discontinuous permafrost in Alaska. *Permafr. Periglac. Process.,* **1999**, *10*(1), 17-37.
[http://dx.doi.org/10.1002/(SICI)1099-1530(199901/03)10:1<17::AID-PPP303>3.0.CO;2-4]

[42]   Slater, A.G.; Lawrence, D.M. Diagnosing Present and Future Permafrost from Climate Models. *J. Clim.,* **2013**, *26*, 5608-5623.
[http://dx.doi.org/10.1175/JCLI-D-12-00341.1]

[43]   Anthony, K.M.; Zimov, S.A.; Grosse, G.; Jones, M.C.; Anthony, P.M.; Chapin, F.S., III; Finlay, J.C.; Mack, M.C.; Davydov, S.; Frenzel, P.; Frolking, S. A shift of thermokarst lakes from carbon sources to sinks during the Holocene epoch. *Nature,* **2014**, *511*(7510), 452-456.
[http://dx.doi.org/10.1038/nature13560] [PMID: 25043014]

[44]   Laurion, I.; Vincent, W.F.; MacIntyre, S. Variability in greenhouse gas emissions from permafrost thaw ponds. *Limnol. Oceanogr.,* **2010**, *55*(1), 115-133.
[http://dx.doi.org/10.4319/lo.2010.55.1.0115]

[45]   Guo, D.; Sun, J. Permafrost Thaw and Associated Settlement Hazard Onset Timing over the Qinghai-Tibet Engineering Corridor. *Int. J. Disaster Risk Sci.,* **2015**, *6*, 347-358.

[http://dx.doi.org/10.1007/s13753-015-0072-3]

[46]   Bjella, K. Thule Air Base Airfield White Painting and Permafrost Investigation, Phases I-IV. *Cold Regions Research and Engineering Laboratory, Final report, ERDC/CRREL TR-13-8, Hanover, NH, USA*, **2013**. http://acwc.sdp.sirsi.net/client/search/asset/1027825

[47]   Colgan, W.; Machguth, H.; MacFerrin, H. The abandoned ice sheet base at Camp Century, Greenland, in a warming climate. *Geophys. Res. Lett.,* **2016**, *43*(15), 8091-8096.
[http://dx.doi.org/10.1002/2016GL069688]

[48]   Filler, D.M.; van Stempvoort, D.R.; Leigh, M.B. *Permafrost Soils, Soil Biology*; Margesin, R., Ed.; Springer: Berlin, Heidelberg, **2009**, Vol. 16, pp. 279-301.

[49]   Mackay, D.; Wania, F. Transport of contaminants to the Arctic: Partitioning, processes and models. *Sci. Total Environ.,* **1995**, *160*(161), 25-38.
[http://dx.doi.org/10.1016/0048-9697(95)04342-X]

[50]   Zhang, L.; Ma, J.; Tian, C. Atmospheric transport of persistent semi-volatile organic chemicals to the Arctic and cold condensation in the mid-troposphere – Part 2: 3-D modeling of episodic atmospheric transport. *Atmos. Chem. Phys.,* **2010**, *10*, 7315-7324.

[51]   Schuster, P.F.; Krabbenhoft, D.P.; Naftz, D.L.; Cecil, L.D.; Olson, M.L.; Dewild, J.F.; Susong, D.D.; Green, J.R.; Abbott, M.L. Atmospherc mercury deposition during the last 270 years: a glacial ice core record of natural and anthropogenic sources. *Environ. Sci. Technol.,* **2002**, *36*(11), 2303-2310.
[http://dx.doi.org/10.1021/es0157503] [PMID: 12075781]

[52]   Johnson, K.P.; Blum, J.D.; Keeler, G.J. Investigation of the deposition and emission of mercury in arctic snow during an atmospheric mercury depletion event. *J. Geophys. Res.,* **2008**, *113*, D17304.

[53]   Holmes, C.D.; Jacob, D.J.; Corbitt, E.S. Global atmospheric model for mercury including oxidation by bromine atoms. *Atmos. Chem. Phys.,* **2010**, *10*, 12037-12057.
[http://dx.doi.org/10.5194/acp-10-12037-2010]

[54]   Obbard, R.W.; Sadri, S.; Qi Wong, Y. Global warming releases microplastic legacy frozen in Arctic Sea ice. *Earths Futur.,* **2014**, *2*, 315-320.

[55]   Stubbins, A.; Hood, E.; Raymond, P.A. Anthropogenic aerosols as a source of ancient dissolved organic matter in glaciers. *Nat. Geosci.,* **2012**, *5*, 198-201.
[http://dx.doi.org/10.1038/ngeo1403]

[56]   Rosen, J. Cold truths at the top of the world. *Nature,* **2016**, *532*(7599), 296-299.
[http://dx.doi.org/10.1038/532296a] [PMID: 27111613]

[57]   Melia, N.; Haines, K.; Hawkins, E. Sea ice decline and 21st century trans-Arctic shipping routes. *Geophys. Res. Lett.,* **2016**, *43*(18), 9720-9728.
[http://dx.doi.org/10.1002/2016GL069315]

[58]   Chu, P.-Y. Mapping permafrost country: Creating an environmental object in the Soviet Union, 1920s–1940s. *Environ. Hist.,* **2015**, *20*(3), 396-421.
[http://dx.doi.org/10.1093/envhis/emv050]

[59]   Zhukov, V.F. Mikhail Ivanovich Sumgin (on the centennial of his birth). *Soil Mech. Found. Eng.,* **1973**, *10*(3), 160-162.
[http://dx.doi.org/10.1007/BF01706676]

[60]   Margesin, R., Ed. *Permafrost Soils*; Springer Verlag: Berlin, Heidelberg, **2009**.
[http://dx.doi.org/10.1007/978-3-540-69371-0]

[61]   South Tyrol Museum of Archaeology. *Ötzi the Iceman,* **2016**. http://www.iceman.it/en/the-iceman/

[62]   Mercer, J.H. West Antarctic ice sheet and CO_2 greenhouse effect: a threat of disaster. *Nature,* **1978**, *271*, 321-325.
[http://dx.doi.org/10.1038/271321a0]

CHAPTER 9

Groundwater

Abstract: Due to technical development and sharp population increase, the global groundwater balance has been negative since at least 1900. Groundwater abstraction between 1965-2010 tripled to 986 km^3. As a consequence, resources like the Australian Artesian Basin were depleted. Groundwater must be seen as resource in transition. In general, its level fell due to pumping and the sealing of large surfaces, causing them to become impervious to infiltrating water. Groundwater deterioration occurred due to excess pumping, resulting in salinisation. Groundwater contamination occurred because of leaking toxic liquid waste at nuclear repositories; leaking unconfined, tailings containing residual heavy metals, sulphuric acid, and cyanide; dispersed PCBs; industrial waste dumped in abandoned mine workings; injection of residual salt brines and liquid radioactive waste underground; and excess utilisation of artificial fertilisers and manure in industrial agriculture, resulting in elevated NO_3^- -concentrations in groundwater wells. Example: Widespread decreasing groundwater availability and quality, as well as groundwater level fall, in China due to overuse for domestic, agricultural, and industrial demand. Groundwater protection measures improved the situation locally. In general, climate change-induced water scarcity can only be temporarily compensated by technology.

Keywords: Aquifer, Brine injection, Contaminant transport, Contamination, Groundwater abstraction, Groundwater protection, Pollution, Quality standards, Reactive nitrogen, Resource depletion, Salinisation, Transitional resource.

The global, total volume of groundwater present in the upper 2 km of exposed continental crust is estimated at 22.6 Mkm^3, and groundwater rechargeable within 50 years - a resource highly vulnerable to global change - is estimated at 0.1-5.0 Mkm^3 [1]. Driven by exponential population growth, and techno-scientific and economic development, groundwater abstraction globally has accelerated very quickly since the beginning of the 20th century. The cumulative amount of global ground water abstraction has tripled since 1965 and was estimated at ca. 986 km^3 in 2010 (household: 212 km^3, virtual water for irrigation: 666 km^3 and for industry: 108 km^3) [2].

Global groundwater net depletion between 1900 and 2008, is estimated to be ca. 4500 km^3 [3]. Example: Discharge of groundwater in the Australian Great Artesian Basin diminished from a maximum of 750 million litres/a in 1915 to 330 million litres/a in 2000 due to the implementation of ca. 4800 artesian wells between 1890 and 2000 and their operation [4]. Salinification of groundwater was caused by several types of utilisation or interferences, *e.g.* incongruous abstraction, which generated ingression of saline fluids [5]. Groundwater levels dropped due to excess pumping [2, 6, 7] and below large surfaces being made impervious by activities of the building industry: covering with asphalt, concrete, dimension stones, and implementation of settlements [5]. Much of the global road network, which, in 2007, had a cumulative length of ca. 50 million km, is covered with tarmacadam and asphalt concrete [8], which are impervious to percolating water[1]. The cumulative urban area of all 1022 cities with populations over 500,000 persons (cumulative number of inhabitants: ca. 2.12 billion), is ca. 481,601 km^2 [9], of which large parts are made impervious to infiltration water. The area occupied by all settlements globally is estimated to be ca. 1.5 million km^2.

The water quality of aquifers has decreased where infiltered with contaminated, polluted or nutrient-enriched waters from point- or diffuse sources [5]. Several examples follow:

At Hanford Nuclear Reservation Site - established as component of the Manhatten Project -, which is situated close to the Columbia River in Washington State (USA), where 177 underground tanks containing 200 million litres of high-level radioactive waste, were implemented. 60 of them began to leak accidentally and caused contamination plumes percolating through the unsaturated zone and through fractured rock, seeping into the groundwater and migrating along its flow paths, carrying tritium, ^{99}Tc, ^{137}Cs, ^{60}Co and other toxic chemicals, towards the receiving stream. In the meantime, 520 km^2 of that aquifer has been contaminated. A clean-up priority act was initiated in 2004 [10, 11].

Hydrogeological investigations at Lake Karachay (south Urals), which was contaminated with liquid radioactive waste (activity in 1999: 4.44×10^8 TBq), revealed that radionuclides (*e.g.* ^{90}Sr, ^{60}Co, ^{137}Cs, ^{106}Ru, *etc.*) have migrated through fissured bedrock metavolcanics into the aquifer below, forming contamination plumes and aureoles spreading according to hydrodynamic gradients. The amount of contaminated groundwater was estimated to be ca. 5 million m^3 [12, 13].

Intermediate and deep well injections of all kinds of liquid radioactive waste were carried out in the vicinity of several reprocessing plants (*e.g.* Hanford, Oak Ridge, Mayak, Tomsk-2, Krasnoyarsk-26), relying on the isolation capacity of confined aquifers, on the duration of migration in pore aquifers, as well as on their retention time [14].

Dispersed solid radioactive material fragments, ejected during the Chernobyl disaster, made safe clean-up necessary; but the matter was dumped in several adjacent unsealed sites, which have contaminated ambient groundwater [15]. Toxic seepage from many undrained and unconfined tailings dumps, containing 1.3 GT of waste rock milled to powder during 120 years of gold mining[2] in the South African Witwatersrand Basin leaked into groundwater, got dispersed by its flowpaths, and entered, in diluted form, perennial rivers and lakes. The seepage was produced by rainwater entering the dumps and decomposed pyrite in several consecutive chemical reactions, which resulted in precipitated $Fe(OH)_3$ and sulphuric acid. In this way, low pH percolating waters dissolved toxic heavy metals, which were not extracted and remained in the slurry[3]. The contamination of ground and surface waters manifested itself in low pH-values and elevated SO_4^{2-}-concentrations, which turned out to increase regularly with the rising ground water table. Depending on alkalinity and distance from the point sources, elevated concentrations of Al, Cu, Zn, Co, Mn, Ni, and U were detected [16, 17]. The duration of generation of acid and toxic metals from such tailings deposits depends on the concentration of pyrite and other sulphides, which were not extracted and left in the slurries, as well as on geochemical and ecological parameters; acid and metal generation and their release into the environment may take hundreds of years [18].

Export of PCB, dispersed between 1977-1983 in the workings of Ruhr mining area (Germany), into the biosphere occurred during extraction of coal and during pumping of acid mine drainage into receiving surface water courses [19, 20].

During potassium mining in the Werra Valley (Germany) between 1928-1992, ca. 625.5 million m^3 of salt brines were injected into the subsurface and caused salinisation of groundwater and drinking water wells [21].

Hydrochemical analyses conducted between 1926 and 1989 by German water supply companies, which predominantly use ground water for drinking water purposes, revealed an overall increase in nitrate concentrations: 78.1% of hydrochemical analyses carried out in 1926 contained less than 10 mg nitrate per liter; until 1989, the percentage of analytic results with less than 10 mg/L dropped continually to only 46.1%. This fact has been attributed to anthropogenic activities, which have saturated the biosphere beyond its capacity to mineralise N through intensified industrial agriculture, domestic and industrial waste, sewage, and atmospheric depositions. Nitrogen balances in agricultural areas in 15 European countries, conducted between 1976 and 1995, resulted in each case in surpluses varying between 16 and 367 kgN/a/ha. Excess nitrogen originates when more N is brought into soil by fertiliser, fodder and atmospheric deposition, as can aggregate in animals and plants. The difference accumulates in the environment, because the N-cycle is open [22].

According to the evaluation of 33,493 survey stations in the EU, ca. 35% of the groundwater bodies contain more than 25 mg NO_3^-/L [23]. Inappropriate industrial agricultural management resulted in application of excessive quantities of fertilisers and the generation of too many food animals. Global fertiliser consumption rose from ca. 4.5 MT in 1950 to 141 MT in 2000 [5, 24]. Considering the groundwater situation in Germany for instance, selective monitoring of near-surface aquifers, not confined by aquitards, between 1992 and 2010 at 186 points known for elevated inputs from agricultural activity revealed that 49% of the groundwater check points was polluted with > 50 mg NO_3^-/L nitrate, and that the general trend consists of 25.9% stable monitoring points, while at the others a slight decrease of NO_3^--values prevails. However, the overall shallow groundwater situation, derived from 739 common monitoring stations of the European Environment Agency Network, reveals that 14.3% exceeded the quality standard (> 50 mg/L NO_3^-) in 2010. Considering the subset (n = 342) present in regions influenced by agricultural activity, the share of > 50 mg/L is 23.1% [25]. The main contribution resulting in cascading concentrations of NO_3^- load in surface- and groundwater, is from industrial agricultural activity[4]. Subsequent to the creation of groundwater protection areas, as well as temporal and spatial limitation of land use and soil fertilisation in their vicinities, the situation improved locally. Reduction of NO_3^- contamination was effected by reducing the application of fertilisers and the implementation of water retention ponds [26].

Monitoring of groundwater quality in China, based on hydrochemical analysis of 641 wells, revealed that, in 2009, 73.8% failed drinking water quality standard, because of pollution with ammonia, nitrite, nitrate, metal ions, *etc.*, while 27% of 397 centralised drinking water sources for cities yielded substandard water [27].

The Monthly Groundwater Report, published in 01/2016 by the Chinese Ministry of Water Resources, declared that a hydrochemical survey of 2103 wells found the quality of 80.2% of shallow groundwater is below that of drinking water standards, because of pollution with discharge from industry and washout from agriculture. 73% of the deep ground water below the North China Plain failed drinking water standards [28, 29]. Another source, however, stated that 38% of the groundwater is of satisfactory, good, and excellent quality in 2014 and that its state has deteriorated gradually by 5% since 2011 [30].

Total water availability declined, between 1956 and 2013, from 65 billion m^3 to 30 billion m^3, and rising total water demand in Beijing, due to fast population growth and increasing domestic, agricultural, and industrial demand, resulted in overexploitation of groundwater, land subsidence and ground water quality degradation [31].

For 2000 years, transmission of meltwater from glaciated mountain ranges *via* subterranean aqueducts - so called karezes - has provided safe and reliable water for domestic purposes and agricultural irrigation in the semiarid climate of Turpan Basin (Xinjiang Province, northwest China). It has decreased because of recent overexploitation of deep groundwater wells, construction of river dams, and oil field development, resulting in reduction of the shallow aquifer feeding the karez system. It was built to prevent high evaporation loss, consisted of ca. 1800 karezes in 2003 and measured in sum 5300 km long, though only 614 were active. Groundwater level fell during the past 10 years by 25 m [32].

Due to numerous groundwater protection measures, the situations improved locally. Climate change-induced water scarcity affecting societies can be compensated only temporarily by technological development [33]. Because of progressive pollution and excess abstraction of groundwater, it must be recognised as "resource in transition" [2].

NOTES

[1] Assuming that 50% of the global road network is asphaltised and its average width is only 6 m, a surface summing up to 150,000 km^2 is made impervious.

[2] It started in 1884 in Witwatersrand Basin by extracting Au from crushed and milled quartzconglomerate and quartzite – consisting of meso- to late Archaean, ca. 11 km thick distal fluvial sediments deposited between 2.9-2.7 Ga ago on the Kaapvaal-Craton - by applying Hg-containing amalgam and since 1915 the MacArthur-Forrest cyanidation process. Depending on the gold price per ounce, profitable concentrations of detrital Au-grains (palaeoplacer gold) vary around a yield of ten grams per 1000 kg host rock [34]. In sum, 90,000 T of gold were recovered [35].

[3] A part of the dissolved heavy metals like U (and all transmutation products of the nuclear decay series), As, Cu, Ni, Pb, Co, and Zn is co-precipitated and adsorbed by $Fe(OH)_3$ [16].

[4] Data (2003-2010) from the official agricultural statistics of industrial agriculture: Area under consideration: 17 million ha; N applications from artificial fertilisers: 1.8-1.6 MT; 105-94 kg/ha; N applications from manure: 1.3-1.3 MT; 75-75 kg/ha; number of food animals: cattle: 13.4- 12.7 million; pigs: 26.5-26.9 million;

poultry: 123.4 -128.9 million; other animals: 3.3-2.7 million [24, 25].

CONFLICT OF INTEREST

The author (editor) declares no conflict of interest, financial or otherwise.

ACKNOWLEDGEMENTS

Declare none.

REFERENCES

[1] Gleeson, T.; Befus, K.M.; Jasechko, S. The global volume and distribution of modern groundwater. *Nat. Geosci.,* **2015**, *9*, 161-167.
[http://dx.doi.org/10.1038/ngeo2590]

[2] Van der Gun, J. *Groundwater and Global Change: Trends, Opportunities and Challenges,* **2012**. http://www.unesco.org/water/wwap

[3] Konikow, L. Contribution of global groundwater depletion since 1900 to sea-level rise. *Geophys. Res. Lett.,* **2011**, *38*, 1-5.
[http://dx.doi.org/10.1029/2011GL048604]

[4] Habermehl, M. *Non-Renewable Groundwater Resources: A Guidebook on Socially-Sustainable Management for Water-policy Makers*; Foster, S.; Loucks, D.P., Eds.; UNESCO-IHP, IHPVI, Paris/Reading, UNESCO/IAH, Series on Groundwater, **2006**, Vol. 10, pp. 82-86.

[5] Goudie, A.S. *The Human Impact on the Natural Environment. Past, Present and Future,* 7th ed; Wiley-Blackwell: Oxford, **2013**.

[6] Wada, Y. Heinrich, L. Assessment of transboundary aquifers of the world - vulnerability arising from human water use. *Environ. Res. Lett.,* **2013**, *8*, 024003.
[http://dx.doi.org/10.1088/1748-9326/8/2/024003]

[7] IGRAC. *Global Groundwater Information System (GGIS). Delft, the Netherlands, IGRAC,* **2010**. http://www.igrac.net

[8] Zalasiewicz, J.; Waters, C. *The Anthropocene. Environmental Science, Oxford Research Encyclopedias,* **2015**.

[9] Demographia World Urban Areas. *Built up urban areas by land area (urban footprint), 12th Annual Edition,* **2016**. http://www.demographia.com/db-worldua.pdf

[10] Wilshire, H.; Nielson, J.E.; Hazlett, R.W. *The American West at Risk. Science, Myths and Politics of Land Abuse and Recovery*; Oxford University Press: New York, **2008**.

[11] Physicians for Social Responsibility. *Hanford facts,* **2015**. http://www.psr.org/chapters/washington/hanford/hanford-facts.html

[12] Merkushkin, A.O. *Karachay Lake is the storage of the radioactive wastes under open sky,* http://www.iaea.org/inis/collection/NCLCollectionStore/_Public/33/011/33011239.pdf**1999**.

[13] Solodov, I.N.; Zotov, A.V.; Khoteev, A.D. Geochemistry of natural and contaminated subsurface waters in fissured bed rocks of the Lake Karachai area, Southern Urals, Russia. *Appl. Geochem.,* **1998**, *13*(8), 921-939.
[http://dx.doi.org/10.1016/S0883-2927(98)00025-0]

[14] Solodov, I.N.; Zotov, A.V.; Khoteev, A.D. Geochemistry of natural and contaminated subsurface waters in fissured bed rocks of the Lake Karachai area, Southern Urals, Russia. *Appl. Geochem.,* **1998**, *13*(8), 921-939.

[http://dx.doi.org/10.1016/S0883-2927(98)00025-0]

[15]   Deutsches Atomforum, V. *Der Reaktorunfall in Tschernobyl - Unfallursachen, Unfallfolgen und deren Bewältigung; Sicherung und Entsorgung des Kernkraftwerks Tschernobyl, Berlin*, **2011**. http://www.kernenergie.de/kernenergie-wAssets/docs/service/025reaktorunfall_tschernobyl2011.pdf

[16]   Tutu, H.; McCarthy, T.S.; Cukrowska, E. The chemical characteristics of acid mine drainage with particular reference to sources, distribution and remediation: The Witwatersrand Basin, South Africa as a case study. *Appl. Geochem.*, **2008**, *23*, 3666-3684. [http://dx.doi.org/10.1016/j.apgeochem.2008.09.002]

[17]   McCarthy, T.S. The impact of acid mine drainage in South Africa. *S. Afr. J. Sci.*, **2011**, *107*(5-6), 7 pages.

[18]   Kalin, M. *Post-mining Landscapes. Reclamation, Ecology, Nature Conservation and Socio-economy in Practice*; Xylander, W.E., Ed.; Peckiana: Görlitz, Germany, **2004**, Vol. 3, pp. 101-112.

[19]   Duin, G. *Bericht der Landesregierung an die Präsidentin des Landtags*, **2015**. https://www.landtag.nrw.de/portal/WWW/dokumentenarchiv/Dokument/MMV16-2631.pdf

[20]   Rahm, H. *Schwebstoffuntersuchungen des LANUV zu bergbaubürtigen Indikator-PCB in Grubenwasser und Oberflächengewässern 2015*, **2015**. http://www.bund-nrw.de/fileadmin/ bundgruppen/bcmslvnrw/PDF_Dateien/Themen_und_Projekte/Energie_und_Klima/Altlasten_Steinko hle/PCB_-_RAG_LANUV_2015-07-28_Zwischenbericht_Ergebnisse.pdf

[21]   Hübner, G. *Ökologisch-faunistische Fließgewässerbewertung am Beispiel der salzbelasteten unteren Werra und ausgewählter Zuflüsse. Ökologie und Umweltsicherung*, **2007**. 27, Diss. Univ. Kassel, Germany

[22]   WHO. *Chapter 1 - Origin of nitrate in drinking water*, **2001**. https://www.yumpu.com/ en/document/view/10554959/chapter-1-origin-of-nitrate-in-drinking-water-world-health-

[23]   Europäische Kommission. *Bericht der Kommission an den Rat und das Europäische Parlament*, **2013**. http://eur-lex.europa.eu/LexUriServ/LexUriServ.do?uri=COM:2013:0683:FIN:DE:PDF

[24]   Conway, G.R.; Pretty, J.N. *Unwelcome harvest: agriculture and pollution*; Earthscan: London, **1991**.

[25]   Keppner, L.; Rohrmoser, W.; Wendang, J. *Nitrates Report 2012*, **2012**. http://www.bmel.de/ SharedDocs/Downloads/EN/Agriculture/OrganicFarming/Nitratbericht-2012.pdf?__blob=publicationFile

[26]   Garnier, J.; Billen, G.; Vilain, G.; Benoit, M.; Passy, P.; Tallec, G.; Tournebize, J.; Anglade, J.; Billy, C.; Mercier, B.; Ansart, P.; Azougui, A.; Sebilo, M.; Kao, C. Curative *vs.* preventive management of nitrogen transfers in rural areas: lessons from the case of the Orgeval watershed (Seine River basin, France). *J. Environ. Manage.*, **2014**, *144*, 125-134. [http://dx.doi.org/10.1016/j.jenvman.2014.04.030] [PMID: 24935024]

[27]   *China Report 2009, Ministry of Environmental Protection of the People's Republic of China. Report on the State of the Environment of China, Water Environment*, **2009**. http://english.mep. gov.cn/down_load/Documents/201104/P020110411532104009882.pdf

[28]   Yi, J. *More than 80 percent of China's groundwater polluted*, **2016**. http://en.tuidang.org/news/ environment/2016/04/more-than-80-percent-of-chinas-groundwater-polluted.html

[29]   MWR. *Groundwater Monitoring Monthly Report, China Government Reports*, **2016**. http://www.mwr.gov.cn/zwzc/hygb/dxsdtyb/201604/P020160405539942030096.pdf

[30]   China Water Risk. *State of Environment Report Review*, **2014**. http://chinawaterrisk.org/resources/ analysis-reviews/2014-state-of-environment-report-review/

[31]   Global Water Partnership, Technical Focus Paper. *China's water resources management challenge: The 'three red lines'*, **2015**. http://www.gwp.org/Global/ToolBox/Publications/Technical%20 Focus%20Papers/TFPChina_2015.pdf

[32]   Abudu, S.; Sheng, Z.; Cui, C. *In: Harvesting Water and Harnessing Cooperation: Qanat Systems in the Middle East and Asia, Middle East Institute,* **2014**. http://www.mei.edu/content/harvesting- water-and-harnessing-cooperation-qanat-systems-middle-east-and-asia

[33]   Pande, S.; Ertsen, T.; Sivapalan, M. Endogenous technological and population change under increasing water scarcity. *Hydrol. Earth Syst. Sci.,* **2014**, *18*, 3239-3258.
[http://dx.doi.org/10.5194/hess-18-3239-2014]

[34]   Janisch, P.R. Gold in South Africa. *J. S. Afr. Inst. Min. Metall.,* **1986**, *86*(8), 273-316.

[35]   Frimmel, H.E. Von den Frühstadien des Lebens zur Bildung der weltweit größten krustalen Goldanreicherung. *Geowiss. Mitt.,* **2015**, *62*, 6-17.

CHAPTER 10

Rivers, Lakes, Reservoirs, and Inland Seas

Abstract: Intensified agriculture, input of sewage and waste, as well as deforestation in the 20th century in Europe caused a decline of river water quality and an increase in its turbidity, and rapid terrestrialisation of lakes. The situation was improved by sewerages and water purification plants, which, however, cannot retain all pollutants and toxins. Among the problems are elevated nitrate concentrations, Hg, microplastics and waste heat from power plants. Generally, immissions from mines, industry, and urban areas have impaired hydrochemistry of rivers and lakes. Examples: Potassium mining salinised the river Werra; unconventional oil development at Athabasca impaired ambient catchments with toxic aerosols. Ecological and hydrological degradation of very large lakes occurred due to the introduction of xenospecies to foster industrial fish catch, input of acid mine drainage, poor sewage treatment, nearby overgrazing and deforestation, and excess water abstraction from affluents. Washout of agricultural nutrients and inappropriate disposal of sewage effected vertical shoaling of the world's largest anoxic, sulphidic water body in the Black Sea. The global trend of lakes towards CO_2 supersaturation results from enhanced input of dissolved organic carbon (DOC), promoting microbial respiration. In remote glacial lakes, climate warming has caused reduction of ice coverage and of permafrost, and effected substantial alteration of lacustrine hydrochemistries. A generalisation of effects of anthropogenic activities to the flow variability of large rivers is not possible. However, engineering, freshwater withdrawal, and extraction of sand and gravel fostered river bed erosion and caused delta starvation, subsidence, and coastal erosion. Hydrochemical studies recognised severe pollution of parts of surface waters in China. Radioactive contamination of surface waters occurred near nuclear plants and repositories.

Keywords: Anoxia, Contamination, Desiccation, Ecosystem change, Erosion, Eutrophication, Immissions, Irrigation, Reactive nitrogen, River engineering, Surface water, Terrestrialisation, Warming, Water abstraction, Water balance, Water demand, Water pollution, Water protection, Water quality, Xenospecies.

World water use, composed of reservoir losses, as well as industrial, municipial, and agricultural uses, increased from ca. 600 km^3 in 1900 to 3700 km^3 in 2000; the largest increases occurred between 1940 and 1980 [1]. Total global freshwater withdrawal in the year 2000 amounted to ca. 2105 km^3 [2] and rose in 2010 to ca. 2845 km^3 [3]. But improved accounting of effects of flow regulation and irrigation on evapotransration and temporal runoff variability of surface waters according to

Hubert Engelbrecht

hydroclimatic data - dating back to 1901 - from 100 large basins revealed that the global fresh water footprint of humanity has currently risen to $10,688 \pm 979$ km^3/a [4].

Fresh water, which is only 0.01% of global water and which is an important environment for ca. 100,000 species (ca. 6% of all described globally), is loosing biodiversity because of the combined and interacting influences of overexploitation, pollution, flow modification, destruction or degradation of habitats, and invasion of exotic species [5]. Turbid waters from washed-out agricultural land and impaired river water chemistry from discharged pollutants caused decline of the pearl producing bivalve, *Margaritifera margaritifera* L., population in Central Europe by 95%, thus putting an end to pearl fisheries. Juveniles of this bivalve can be regarded as biosensors, which live in symbiosis with host fishes (trout, salmon) and only tolerate, when settled, sandy/cobbly substrata, with quick flowing, shadowed, clear water; a pH of 6.2-7.5, a temperature of < 25 °C, and low contents of aluminium, total phosphorus, and nitrate; and without silt or mud [6, 7]. But, predominantly due to intensified industrial agriculture and forestry, the increase of fluvial sediment transport and its deposition in European lakes is estimated to have increased by factor 100 since 1900 [8], causing accelerated erosion of the diffuse sources and terrestrialisation of these sinks. The latter was first described in 1953 in Germany as "rapid lake ageing" [9].

In the following, several examples detail the degradation of lake hydrology and ecology caused by anthropogenic activities:

The introduction in the 1950s of the Nile perch (*Lates niloticus* Linnaeus 1758) into Lake Victoria to foster industrial fisheries, as well as intensified agriculture in the lake's catchment area, caused a fundamental modification of its ecosystem. The development of primary production and chlorophyll concentration, as well as the increase in volume and duration of bottom anoxia point to transition to eutrophication. In the time considered, input of total phosphorus increased twentyfold to 400 kg/km^2/a, and fish catch increased from 25,000 T in 1960 to 500,000 T in 1990 [10].

Lake Titicaca - a former Incan sanctuary - and its neighbouring lakes are situated in a low humidity to arid basin of 57,293 km^2, established on the Altiplano (3600-4500 m asl) within the Andean Mountain Range. Their water quality deteriorated due to anthropogenic activities. Discarded acid mine drainage from heavy metal mining in vicinity of the lakes caused elevated levels of lead, zinc, cadmium, nickel, cobalt, manganese, mercury, and arsenic in the water and limnic sediments. Poor communal waste disposal and discharge of predominantly untreated sewage from nearby urban centres into the lakes resulted in accumulation of organic pollutants and the presence of elevated levels of coliforms. Nitrogen compounds from livestock raising and agricultural activity in the surrounding areas contributed to a higher load of organic pollutants and partial eutrophication. Water abstraction from tributaries for irrigation purposes (in sum 7379 L/sec.) reduces freshwater influx into the lakes. In addition, the high evaporation rate increases the lake ecosystem's sensitivity concerning pollution [11, 12].

Superimposed on a nonlinear response to orbital forcing[1], the shrinkage speed of Lake Chad's area doubled due to anthropogenic activities like abstraction of irrigation water at inflows, and overgrazing and deforestation in nearby areas. The lake's size has diminished to one twentieth - 1350 km^2 - of its size in 1963 [13, 14].

Saline water abstraction since 1929, necessary for industrial mineral production (*e.g.* potash, sylvite, bromine, carnallite) in evaporation ponds, and diversion of fluvial waters from the main tributary Jordan River since the 1960s, serving for supply of cities and industrial scale irrigation, caused a mass imbalance of the hypersaline terminal lake, the Dead Sea. The annual water deficit runs up to ca. 650 million m^3 and the annual inflow of the River Jordan has diminished from 1600 million m^3 in 1948 to only 25 million m^3 in 2010. The continuous drop in water level and diminution of its surface from 1650 km^2 in 1930 to 600 km^2 in 2014 have been the consequences [15 - 17].

Since ca. 1960, the Aral Sea's water balance has been in disequilibrium due to unsustainable water abstraction to irrigate cotton plantations. Thus its water volume decreased from initially 1070 km^3 by at least 85%. During this process, Aral Sea disintegrated into three separate residual water bodies, in which differing hydrological and limnobiological developments proceeded. The Large Aral Sea and Lake Thchebas became hypersaline, but the Small Aral Lake retained its brackish milieu due to resumed input of fluvial freshwater. The waters of Small Aral Sea and Lake Thchebas fully mix, but those of Large Aral Sea are stratified by salinity and temperature. The highly diverse salinities (11.1 g/kg - 133.8 g/kg) in these water bodies have developed different microbial communities (*e.g.* diatom microphytobenthos) [18]. Similar cases of detrimental anthropogenic influence - *e.g.* overuse of water resources, contamination - were reported from *e.g.* Lop Nor, Caspian Sea, Issyk-Kul, Lake Corangamite, Lake Chad, Lake Urmia, and Lake Titicaca [19].

Monitoring of the O_2-inventory of the Black Sea between 1955 and 2015 (4385 ship-based vertical profiles measured O_2, salinity, and temperature) indicated that its O_2 content has decreased by 44%. As a result, the permanent chemocline separating the oxygenated, low salinity surface layer from the hypoxic and anoxic, sulphidic and high salinity bottom layer shoaled in that time interval vertically upward by 50 m to 90 m water depth. In the surface layer, fed by inflowing fluvial waters, mixing occurs and a seasonal thermocline is present. The stratified and stagnating, euxinic[2] lower layer is fed by sea water from the Bosporus and contains the world's largest reservoir of hydrogen sulphide. The upward excursion of the chemocline is explained by restricted connection to open sea waters; by eutrophication of the lower part of the oxic surface layer due to fluvial input of agricultural nutrients, pesticides, and sewage; as well as by reduced ventilation of the surface layer resulting from climate change-induced warming [20].

Because high quality freshwater in sufficient quantity is essential in all dimensions for global human well-being, economic prosperity, and working ecosystem services, the "UN International Decade for Action - Water for Life 2005-2015" recommended to protect freshwater from pollution and excess abstraction, to prevent water-related disasters, and to improve water efficiency [21]. The share of total water use between 1998-2007 is 70% for agriculture, 20% for industry and 10% for domestic use [22]. However, freshwater quality degraded due to acid rain, meltwaters loaded with particulate atmospheric contaminants, deposition of toxic dust, input of trash, communal sewage, acid mine drainage, washed mud, artificial fertilisers, manure, and biocides from farmland, as well as waste waters and waste heat from industrial plants. Drained cooling water from power plants was identified as serving as an artificial habitat for self-sustaining tropical fish and plant species, as well as sources of novel pests in European waters: released pet fish *Amatitlania nigrofasciata* was found to host a non-indigenous parasite fauna (the tropical nematode *Camallanus cotti*), which has already spread to native fish species [23].

Monitoring of 235 lake summer surface water temperatures, conducted between 1985 and 2009, points to a global warming rate of averaged 0.34 °C/10a, which results from climate change-induced atmospheric heat transfer [24].

Considering differing production systems and feed composition per food animal species between 1996 and 2005, it has been estimated that more than one third (2.422 Gm^3/a) of global water footprint of agricultural production (8.363 Gm^3/a) is related to generation of animal products [25].

According to the evaluation of 29,018 survey stations in the EU, ca. 15% of the surface waters contain more than 25 mg NO_3/L [26].

Rapid industrialisation since 1950 and population growth in China effected that on average 42.7% of 203 rivers monitored failed to reach drinking water quality standard, predominantly because of piped-in untreated wastewater containing organic pollutants, ammonia, *etc*. A hydrochemical study of 26 key reservoirs and lakes revealed that the eutrophication level of 76.9% was intermediate to heavy due to agrochemicals and wastewater containing ammonia and phosphorus [27].

Locally, situations improved depending on the time of implementation of drains, filters, water purification plants, sewerages[3], and water protection laws. An important method to evaluate degradation of micropollutants (ng/L, µg/L) like washed out pesticides in catchments or incompletely removed pharmaceuticals from water treatment plants, and to identify micropollutant sources, is compound-specific stable-isotope analysis [28].

More than 1000 long time series analyses (> 40 a) of runoff data of 195 rivers reveal a complex correlation between anthropogenic activity (deforestation in catchment areas, urbanisation, reservoir construction, water abstraction, river engineering, and river deepening), natural flow variability, and recent climatically-forced hydrological changes: annual maximum flows changed insignificantly in 137 cases, decreased in 31 cases, and increased in 27 cases [29]. Examples: Annual flow rates of the Colorado River at its estuary decreased from averaged 27 billion m^3 in 1905 to almost zero in 2005 due to rising anthropogenic water withdrawal. Similar developments towards the limits of peak surface water were observed at the Yellow, Nile, Amu Darya and Jordan Rivers [30].

Microplastic profile measurements conducted in lakes and rivers revealed considerable loading with microplastic fragments, spherules, and fibres, *e.g.* the surface water of river Rhine yielded an average of 892,777 pieces of plastic debris per km^2 [31]. Trawler-based sampling with a 333 µm mesh net over a distance of 1300 km in the surface waters of the Great Lakes resulted in an average abundance of ca. 43,000 microplastic particles/km^2 and maximum concentration of 466,000 fragments/km^2 [32].

Straightening of river courses and construction of artificial levees caused intensified erosion and more frequent inundation events [15]. More conferred building permits - due to increase in population and growth of cities - in areas close to rivers enhances the risk of flooding and loss of material and lives.

Drainage, dredging, and inappropriate water management in wetland areas developed at lakes reduced their ecological functions as habitats for wildlife and plants, as well as their function as biochemical sinks for organic nitrogen and phosphorus [33].

The global environmental legacy of industrial mining, especially coal, gold, and platinum, is - among other - acid mine drainage (AMD), which contains sulphuric acid and heavy metal ions, and contaminates surface waters if dumped into receiving streams, and which contaminates groundwater by advection and mixing if pumping of AMD in decommissioned mines has been stopped and the workings are flooded with rising groundwater and mix with acid mine drainage. Decant from post-closure gold and coal mines in the Republic of South Africa is estimated to be more than 62 million L/day, which degrades surface water quality on a regional scale. Thus, expensive treatment methods are unavoidable to maintain drinking water quality and to prevent further contamination of ambient fauna and flora[4] [34].

Unconventional oil development in the Athabasca River Basin (Canada), which grew from 1995 to 2008 by ca. 200% to 1.3 million barrels/d, resulted in the devastation of 530 km^2 of boreal forest landscape, in the creation of 130 km^2 of tailings ponds and in loading of ambient surface waters with trace metal elements toxic at low concentrations. Comparison of hydrogeochemical analyses from upstream, near-development, and downstream sites, as well as from undeveloped reference sites, revealed the clearly detrimental impacts of immissions - from oil sand production and further industrial processing - on surface waters. Priority pollutants like antimony, arsenic, beryllium, cadmium, chromium, copper, lead, mercury, nickel, selenium, silver, tellurium, and zinc, contained in those immissions have been proven to be present in ambient fluvial and lacustrine environments above natural background concentrations [35].

Reported spills from 31,481 unconventional oil and gas developments, carried out from 2005-2014 in four US-states, amounted to 6648 incidents, which impaired - beside soil and groundwater - also surface freshwater. Spill volumes varied between 0.004 m^3 and 3756 m^3 and cumulated volume of reported spills came up to 46,755 m^3 [36].

Potassium mining in Werra Valley (Germany) since 1901 has resulted in the discharge of considerable amounts of brines into ambient fluvial systems and salinisation of the river Werra. Ion concentrations temporarily exceeded 2251 mg/L Cl^- (estimated maximum: 40,000 mg/L Cl^- in 1976), 104 mg/L Ca^{2+} and 127 mg/L Mg^{2+}, which were measured on 30/08/1928 [37].

Between 1949 and 1956 76×10^6 m^3 of liquid radioactive waste (activity 10^5 TBq) from Mayak nuclear weapon factory was discharged into the Techa River (Ural). The storage of liquid radioactive waste (activity 22 TBq) in Lake Karachai (Ural) caused contamination of the sediment, which was dispersed in 1967 by wind over an area of 1800 km^2 [38, 39].

Aerosol depositions - sulphuric acid, heavy metal sulphides - from smelters have caused acidification and toxic metal concentrations in adjacent lakes, resulting in adaptive stress to original lacustrine biota and, consequently, in compositional changes of indigenous fauna and flora [40].

Geochemical analytics of dust, emitted by vehicles and deposited as mud in drainage-containments of superhighways revealed elevated concentration of particulate platinum and rhodium (several 10s of µg/kg) and that the sewage is enriched in these rare earth elements by factors between 4 and 10 (maximum 196 µg/l); the anthropogenic sources were identified as catalytic converters [41].

Averaged annual fluvial runoff of mercury from continents to oceans has increased from 40 T/a in preindustrial time to 200 T/a [42], while other calculations estimate as much as 380 T/a [43].

Geochemical analysis of globally distributed lacustrine deposits reveal that a preindustrial background sedimentation rate of 3-3.5 µg $Hg/m^2/a$ increased during industrialisation and attained up to 20 µg $Hg/m^2/a$ in the 1980s [44]. Bioavailability and bioaccumulation studies of total Hg and methyl Hg revealed that, downstream of an industrial Hg polluted point source, the contamination levels in fish increase [45].

Pre-human suspended sediment fluvial transport to the marine coasts has been estimated to be ca. 15.1 GT/a. From the beginning of industrialisation to the first half of the 20th century, machine-based soil moving, agriculture, mining, terracing, and forest clearance increased the amount of global fluvial sediment load. But during the second half of the 20th century, global fluvial sediment discharge diminished by ca. 15% to 12.8 GT/a due to numerous dam constructions - globally ca. 50,000 large dams - and reservoirs. The resulting cumulative sediment deficit at the coasts - causing subsidence of deltas - is calculated to be ca. 73 km^3, represented by an area of 7300 km^2 covered by a progradational sediment wedge of 10 m thickness [46].

Extraction of fluvial sand and gravel - used in the building industry - has deteriorated habitat conditions for flora and fauna, triggered landslides, increased riverbank erosion, and labilised channels, and the course of streams [47, 48].

Global monitoring of 2898 CO_2 data points from 4902 lakes resulted in the fact that the state of many of them (93%) are close to CO_2 supersaturation. This can be explained by climate change-induced increase in the concentrations of predominantly fluvially imported dissolved organic carbon from soils, which impedes photosynthesis due to light absorption, fosters microbial respiration, and consequently contributes to the CO_2 partial pressures in the lakes [49]. This has been confirmed by outdoor stream mesocosm experiments in streams draining agricultural catchments: simulation of climate change conditions - impacts of stressors like rise in water temperature, nutrient concentration, and fine sediment load - result in complex ecosystem responses, *e.g.* enhanced decomposition processes of organic matter *via* aerobic and anaerobic metabolic activity of microbes and fungi [50], resulting in deoxygenation of streams and the release of more CO_2 and N_2O.

Global fluvial export of dissolved organic carbon from the soils in drainage areas to the ocean basins is estimated to be 400-900 MT/a. Due to climate change-induced global warming, dissolved organic carbon loads in Arctic rivers have risen because they drain the active layers of permafrost areas, which have increased in depth, temperature, and seasonal duration [51].

Other hydrological consequences of climate change are as follows:

Dry river beds and impounded waters of fluvial networks, which form temporarily during more frequent summer droughts, turn into sources emitting additional CO_2 and CH_4 [52].

The vertical distribution of organic C and N concentrations, as well as planctonic diatom, chrysophyte, and chironomid assemblages in radiometrically dated sediment cores of seven remote arctic and alpine lakes of similar basic limnologic characteristics (oligotrophic, dimictic) revealed relatively unambiguous temporal correspondence with lowland instrumental meteorological time series data, thus confirming responses of lake ecosystems to rising anthropogenic contribution of climate forcing since ca. 1800 [53]. The investigation of diversity development of the chironomid midge assemblages present in three remote, high alpine lakes of the Eastern Alps resulted in the recognition of substantial shifts since 1850 because the annual duration of lake ice coverage has diminished [54].

20 years of monitoring of both the hydrochemistry of a Himalayan glacial lake, and the retreat of glacial ice in its catchment area, revealed that the retreat of ice and permafrost increased the volume of rock exposed to atmospheric oxygen and enhanced sulphide oxidation, resulting in elevated sulphate concentrations in the lake [55].

Monitoring of meteorological parameters at the source regions of the waters accumulating at the Three Rivers Dam in China (1960-2009) revealed unstable, regionally differentiated, temporal transitions to dryer and warmer, but in some places more humid, conditions [56].

China's water demand in 2008 - 591 billion m^3 - exceeded by 50 billion m^3 the regional water availability because annual rainfall and surface water flows have declined considerably. 15 million ha of farmland were destroyed by drought [57].

NOTES

[1] Pluvial palaeolake Mega-Chad, which developed during the African Humid Period 15,000-5000 a ago, began to shrink with the onset of aridification. This transition was caused by an orbitally controlled decrease of insolation, confining the intensity of the West African Monsoon. When the northern part of that lake dried out ca. 1000 a ago, the world's largest single dust source originated, which began to supply, *via* aeolian transport, the Central Atlantic and Amazonian ecosystems with nutrients, previously deposited on the former lake floor [14].

[2] Pontus Euxinus (Latin and Ancient Greek) = Black Sea. Due to the anaerobe milieu present in the 2150 m thick lower water layer of the Black Sea, organic matter accumulates at the bottom of the water column to form sapropelite containing up to 10% organic matter. Hydrogen sulphide originates from anaerobe microbial respiration and dissolves in the water column. Lithified sapropelite containing more than 5% C_{org} is defined as black shale: the source matter of natural gas, petroleum, tar and coal.

[3] However, these installations fail to filter micro- and nanoplastics, to retain many sorts of chemicals and total phosphate and fail to decompose metabolic products of pharmaceuticals. Downstream aquatic flora and fauna is exposed to these pollutants. Utilisation of many T of sludge-masses originating in sewerages (biomass-burning to generate electricity, component in noise-protection embankment, fertiliser in agroindustry) depends from the amount of chemical contaminants and infectious microbes analysed and from established legislative threshold values.

[4] Allied mining companies engaged themselves in water stewardship: application of strong and transparent corporate water governance, effective management of water at mining operations, and collaboration to achieve responsible and sustainable water use [58].

CONFLICT OF INTEREST

The author (editor) declares no conflict of interest, financial or otherwise.

ACKNOWLEDGEMENTS

Declare none.

REFERENCES

[1]   Chenoweth, J. Water, water everywhere. *New Sci.,* **2008**, *199*(2670), 28-32.
        [http://dx.doi.org/10.1016/S0262-4079(08)62124-7]

[2]   Dittrich, M.; Giljum, S.; Lutter, S. *Green economies around the world? Implications of resource use for development and the environment.,* **2012**. http://seri.at/wp-content/uploads/2012/06/green_economies_around_the_world.pdf

[3]   Van der Gun, J. *Groundwater and Global Change: Trends, Opportunities and Challenges,* **2012**. http://www.unesco.org/water/wwap

[4]   Jaramillo, F.; Destouni, G. Local flow regulation and irrigation raise global human water consumption and footprint. *Science,* **2015**, *350*(6265), 1248-1251.
        [http://dx.doi.org/10.1126/science.aad1010] [PMID: 26785489]

[5]   Dudgeon, D.; Arthington, A.H.; Gessner, M.O.; Kawabata, Z.; Knowler, D.J.; Lévêque, C.; Naiman, R.J.; Prieur-Richard, A.H.; Soto, D.; Stiassny, M.L.; Sullivan, C.A. Freshwater biodiversity: importance, threats, status and conservation challenges. *Biol. Rev. Camb. Philos. Soc.,* **2006**, *81*(2), 163-182.
        [http://dx.doi.org/10.1017/S1464793105006950] [PMID: 16336747]

[6]   Young, M.R. Conserving the freshwater pearl mussel (*Margaritifera margaritifera* L.) in the British Isles and Continental Europe. *Aquat. Conserv. Mar. Freshwater Ecosyst.,* **1991**, *1*(1), 73-77.
        [http://dx.doi.org/10.1002/aqc.3270010106]

[7]   Degerman, E.; Alexanderson, S.; Bergengren, J. *Restoration of freshwater pearl mussel streams,* **2009**. http://www.wwf.se/source.php/1257735/Restoration%20of%20FPM%20streams.pdf

[8]   Ripl, W.; Wolter, K-D. Lake restoration and rehabilitation In: *The Lakes Handbook*; O'Sullivan, P.E.; Reynolds, C.S., Eds.; Wiley-Blackwell Science Ltd.: Oxford, UK, **2005**; Vol. 2, pp. 25-64.

[9]   Von Ohle, W. Der Vorgang rasanter Seenalterung in Holstein. *Naturwissenschaften,* **1953**, *40*(5), 153-162.
        [http://dx.doi.org/10.1007/BF00639939]

[10]   Gophen, M. Ecological devastation in Lake Victoria: Part B: Plankton and fish communities. *O. J. E.,* **2015**, *5*, 315-325.

[11]   UNESCO. *World Water Assessment Programme, Pilot case studies: a focus on real-world examples: The Lake Titicaca Basin, Bolivia and Peru,* **2011**. http://webworld.unesco.org/water/wwap/case_studies/titicaca_lake/titicaca_lake.pdf

[12]   UNEP. Diagnostico Ambiental del Sistema Titicaca-Desaguadero-Poopo-Salar de Coipasa (Sistema TDPS) Bolivia-Perú *Comité Ad-Hoc de Transición de la Autoridad Autónoma Binacional del Sistema TDPS, Washington, 1996,* **1996**. https://www.oas.org/DSD/publications/Unit/oea31s/begin.htm#Contents

[13]   UNEP. *Lake Chad: almost gone,* **2008**. http://www.unep.org/dewa/vitalwater/article116.html

[14]   Armitage, S.J.; Bristow, C.S.; Drake, N.A. West African monsoon dynamics inferred from abrupt fluctuations of Lake Mega-Chad. *Proc. Natl. Acad. Sci. USA,* **2015**, *112*(28), 8543-8548.
        [http://dx.doi.org/10.1073/pnas.1417655112] [PMID: 26124133]

[15]   Goudie, A.S. *The Human Impact on the Natural Environment. Past, Present and Future,* 7th ed; Wiley-Blackwell: Oxford, **2013**.

[16]   Saab, M.A. Environmental impacts on the Dead Sea, sustainability cost estimates. *Environ. Ecol.,* **2010**, *1*(1), 113-124.

[17]   Hammour, S.A. *River Basin Management,* **2014**. http://www.sesrtcic.org/Presentations/Water_Management_Symposium/Jordan/Jordan.pdf

[18]   Izhitskiy, A.S.; Zavialov, P.O.; Sapozhnikov, P.V.; Kirillin, G.B.; Grossart, H.P.; Kalinina, O.Y.;

Zalota, A.K.; Goncharenko, I.V.; Kurbaniyazov, A.K. Present state of the Aral Sea: diverging physical and biological characteristics of the residual basins. *Sci. Rep.,* **2016**, *6*, 23906.
[http://dx.doi.org/10.1038/srep23906] [PMID: 27032513]

[19] Nihoul, J.C.; Zavialov, P.O.; Micklin, P.P. Dying and dead seas: climatic *versus* anthropic causes. **2004**.
[http://dx.doi.org/10.1007/978-94-007-0967-6]

[20] Capet, A.; Stanev, E.V.; Beckers, J.-M. Decline of the Black Sea oxygen inventory. *Biogeosciences,* **2016**, *13*, 1287-1297.
[http://dx.doi.org/10.5194/bg-13-1287-2016]

[21] Water, U.N. *A Post 2015 Global Goal for Water: Synthesis of key findings and recommendations.,* **2014**. http://www.unwater.org/fileadmin/user_upload/unwater_new/docs/Topics/UN-Water_paper_on_a_Post-2015_Global_Goal_for_Water.pdf

[22] FAO Statistics. *Resources, Land use,* **2010**. http://www.fao.org/fileadmin/templates/ess/ess_test_folder/Publications/yearbook_2010/a04.xls

[23] Emde, S.; Kochmann, J.; Kuhn, T.; Dörge, D.D.; Plath, M.; Miesen, F.W.; Klimpel, S. Cooling water of power plant creates "hot spots" for tropical fishes and parasites. *Parasitol. Res.,* **2016**, *115*(1), 85-98.
[http://dx.doi.org/10.1007/s00436-015-4724-4] [PMID: 26374537]

[24] O'Reilly, C.M.; Sharma, S.; Gray, D.K. Rapid and highly variable warming of lake surface waters around the globe. *Geophys. Res. Lett.,* **2015**, *42*(24), 10,773-10,781.
[http://dx.doi.org/10.1002/2015GL066235]

[25] Mekonnen, M.M.; Hoekstra, A.Y. A global assessment of the water footprint of farm animal products. *Ecosystems (N. Y.),* **2012**, *15*(3), 401-415.
[http://dx.doi.org/10.1007/s10021-011-9517-8]

[26] Europäische Kommission. *Bericht der Kommission an den Rat und das Europäische Parlament,* **2013**. http://eur-lex.europa.eu/LexUriServ/LexUriServ.do?uri=COM:2013:0683:FIN:DE:PDF

[27] *China Report 2009, Ministry of Environmental Protection of the People's Republic of China. Report on the State of the Environment of China, Water Environment,* **2009**. http://english.mep.gov.cn/down_load/Documents/201104/P020110411532104009882.pdf

[28] Elsner, M.; Imfeld, G. Compound-specific isotope analysis (CSIA) of micropollutants in the environment - current developments and future challenges. *Curr. Opin. Biotechnol.,* **2016**, *41*, 60-72.

[29] Kundzewicz, Z.W.; Graczyk, D.; Maurer, T. Trend detection in river flow series: 1. Annual maximum flow. *Hydrol. Sci.,* **2005**, *50*(5), 797-810.

[30] Gleick, P.H.; Palaniappan, M. Peak water limits to freshwater withdrawal and use. *Proc. Natl. Acad. Sci. USA,* **2010**, *107*(25), 11155-11162.
[http://dx.doi.org/10.1073/pnas.1004812107] [PMID: 20498082]

[31] Mani, T.; Hauk, A.; Walter, U.; Burkhardt-Holm, P. Microplastics profile along the Rhine River. *Sci. Rep.,* **2015**, *5*, 17988.
[http://dx.doi.org/10.1038/srep17988] [PMID: 26644346]

[32] Eriksen, M.; Mason, S.; Wilson, S.; Box, C.; Zellers, A.; Edwards, W.; Farley, H.; Amato, S. Microplastic pollution in the surface waters of the Laurentian Great Lakes. *Mar. Pollut. Bull.,* **2013**, *77*(1-2), 177-182.
[http://dx.doi.org/10.1016/j.marpolbul.2013.10.007] [PMID: 24449922]

[33] Grosshans, R.E.; Venema, H.D.; Oborne, B. *Advancing Netley-Libau Marsh Restoration Efforts, The International Institute for Sustainable Development, Winnipeg, Manitoba, Canada,* **2012**. https://www.iisd.org/sites/default/files/publications/advancing-netley-libau-ma-sh-restoration-efforts.pdf

[34] Ochieng, G.; Seanego, E.S.; Nkwonta, O. Impacts of mining on water resources in South Africa: A review. *Sci. Res. Essays,* **2010**, *5*(22), 3351-3357.

[35] Kelly, E.N.; Schindler, D.W.; Hodson, P.V.; Short, J.W.; Radmanovich, R.; Nielsen, C.C. Oil sands development contributes elements toxic at low concentrations to the Athabasca River and its tributaries. *Proc. Natl. Acad. Sci. USA,* **2010**, *107*(37), 16178-16183.
[http://dx.doi.org/10.1073/pnas.1008754107] [PMID: 20805486]

[36] Patterson, L.A.; Konschnik, K.E.; Wiseman, H.; Fargione, J.; Maloney, K.O.; Kiesecker, J.; Nicot, J.P.; Baruch-Mordo, S.; Entrekin, S.; Trainor, A.; Saiers, J.E. Unconventional oil and gas spills: Risks, mitigation priorities, and state reporting requirements. *Environ. Sci. Technol.,* **2017**, *51*(5), 2563-2573.
[http://dx.doi.org/10.1021/acs.est.6b05749] [PMID: 28220696]

[37] Hübner, G. Ökologisch-faunistische Fließgewässerbewertung am Beispiel der salzbelasteten unteren Werra und ausgewählter Zuflüsse. *Ökologie und Umweltsicherung 27, Diss. Univ. Kassel, Germany,* **2007**.

[38] International Atomic Energy Agency. *Estimation of global inventories of radioactive waste and other radioactive materials,* **2009**. http://www-pub.iaea.org/MTCD/publications/PDF/te_1591_web.pdf

[39] Trapeznikov, A.V.; Pozolotina, V.N.; Chebotina M.Ya. ; Chukanov, V.N.; Trapeznikova, V.N.; Kulikov, N.V.; Nielsen, S.P.; Aarkrog, A. Radioactive contamination of the Techa River, the Urals. *Health Phys.,* **1993**, *65*(5), 481-488.
[http://dx.doi.org/10.1097/00004032-199311000-00002] [PMID: 8225983]

[40] Dixit, S.S.; Dixit, A.S.; Smol, J.P. *Restoration and Recovery of an Industrial Region - Progress in Restoring the Smelter-Damaged Landscape Near Sudbury, Canada*; Gunn, J.M., Ed.; Springer: New York, **1995**, pp. 33-44.
[http://dx.doi.org/10.1007/978-1-4612-2520-1_3]

[41] Zereini, F.; Urban, H. *Geowissenschaften und Umwelt*; Matschullat, J.; Müller, G., Eds.; Springer: Heidelberg, **1994**, pp. 185-192.

[42] Selin, N.E.; Jacob, D.J.; Yantosca, R.M. Global 3-D land-ocean-atmosphere model for mercury: Present-day *versus* preindustrial cycles and anthropogenic enrichment factors for deposition. *Global Biogeochem. Cycles,* **2008**, *22*, GB2011.

[43] Amos, H.M.; Jacob, D.J.; Streets, D.G. Legacy impacts of all-time anthropogenic emissions on the global mercury cycle. *Global Biogeochem. Cycles,* **2013**, *27*, 410-421.
[http://dx.doi.org/10.1002/gbc.20040]

[44] Biester, H.; Bindler, R.; Martinez-Cortizas, A.; Engstrom, D.R. Modeling the past atmospheric deposition of mercury using natural archives. *Environ. Sci. Technol.,* **2007**, *41*(14), 4851-4860.
[http://dx.doi.org/10.1021/es0704232] [PMID: 17711193]

[45] Carrasco, L.; Barata, C.; García-Berthou, E.; Tobias, A.; Bayona, J.M.; Díez, S. Patterns of mercury and methylmercury bioaccumulation in fish species downstream of a long-term mercury-contaminated site in the lower Ebro River (NE Spain). *Chemosphere,* **2011**, *84*(11), 1642-1649.
[http://dx.doi.org/10.1016/j.chemosphere.2011.05.022] [PMID: 21663932]

[46] Syvitski, J.P.; Kettner, A. Sediment flux and the Anthropocene. *Philos. Trans. A Math. Phys. Eng. Sci.,* **2011**, *369*(1938), 957-975.
[http://dx.doi.org/10.1098/rsta.2010.0329] [PMID: 21282156]

[47] Peduzzi, P. *Sand, rarer than one thinks,* **2014**. http://www.grid.unep.ch/products/3_Reports/GEAS_Mar2014_Sand_Mining.pdf

[48] Kondolf, G.M. Geomorphic and environmental effects of instream gravel mining. *Landsc. Urban Plan.,* **1994**, *28*, 225-243.
[http://dx.doi.org/10.1016/0169-2046(94)90010-8]

[49] Sobek, S.; Tranvic, L. J.; Cole, J. J. Temperature independence of carbon dioxide supersaturation in

global lakes. *Global Biochem. Cycles,* **2005**, *19*(2), GB2003.

[50]   Piggott, J.J.; Niyogi, D.K.; Townsend, C.R. Multiple stressors and stream ecosystem functioning: climate warming and agricultural stressors interact to affect processing of organic matter. *J. Appl. Ecol.,* **2015**, *52*(5), 1126-1134.
[http://dx.doi.org/10.1111/1365-2664.12480]

[51]   Prokushkin, A.S.; Kawahigashi, M.; Tokareva, I.V. *Permafrost soils, Soil Biology*; Margesin, R., Ed.; Springer: Berlin, Heidelberg, **2009**, Vol. 16, pp. 237-250.
[http://dx.doi.org/10.1007/978-3-540-69371-0_16]

[52]   Gómez-Gener, L.; Obrador, B.; von Schiller, D. Hot spots for carbon emissions from Mediterranean fluvial networks during summer drought. *Biogeochemistry,* **2015**, *125*(3), 409-426.
[http://dx.doi.org/10.1007/s10533-015-0139-7]

[53]   Battarbee, R.W.; Grytnes, J-A.; Thompson, R. Comparing palaeolimnological and instrumental evidence of climate change for remote mountain lakes over the last 200 years. *J. Paleolimnol.,* **2002**, *28*(1), 161-179.
[http://dx.doi.org/10.1023/A:1020384204940]

[54]   Ilyashuk, E.A.; Ilyashuk, B.P.; Tylmann, W. Biodiversity dynamics of chironomid midges in high-altitude lakes of the Alps over the past two millennia. *Insect Conserv. Divers.,* **2015**, *8*(6), 547-561.
[http://dx.doi.org/10.1111/icad.12137]

[55]   Salerno, F.; Rogora, M.; Balestrini, R.; Lami, A.; Tartari, G.A.; Thakuri, S.; Godone, D.; Freppaz, M.; Tartari, G. Glacier melting increases the solute concentrations of Himalayan glacial lakes. *Environ. Sci. Technol.,* **2016**, *50*(17), 9150-9160.
[http://dx.doi.org/10.1021/acs.est.6b02735] [PMID: 27466701]

[56]   Liang, L.; Li, L.; Liu, C. Climate change in the Tibetan Plateau Three Rivers Source Region: 1960-2009. *Int. J. Climatol.,* **2013**, *33*(13), 2900-2916.
[http://dx.doi.org/10.1002/joc.3642]

[57]   Global Water Partnership, Technical Focus Paper. **2015**. http://www.gwp.org/Global/ToolBox/Publications/Technical%20Focus%20Papers/TFPChina_2015.pdf

[58]   International Council on Mining and Metals. *Position statement on water stewardship,* **2017**. http://www.icmm.com/water-ps

CHAPTER 11

Marine Shores

Abstract: Interferences of elevated human pressure at marine coasts, and contaminants transported by marine currents and rivers contribute to forced degradation of littoral environments. Additional impairment has occurred due to inappropriate technical constructions to protect built-up areas against floods caused by sea level rise. Delta subsidence due to upriver damming, on and offshore hydrocarbon extraction, beach sand mining, onshore ground water pumping, and soil compaction and loading because of coastal urbanisation has resulted in accelerated sea level rise, marine ingressions, and considerable land loss. Nutrient-rich runoffs, produced by intensified agriculture, effected the origination of - partially temporary - hypoxic coastal water bodies. Eutrophication and degradation of littoral waters also result from industrial aquafarming. Harbour mud dredged for navigational purposes has been contaminated with toxic metals. Oil and gas developments and havaries have caused severe hydrocarbon pollution in littoral areas several times. Toxic legacies exist in the form of waste-filled river and coastal impoundments, now exposed to sea level rise. Because of the mobilisation of toxic leachates and submarine discharge of contaminated plumes (*e.g.* in estuaries), redistribution of toxic metal species and their bioaccumulation occurs. Saline brines, discharged from desalination plants in arid climate zones to compensate for water scarcity, negatively impact water quality of ambient littoral regions. Accidentally dispersed nuclear weapon material contaminated the inner shelf area off Thule Airbase.

Keywords: Aquafarming, Coastal engineering, Coastal urbanisation, Contamination, Delta subsidence, Desalination plants, Erosion, Estuaries, Eutrophication, Flooding, Harbour mud, Hypoxia, Littoral, Marine ingression, Oil spills, Pollution, Sand mining, Sea level rise, Tailings, Waste, Wetlands.

One third of global human population lives in coastal regions, which include only ca. 4% of global land area. This leads, in addition to climate change-induced sea level rise, to increased anthropogenic pressure on epilittoral ecosystems [1]. At marine coasts, where material flows from the open seas interact and merge with those from the continents, the environments are exposed to stronger and more diverse effects of anthropogenic pollutants.

Global sea level rise has been caused by thermosteric expansion, by influx of meltwaters from glaciated land areas, and by diminution of liquid water storage

Hubert Engelbrecht

capacity on land [2]. According to coastal tide gauge data, global sea level rose between 1870 and 2000 by ca. 20 cm on average. Since 1993, satellite data revealed ongoing rise by 6.5 cm on average. The average sea level rise is 3.4 ± 0.4 mm/a[1] [3]. Several examples detail the consequences arising from coastal/inland industrial development and sea level rise:

• Sea level rise, sediment starvation due to upstream dams, as well as natural and induced subsidence (the latter due to long-term on and offshore hydrocarbon extraction) has caused land surface loss - containing *e.g.* ecologically valuable salt-marshes - of ca. 5000 km^2 within 80 years at the Mississippi Delta and Northern Gulf of Mexico coast. Protection and restoration measures are planned to mitigate the additional loss of terrain predicted by simulated scenarios [4, 5].
• Oil and gas extractive business at the Caspian Sea plays - with an estimated hydrocarbon potential of ca. 24 billion barrels - caused pollution of recent littoral, delta, and inner shelf sediments with aliphatic and polycyclic hydrocarbons; additional fluvial import of hydrocarbon pollutants caused detriments to bioproductivity and fishery [6].
• Satellite imagery data over four decades (1973-2013) from the Yellow River Delta area revealed that urbanisation, oil and natural gas development, as well as geotechnical river bed regulations and diversions induced a shoreline retreat of 13 km and a subsequent shoreline advance of 21 km. Hydrocarbon extraction also caused impingement of protected nature reserves, declared as Ramsar[2] wetland sites [7].
• Coastal erosion and flooding is enhanced by sea level rise (4-5 cm/a), inappropriate ground water abstraction, and soil compaction from the load of numerous constructions implemented during the last three decades, which were characteristic of rapid urban expansion: Long-term satellite monitoring of relative vertical motion of land area at the coastal city of Jakarta (9.6 million inhabitants, 660 km^2) revealed that the subsidence rate amounts to an average of 1-15 cm/a (maximum 20-25 cm/a) [8].
• Groundwater pumping in Eocene artesian aquifers and the construction of navigation channels contributed to relative subsidence of the Tangier Islands (Chesapeak Bay, USA), which lost 66.75% of their terrain since 1850 [9].
• Exploitation of coastal oil and gas reservoirs caused subsidence, locally increased sea level rise [10], and induced earthquakes [11].

Coastal pollution due to anthropogenic activities are described in the following examples:

• The nitrogen fertiliser consumption rate in industrial agronomy of the USA increased fourfold, from ca. 3 million T/a in 1960 to 12 million T/a in 2009, resulting in a corn yield gain of 5 T/ha. Despite improved nitrogen use efficacy, nutrient-loaded runoffs from agrarian land caused the formation of ca. 300 hypoxic zones in marine coastal water bodies along the shores of the USA [12].

- 14 years of water quality monitoring (1980-1993) of the 384 km^2 Guanabara Bay close to the megacity Rio de Janeiro revealed that this shallow, tidally influenced estuarine system, which receives large quantities of untreated domestic and industrial sewage from ca. 7.6 million inhabitants, is heavily loaded with nutrients, hydrocarbons, heavy metals, and organic pollutants. Influx of total phosphorus (ca. 19 T/day), total nitrogen (ca. 112 T/day), and BOD (ca. 362 T/day) caused hypertrophic and temporarily hypoxic conditions, as well as high chlorophyll-a concentrations, impairing original planktonic and benthic communities. Faecal coliform concentrations were measured at far above tolerable values [13]. This state of water quality has developed since 1930, because of industrialisation and a sharp population increase. Analyses of Guanabara Bay waters, conducted between 2010 and 2015, revealed that they are loaded each day with 470 T BOD, 150 T of industrial waste water, 18 T of hydrocarbons, and 813 T of solid waste. In the inner portion of the bay, eutrophication caused a shift in microbial communities: methanogenic archaea and opportunistic pathogenic vibrio species developed. In addition, antibiotic resistance genes were commonly found in metagenomes of planktonic microorganisms [14].
- Biochemical analyses of 90 estuarine sediment samples recovered from 18 river deltas at the Chinese coast revealed the problematic omnipresence of antibiotic resistance genes, whose abundance and biodiversity appeared to be positively correlated with agri- and aquacultural activity as well as intense urban littoral development [15].
- Dissipated nutrients and metabolic products from aquacultures[3] contributed to eutrophication of ambient water bodies [16].
- The Deepwater Horizon havary on 04/20/2010 resulted in shoreward moving oil carpets and contamination of Texas's, Louisiana's Mississippi's, Alabama's, and Florida's beaches, marshes, wetlands, mangroves, estuaries, and aquacultures, despite multiple preventative efforts were made using containment booms, dispersants, controlled *in situ* combustion, offshore filtration, skimming, and sorption. Although application of dispersants near the wellhead kept a large part of oil in the bathypelagic zone, oil came to the surface in May 2010 and formed a patchwork of oil slicks measuring ca. 17,725 km^2 [17, 18].
- Accumulation of contaminants in deposits of marine deltas and harbour basins occurred due to upstream disposal and dispersal during charging and discharging of commodities on containerships at wharfs [10]. For example, the distal fluvial and deltaic mud deposits of Lower Weser and at Bremen industrial harbour contain elevated concentrations of toxic heavy metal species Cd, Cr, Cu, Ni, Pb, and Zn [19]. In northwestern European harbours, ca. 200 million m^3 of

contaminated mud is dredged annually for navigational purposes. Experimental biomonitoring of microbial diversity in sand, covered with that kind of dredged mud revealed that, in the pore water of the sand, significant community changes, including temporary denaturation of archaea, have occurred [20]. Another point is ship scrapping at marine shores of Bangladesh and Indonesia.

- Proximal glaciomarine sediments off of Thule Airbase (Greenland) have been found to contain elevated levels of the transuranic elements Pu and Am. Identified sources were fallout from numerous above ground nuclear tests and from dispersed nuclear weapon material, *e.g.* a B52-bomber crashed in 1968 on sea ice near Thule Airbase [21].

- In the past, low lying estuaries and coastal regions, not utilisable for agricultural purposes, were regarded as suitable waste disposal areas. Waste was also used as a fill material for river or coast embankment constructions. It is estimated that in England and Wales ca. 20,000 embankments of these types exist. Due to climate change-induced sea level rise, about 25% of them are considered at risk of flooding and erosion, causing remobilisation of toxic matter. Contamination of the ambient environment has already occurred by toxic landfill leachate entering surface waters and groundwater. Access to the latter has been effected by percolation into littoral deposits and mixing. The contaminated plumes moved - depending on the geochemical mobility of the contaminants - *via* groundwater flow paths towards submarine discharge sites, where mixing with sea water occurred [22, 23]. The dissemination and dispersion, but also secondary concentration, of anthropogenic trace metals (Cr, Cu, Zn, Pb, Cd, Pt, Ag, Co, Ni, and Hg) - in solution, as well as particulate - in estuaries depends on several hydro- and biochemical factors as follows [23]:

- Metal desorption from sediments and suspended matter increases proportionally with salinity because of the formation of inorganic complexes.

- Metal ions can be absorbed by particulate matter, depositing out of turbid zones onto the sediment of the estuary and thus diminishing the residence time of the contaminant in the water column.

- Chemical and physical remobilisation of metals absorbed in sediment - often storing up to 10^5 times more metals than the water column above - is mediated *via* redox conditions in sediment pore water, which cyclically changes with the tide, and by erosive forces of coastal currents and sea surface oscillations (waves, floods).

- Rising salinity reduces bioavailability of metals because of the introduction of cations, causing increased tendency to metal complexation.

- Bioaccumulation of trace metals adsorbed in sediment occurs *via* interaction of macrophyte roots and their symbiotic bacteria, causing transfer of the metals to above ground biomass. Phytoplankton species, which take up dissolved trace metals, bioaccumulate as follows: volume-based bioconcentration factors of *e.g.*

Thalassiosira pseudonana, Dunaliella tertiolecta, and *Emiliana huxleyi* were found to be 10^4 for Ag, 10^3 for Cd, 10^4-10^3 for Hg, and 10^2-10^3 for Zn. Phytoplankton adsorb metals at the cell walls and also absorbe them into the cells. Active and passive absorption is performed by diffusion according to gradients at membranes or by ion exchange pumps. In this way, transfer of metals to higher trophic levels occurs. Phytoplankton growth alters metal hydrochemistry and contaminate other biota *via* diet-borne and waterborne exposures.

- Metabolic products of benthic fauna like CO_2, CH_4, NH_3, and PO_4^{3-} increase metal mobility and effect methylation processes.
- Bioturbation in sediment promotes trace metal advection and transports O_2-containing water into deeper sediment zones. For example, Pt is increasingly absorbed if sediment is burrowed by *Arenicola marina*. Ingestion by deposit feeders can relocate metals *via* faecal pellets.
- Decomposing phytoplankton masses increase the dissolved organic carbon in the water column and, consequently, metal complexation and solubilisation.
- Phytoplankton growth decreases the transparency of the water column, limiting growth of submerged plants and their capacity to absorb metals.
- Phytoplankton communities exposed to metals reduce their bioproductivity up to 50% and alter their species compositions [23].
- Porphyry copper ore mining on Marinduque Island (Philippines) has resulted in severe degradation of inner shelf regions. From 1975 to 1990, 200-300 MT of copper tailings were disposed of into Calancan Bay at the northern coast of the island and were dispersed over an area of ca. 70 km^2. The acid pore waters on the tailings causeway extending 6 to 8 km offshore contain elevated concentrations of aluminium, arsenic, copper, selenium, iron, and molybdenum. At the western coast of the island, acid rock drainage from copper tailings extending into littoral waters at Mogpog River Delta raised substantial concern about adverse effects on ambient sea water quality [24].
- From ca. 1900 onwards, sea water has been processed in industrial desalination plants to supply cities and agricultural land in semiarid and arid coastal regions. Because of the fast-rising world population and its demands, water scarcity has increased, because ground water resources remained constant or depleted. The annual water withdrawal, caused by agricultural, industrial, and municipal demand, increased from ca. 600 km^3/a in 1900 to 5000 km^3/a in 2000 [25]. The global demand in 2005 amounted to 9000 km^3 [26]. In 2014, ca. 2.5 billion persons lacked access to clean sanitation systems, 780 million persons had no access to safe drinking water, and ca. 3.5 million died because of preventable water-related diseases [27]. Ca. 13,000 desalination plants, compensating for depleted groundwater resources, produce 50 million m^3/d. The discharged brines often cause increase in salinity and water temperature, accumulation of heavy

metals, hydrocarbons, and anti fouling material. The ecotoxic effects vary with discharge site selection, flushing degree of brines, and sensitivity of flora and fauna present [28].

Deforestation at marine shores has given rise to wind- and water-driven mass movement of soil and rock [10]. Overgrazing has brought about remobilisation of dunes. Sand mining for building raw material or to exploit placer deposits[4] has intensified erosion because of the reduction of the amount of moving, ductile sand masses absorbing the kinetic energy of water waves. Interruption of fluvial sediment flow by the implementation of numerous dams upriver entailed sediment starvation at deltas [29], intensified coastal erosion, and resulted in recession of the extent of wetland areas because of marine ingressions [10]. In addition to the impairments mentioned, the following detriments caused by anthropogenic activity and utilisation occur: increased erosion and modified sediment flow originated due to inappropriate coastal engineering (groynes, breakwaters, jetties, retaining walls, sea walls, revetments, *etc.*) and building operations to protect urban or industrial areas established close to the sea [10, 30].

NOTES

[1] There exists broad agreement among ca. 1300 independent scientists that accelerated warming trends since ca. 1960 in the troposphere and in the oceans have very probably been caused by anthropogenic greenhouse gases [3].

[2] The Ramsar Convention, signed in 1971 in Ramsar (Iran), consists of an intergovernmental treaty providing the framework for action and cooperation to preserve and intelligently use wetlands and their resources. 169 contracting parties and 2247 Ramsar sites exist. The total area of declared sites measures ca. 215,000,000 ha [31].

[3] World aquaculture production - marine and inland - rose from 0.5 MT in 1950 continually to 66.6 MT in 2012 [32].

[4] They originate, when the forces of wind, flowing water and gravity transport, separate, and concentrate certain mineral grains, characterised by higher specific weight than the main constituents. Such placers form in the alluvium of rivers, in aeolian dunes of deserts, and in beach sands at seashores. The mineral grains concentrated in placers consist of gold, platinum, zircon, monazite, cassiterite, magnetite, chromite, ilmenite, rutile, *etc.*

CONFLICT OF INTEREST

The author (editor) declares no conflict of interest, financial or otherwise.

ACKNOWLEDGEMENTS

Declare none.

REFERENCES

[1]   UNEP. *Marine and coastal ecosystems and human well-being: A synthesis report based on the findings of the Millennium Ecosystem Assessment,* **2006**. http://www.unep.org/pdf/Completev6_LR.pdf

[2]   Gregory, J. *Projections of sea level rise,* **2013**. https://www.ipcc.ch/pdf/unfccc/cop19/3_gregory13sbsta.pdf

[3]   NASA Global Climate Change. *Sea level,* **2016**. http://climate.nasa.gov/vital-signs/sea-level/

[4]   USGS. *Subsidence and Wetland Loss Related to Fluid Energy Production, Gulf Coast Basin,* **2014**.http://coastal.er.usgs.gov/gc-subsidence/induced-subsidence.html

[5]   Couvillion, B.R.; Steyer, G.D.; Wang, H. Forecasting the Effects of Coastal Protection and Restoration Projects on Wetland Morphology in Coastal Louisiana under Multiple Environmental Uncertainty Scenarios. *J. Coast. Res.,* **2013**, *67*, 29-50. [http://dx.doi.org/10.2112/SI_67_3]

[6]   Nemirovskaya, I.A. Hydrocarbons in the modern sediments of the Caspian Sea. *Water Resour.,* **2016**, *43*(1), 111-120. [http://dx.doi.org/10.1134/S009780781506007X]

[7]   Kuenzer, C.; Ottinger, M.; Liu, G. Earth observation-based coastal zone monitoring of the Yellow River Delta: Dynamics in China's second largest oil producing region over four decades. *Appl. Geogr.,* **2014**, *55*, 92-107. [http://dx.doi.org/10.1016/j.apgeog.2014.08.015]

[8]   Abidin, H.Z.; Andreas, H.; Gumilar, I. Land subsidence of Jakarta (Indonesia) and its relation with urban development. *Nat. Hazards,* **2011**, *59*, 1753-1765. [http://dx.doi.org/10.1007/s11069-011-9866-9]

[9]   Schulte, D.M.; Dridge, K.M.; Hudgins, M.H. Climate change and the evolution and fate of the Tangier Islands of Chesapeake Bay, USA. *Sci. Rep.,* **2015**, *5*, 17890. [http://dx.doi.org/10.1038/srep17890] [PMID: 26657975]

[10]   Goudie, A.S. *The Human Impact on the Natural Environment. Past, Present and Future,* 7th ed; Wiley-Blackwell: Oxford, **2013**.

[11]   Bommer, J.J.; Crowley, H.; Pinho, R. A risk-mitigation approach to the management of induced seismicity. *J. Seismol.,* **2015**, *19*(2), 623-646. [http://dx.doi.org/10.1007/s10950-015-9478-z] [PMID: 28190961]

[12]   Davidson, E.A.; David, M.B.; Galloway, J.N. Excess nitrogen in the U.S. environment: Trends, risks, and solutions. *Iss. Ecol.,* **2012**, *15*, 1-16.

[13]   Ribeiro, C.H.; Kjerfve, B. Anthropogenic influence on the water quality in Guanabara Bay, Rio de Janeiro, Brazil. *Reg. Environ. Change,* **2002**, *3*(1), 13-19. [http://dx.doi.org/10.1007/s10113-001-0037-5]

[14]   Fistarol, G.O.; Coutinho, F.H.; Moreira, A.P.; Venas, T.; Cánovas, A.; de Paula, S.E., Jr; Coutinho, R.; de Moura, R.L.; Valentin, J.L.; Tenenbaum, D.R.; Paranhos, R.; do Valle, Rde.A.; Vicente, A.C.; Amado Filho, G.M.; Pereira, R.C.; Kruger, R.; Rezende, C.E.; Thompson, C.C.; Salomon, P.S.;

Thompson, F.L. Environmental and Sanitary Conditions of Guanabara Bay, Rio de Janeiro. *Front. Microbiol.,* **2015**, *6*(6), 1232.
[PMID: 26635734]

[15]   Zhu, Y-G.; Zhao, Y.; Li, B.; Huang, C.L.; Zhang, S.Y.; Yu, S.; Chen, Y.S.; Zhang, T.; Gillings, M.R.; Su, J.Q. Continental-scale pollution of estuaries with antibiotic resistance genes. *Nat. Microbiol.,* **2017**, *2*, 16270.
[http://dx.doi.org/10.1038/nmicrobiol.2016.270] [PMID: 28134918]

[16]   Wang, X.; Olsen, L.M.; Reitan, K.I. Discharge of nutrient wastes from salmon farms: environmental effects, and potential for integrated multi-trophic aquaculture. *Aquacult. Environ. Interact.,* **2012**, *2*, 267-283.
[http://dx.doi.org/10.3354/aei00044]

[17]   Deepwater Horizon Study Group. *Final Report on the Investigation of the Macondo Well Blowout,* **2011**. http://ccrm.berkeley.edu/pdfs_papers/bea_pdfs/dhsgfinalreport-march2011-tag.pdf

[18]   *On Scene Coordinator Report Deepwater Horizon Oil Spill. Submitted to the National Response Team,* **2011**. http://www.uscg.mil/foia/docs/dwh/fosc_dwh_report.pdf

[19]   Kasten, S.; Schulz, H.D. *Geowissenschaften und Umwelt*; Matschullat, J.; Müller, G., Eds.; Springer: Heidelberg, **1994**, pp. 185-192.
[http://dx.doi.org/10.1007/978-3-642-79021-8_23]

[20]   Toes, A.C.; Finke, N.; Kuenen, J.G.; Muyzer, G. Effects of deposition of heavy-metal-polluted harbor mud on microbial diversity and metal resistance in sandy marine sediments. *Arch. Environ. Contam. Toxicol.,* **2008**, *55*(3), 372-385.
[http://dx.doi.org/10.1007/s00244-008-9135-4] [PMID: 18273665]

[21]   Ikäheimonen, T.K.; Ilus, E.; Klemola, S. Plutonium and americium in the sediments off the Thule air base, Greenland. *J. Radioanal. Nucl. Chem.,* **2002**, *252*(2), 339-344.
[http://dx.doi.org/10.1023/A:1015726608302]

[22]   Förstner, U.; Wittmann, G.T.W. *Metal Pollution in the Aquatic Environment*; Springer Berlin Heidelberg, **1981**.

[23]   de Souza Machado, A.A.; Spencer, K.; Kloas, W.; Toffolon, M.; Zarfl, C. Metal fate and effects in estuaries: A review and conceptual model for better understanding of toxicity. *Sci. Total Environ.,* **2016**, *541*, 268-281.
[http://dx.doi.org/10.1016/j.scitotenv.2015.09.045] [PMID: 26410702]

[24]   Plumlee, G.S.; Morton, R.A.; Boyle, T.P. *An Overview of Mining-Related Environmental and Human Health Issues, Marinduque Island: Philippines,* **2000**. https://pubs.usgs.gov/of/2000/ofr-00-0397/o-r-00-0397.pdf

[25]   Obergföll, B.T. *Meerwasserentsalzung, Nachhaltig - Herausforderung für Wissenschaft und Entrepreneurship, Seminararbeit Universität Karlsruhe (TH) Interfakultatives Institut für Entrepreneurship,* **2010**. http://www.nachhaltigkeit-erforschen.de/fileadmin/erforschen/pdf-seminar/Meerwasserentsalzung.pdf

[26]   Hoekstra, A.Y.; Mekonnen, M.M. The water footprint of humanity. *Proc. Natl. Acad. Sci. USA,* **2012**, *109*(9), 3232-3237.
[http://dx.doi.org/10.1073/pnas.1109936109] [PMID: 22331890]

[27]   Cooley, H.; Ajami, N.; Ha, M.-L. *In: World's Water Series, 8: Chapter 1,* **2013**. http://worldwater.org/wp-content/uploads/sites/22/2013/07/ww8-ch1-us-water-policy.pdf

[28]   Roberts, D.A.; Johnston, E.L.; Knott, N.A. Impacts of desalination plant discharges on the marine environment: A critical review of published studies. *Water Res.,* **2010**, *44*(18), 5117-5128.
[http://dx.doi.org/10.1016/j.watres.2010.04.036] [PMID: 20633919]

[29]   Kondolf, G.M. Hungry water: effects of dams and gravel mining on river channels. *Environ. Manage.,* **1997**, *21*(4), 533-551.

[http://dx.doi.org/10.1007/s002679900048] [PMID: 9175542]

[30]   EEA. *Europe's Coasts, reconciling and development,* **2010**. http://www.eea.europa.eu/downloads/
        1346e1f648f7288cd5a3b1972738e4b2/1302718678/europe2019s-coasts-reconciling-developm-
        nt-and-conservation.pdf

[31]   *The Ramsar Convention and its Mission,* **2017**. http://www.ramsar.org/

[32]   Food and Agriculture Organization of the United States. *The State of World Fisheries and
        Aquaculture, Opportunities and Challenges,* **2014**. http://www.fao.org/3/a-i3720e.pdf

CHAPTER 12

Oceans and Seas

Abstract: The surface waters of oceans and seas have been exposed long-term to multiple anthropogenic contaminations, waste heat, and heat transfer from the warming atmosphere. Absorption of ca. 170 GT CO_2 since the beginning of industrialisation has caused on average a reduction of pH by 0.1. Between 1871 and 2005, heat transfer from the atmosphere has resulted in surface water warming of tropical seas by 0.5 °C. Fluvial input of reactive N and P from urban areas and agricultural land has caused partial deoxygenation of the sea waters. In poorly ventilated areas like the Baltic Sea and Gulf of Bengal, hypoxic and anoxic zones developed. Anoxic zones in eddies, observed in the eastern South Atlantic, alter microbial communities and, therefore, primary productivity. Decrease in Atlantic Meridional Overturning Circulation (AMOC) has occurred because of the intensified input of Arctic glacial meltwater, resulting in less storage of atmospheric CO_2 in deeper sea water layers. Average Hg-concentration in the upper seawater layer increased from 0.75 pM to 1.45 pM between 1850 and 2008. Offshore hydrocarbon development, including havaries and oil spills impaired sea water quality. Littering of the seas with plastic particles has occurred since ca. 70 years ago and amounted 2010 to 8-10 MT/a. The fragments act as transport vehicles for pathogens and organotoxins. Radioactive waste was disposed of between 1946 and 1993 at 80 sites in the Atlantic, Arctic, and Pacific Oceans. The Fukushima nuclear havary effected aquatic discharge of minimum 2×10^{16} Bq into the Pacific.

Keywords: Acidification, AMOC, Anoxia, Deoxygenation, Eutrophication, Heat uptake, Meltwater, Oil spills, Plastic, Pollution, Primary production, Revelle Factor, Sea water, Sea water stratification, Thermohaline circulation, Warming.

The oceans' surface near water bodies have decreased their capacity as a sink due to excess anthropogenic discharge: transfer of heat (from atmosphere and rivers), import of numerous sorts of organic and inorganic contaminants (particulate, liquid) and waste[1] dissipated at shores and discharged into rivers, and gaseous and particulate immissions (CO_2, NO_X, aerosols, *etc.*) released into the atmosphere.

Because the volumetric heat capacity ($Jcm^{-3}K^{-1}$) of sea water is ca. 4000 times greater than that of air at sea level, the world oceans[2] have absorbed most of the waste heat and heat produced by anthropogenic greenhouse gases, which have accumulated in the atmosphere (93% of heat is stored in the oceans, 3% in melting ice, 3% in the continents, and 1% in the atmosphere) [1, 2]. Based on the data from many thousand vertical temperature profiles conducted between 1955 and 1998, it appears that the oceans' heat content between 0-3000 m bsl rose, on average, by 0.037 °C, equivalent to a heat import of ca. 14.5×10^{22} J (83.8% of Earth's heat balance for the time period mentioned). The warming of the upper 700 m of the oceans between 1955 and 2008 is, on average, 0.2 °C, equivalent to an increase in heat content by 0.4×10^{22} J/a [3, 4]. Multidecadal datasets from temperature profiles and full-depth transoceanic sections indicate that the industrial era global ocean heat absorption from 1865 to 1997 is equal to the heat input between 1997 and 2015, and that one third of the heat storage occurs below 700 m [5].

Reconstruction of anomalies of tropical sea surface temperature development from 1871 to 2005 resulted in + 0.5°C warming [6].

Sea surface temperature data obtained over the last 20 years from floating buoys, Argo floats, and infrared radiometer-based satellites, indicate an increase by 0.12 °C/decade [7].

Incipient transition to a stratified system of water layers was caused by rising quantities of melt water from Arctic ice, which slowed down the Atlantic Meridional Overturning Circulation (AMOC), driven by thermohaline density gradients [8]. A fully coupled climate simulation tested the sensitivity of the Atlantic Meridional Overturning Circulation. It revealed that the release of 1 million m^3/sec of meltwater between 1965 and 2000 from the Greenland ice shield decreased circulation speed by 13-30%, because of diminished formation of Upper Labrador Sea Water [9]. The resulting reduced exchange with deep waters caused a decline in CO_2 storage in subarctic waters [10] and consequently, accumulation of more CO_2 in the atmosphere [11].

From the onset of industrialisation to 2002, the pH value of ocean surface waters has been reduced by 0.1, equivalent to ca. 30% increase of acid forming H^+ ions in seawaters, predominantly caused by anthropogenic emissions of CO_2 [12]. Absorption and chemical dissolution of CO_2 in the oceans is controlled by water temperature[3] and by the Revelle factor[4]. An inventory of the distribution of dissolved inorganic carbon was carried out, based on a tracer separation technique and on data sets from 9618 hydrographic stations that were collected on 95 cruises as part of the World Ocean Circulation Experiment and the Joint Global Ocean Flux Study. This showed that ca. 48% (118 ± 19 GT) of the anthropogenic CO_2 emitted between 1800 and 1994 has been absorbed by the ocean's upper and middle depths, that the highest vertically integrated CO_2 concentrations (< 80 moles/m^2) are present in the Northern Atlantic, and that about 60% of anthropogenic CO_2 has been stored in the Southern Hemisphere's oceans [13]. Cumulative uptake of anthropogenic CO_2 (produced directly and indirectly) by the world oceans between 1995-2015 is ca. 49.7 ± 0.5 GT CO_2, which is only 24.3% of the anthropogenic CO_2 emitted in that time (ca. 204.8 GT CO_2)[5] [14]. The significant uptake inhibition of CO_2 in that time can be explained by the warming of the oceans and by the fact that the current Revelle factors of the surface waters of the oceans are about one unit higher than at the beginning of industrialisation [15].

Uptake of heat, CO_2 and ammonia, and input of nitrate, caused acidification, eutrophication, and deoxygenation of the oceans [16]. Concerning O_2, coupled climate and ocean models discern between the natural variability of sea water oxygen distribution and anthropogenic forcing of oxygen anomalies in sea water. Simulations of ocean chemistry development between 1920 and 2100 suggest that anthropogenic forcing caused O_2 depletion in the Southern Indian Ocean and in the eastern tropical parts of the Atlantic and Pacific Oceans, and are confirmed by *in situ* observations [17]. It has been observed that, in open ocean eddies of the eastern tropical Atlantic, anoxic zones alter microbial communities and, consequently, primary productivity of open marine waters [18]. A quantitative global assessment of the whole ocean oxygen inventory by *in situ* analyses and supporting literature data revealed that its O_2 content decreased by 2% on average since 1960, due to climate warming, reduced convection and eutrophication [19].

Intensified fishing, oversea transportation, and extraction of mineral raw materials on and below the sea floor gave rise to further environmental deteriorations (dumping at sea; oil, natural gas and tar spills; detergents; hazardous waste; incinerations [20]; turbid water zones; tailings; seepage; *etc.*) [21, 22]. The quality of the ocean hydrochemistry has been deteriorated by contaminants and pollutants from many diffuse and point sources; several examples follow:

The average concentration of Hg in upper ocean waters from 1850 to 2008 rose by 0.75 pM to 1.45 pM. Reservoir mass of early industrial (1840) Hg in the upper ocean waters is estimated at 77,300 T. The corresponding present day amount is 143,100 T [23]. The amount of Hg in global surface ocean waters has increased since the beginning of industrialisation from 2500 T to 7000 T [24]. This is of concern because of the formation of toxic methylmercury in aquatic environments [23].

Disposal of radioactive waste at sea was carried out between 1946 and 1993 at more than 80 sites in the Atlantic, Arctic, and Pacific Oceans. The steel drums, liquids, and components of nuclear submarines and icebreakers contained cumulative radioactivity of ca. 8.5×10^4 TBq [25].

Concerning the aquatic discharge of radioactive matter from the Fukushima havary site, a minimum of 2×10^{16} Bq ^{137}Cs has been released into the ocean. Concentration levels of ^{137}Cs in seawater within a 30 km perimeter of the disaster site exceeded 10^4 Bq/m^3 in early April 2011 [26]. Simulation of long-term dispersal of ^{137}Cs in the Pacific Ocean, according to a global circulation model, indicates that, after 2-3 years, the radionuclide cloud penetrated to depths of 400 m and that the strongly diluted cloud will have reached the North American coast after 5-6 years. Total peak radioactivity levels in the regions affected will then be twice the values prior to Fukushima havary [27].

Since the first oil well in 1858, uncontrolled blowout events have occurred during offshore oil and gas explorations, which impaired the environment. A major disaster came to pass on 04/20/2010 on the Deepwater Horizon drill site in the Macondo prospect in the Gulf of Mexico, 66 km off of Louisiana's coast, with a borehole depth of 4054 m, having worked through Tertiary sediments filling the Mississippi Canyon. The disaster has been attributed to several technical decisions causing deficiencies, the undetected entry of high temperature, high pressure, and highly charged hydrocarbons into the well and multiple failures of blow out prevention systems and well controls. This resulted in the eruption of natural gas, its explosion, and the destruction of the semisubmersible rig. As it sank 1525 m through the water column, the marine riser and the drill pipe separated from it and collapsed. As a consequence, hydrocarbons began to discharge into the seawater. 83 days passed while attempts were made to shut the well. Before the borehole was plugged, ca. 780,000 m^3 of hydrocarbons had poured out into the seawater, causing considerable contamination and deoxygenation [28, 29].

Shallow and/or less ventilated marine areas suffer from eutrophication due to algal blooms, which result in local hypoxic to anoxic sea water conditions. Some examples are given:

Long-term hydrographical and physicochemical monitoring of the exchange of sweet- and saltwater, residence time of nutrients, development of oxygen-deficient zones, and local atmospheric circulations revealed that hypoxic and anoxic zones have spread considerably between 1969 and 2015 in the Baltic Sea. The main factors contributing to the increase in the oxygen-consuming processes were the fluvial input of nutrients (washed out fertilisers from industrial agriculture and sewage from cities), climate change-induced warming of sea water, and decreased frequency of wind driven events that transported cool and oxygen-enriched waters from the North Sea into the Baltic Sea [30].

Since ca. 1975, industrial growth on the Bohai Sea Economic Rim and the origination of a very densely populated coastal zone (230 million persons on

520,000 km^2) caused significant environmental pressure on the littoral, estuarine, and inner shelf areas. Pollution of harbour waters is caused by the exchange of freight on containerships in ports, atmospheric deposition, industrial effluents, and influx of polluted fluvial waters. It is aggravated by a ca. 50% decrease in freshwater input due to upriver water abstraction for urban supply. This has resulted in diminution of marine benthos and phytoplankton biodiversity, as well as increases in algal blooms and deoxygenation. Fishing catch mean trophic level in the Bohai Sea has declined faster than the global average trend, from 4.06 in 1960 to 3.41 in 1999. Trace metals (As, Cd, Cr, Cu, Hg, Ni, Pb, and Zn) concentration values were measured above threshold levels in seawater, sediment and - due to bioaccumulation - in organisms. Primary and metabolised secondary organic micropollutants (polycyclic aromatic hydrocarbons PAHs, DDT, PCBs, *etc.*) from nearshore oil wells, plants, and combustion contribute to major ecological risks. Plastic litter is also of concern (see below). According to the "Bohai Blue Sea Action Plan", the trace metal concentrations have locally decreased since 2001 [31 - 33].

Due to washouts from industrial agricultural cropland and urban effluents into the Mississippi Atchafalaya Basin, a shallow hypoxic zone in the northern Gulf of Mexico has developed seasonally (in summer), which had a five-year average area of 12,500 km^2 from 1988 to 2006 [34].

Unlike other cases, the oxygen minimum zone in the Gulf of Bengal, which originated because of discharge of organic wastes from densely populated coastal and inland urban areas and which has grown in 100-400 m water depth to 60,000 km^2, has not yet developed into a significant source of N_2, because anaerobic microbes - *e.g.* denitrifying and anaerobic ammonium oxidation (anammox) microbial populations - are restricted by nitrite availability due to the presence of trace concentrations of oxygen (10-200 nanomoles) [35].

The littering of the seas with plastic products started with the beginning of industrial mass production of plastic at the end of the first half of the 20th century. In 2015, ca. 50 MT of polyethylene terephthalate was produced. Ca. 80% of macroplastic garbage floating in the oceans was originally dumped at densely populated marine coasts or was imported by waste-loaded rivers. The rest was dumped during economic, tourist, or military activities on the sea. In 2010, 4.8-12.7 MT of land-based plastic waste entered the oceans [36]. The litter follows the courses of the surface waters, which have been explored since the 1950s. Dated and discrete point-source releases of numerous artificial buoyant drifters made it possible to determine structure and orbiting speeds of global oceanic surface gyres and their interconnectedness. This was achieved by long-term tracking of their recovery points and numeric modelling. The results showed that *e.g.* the orbiting period of the 13,500 km measuring North Pacific Gyre takes ca. 3.6 a and that the release of artificial flotsam on 01/10/1992 in the mid-North Pacific drifted into the Bering Strait and was conveyed by pack ice into the North Atlantic Gyre ca. 11 a later [37]. Plastic litter converges and accumulates in the centres of the five largest oceanic gyres: > 200,000 floating pieces/km^2 were counted in their centers. An estimated 8 MT/annum of macroplastic pieces end up in the seas [38]. Seven years of monitoring resulted in the estimation that ca. 5.25 trillion floating plastic particles > 0.33 mm are dispersed in the oceans, with an overall weight of only ca. 269,000 T. Fragmentation, ingestion and biodegradation processes were used in the study and successfully removed a very large portion of floating plastic fragments from the sampling area [39, 40]. Plastic debris present in the wave-dominated zone disintegrates to secondary, meso, and microplastic due to mechanical abrasion, oxidation and exposure to solar irradiation [41]. More than 260 marine species are known to have ingested or have been entangled in drift plastic. It is present in all marine environments and is integrated in all marine depositional systems [42]. Chemical and biological decomposition of plastic litter depends on its original chemical composition, as well as exposure and depositional site. The average decay time of plastic in marine environment is ca. 400 years [38]. Floating plastic litter is also detrimental to overseas traffic, fishing and tourism. Endeavours are being made to clean up the seas [43] and to diminish plastic waste volumes by succeeding with circular plastics economy [44].

Primary microplastic (and also nanoplastic) fragments from clothing, cleansers, cosmetics, *etc.* pass through the filters of waste water treatment plants and enter marine environments *via* fluvial systems [41].

Persistent organic pollutants like PAH and PCB are adsorbed by floating microplastic fragments, which can be mistaken for food by marine species and ingested. In this way, toxic substances bioaccumulate in the food chain [41, 45]. In addition, biofilms grown on polyethylene, polypropylene and polystyrene microparticles drifting in the North and Baltic Seas, contain the potentially pathogenic species *Vibrio parahaemolyticus*. This bacterium positively responds to sea water warming with enhanced population growth above 22 °C [46].

The seas are also exposed to ecological long-term risks and threats, caused by sunk and corroding material; Example: The amount of chemical and explosive munition, dumped during the Second World War in the Baltic Sea, is estimated to ca. 50,000 T. Release of one sixth of the contained toxic matters into the sea water will ruin the Baltic Sea hydrochemistry and ecosystem [47]. Sunken merchant, passenger and war ships are contamination point sources in the seas' ecosystems: an estimate of worldwide wrecks, containing a potential total amount of oil between 2.5 MT and 20.4 MT, amounts to 8569 sunken vessels [48].

Conservation and sustainable use of the oceans, seas, and marine resources has been recognised of prime importance [49].

NOTES

[1] An estimated 73,059 T of waste enters the global marine environment per hour [50].

[2] Carl-Gustav Arvid Rossby - oceanographer and meteorologist - was the first to suggest, in 1959, that ocean heat content may be the dominant component of the variables determining Earth's heat balance [3].

[3] Retrograde solubility of CO_2 effects that less of it is physically absorbed in water of higher temperature.

[4] It functions as buffer factor controlling the CO_2 exchange at the interface between atmosphere and sea water and is expressed as the relation between the fractional change in the partial pressure of CO_2 in sea water and the fractional change in total dissolved inorganic carbon subsequent to re-equilibration. The buffer factor depends also on the relation between sea water alkalinity and dissolved inorganic carbon. Sea waters with lower Revelle factors have larger capacities to buffer CO_2. The Revelle factor varies between 8 at the surfaces of tropical seas and 15 in the seas in high latitudes [15].

[5] Recent sea water acidification caused by dissolution of predominantly anthropogenic CO_2 has mainly affected the surface layers of the oceans, but in the next centuries will migrate deeper and will modify the deep-sea carbonate system and the depths of its critical boundary layers. A shoaling of the carbonate saturation depth, as well as of the carbonate compensation depth will occur, thus implying less geological burial of $CaCO_3$ in pelagic environments [2, 51].

CONFLICT OF INTEREST

The author (editor) declares no conflict of interest, financial or otherwise.

ACKNOWLEDGEMENTS

Declare none.

REFERENCES

[1]   Howes, E.L.; Joos, F.; Eakin, M. The Oceans 2015 Initiative. *The Oceans 2015 Initiative, Part I: An*

updated synthesis of the observed and projected impacts of climate change on physical and biological processes in the oceans. *Studies, N°02/15, IDDRI, Paris, France,* **2015**. http://www.iddri.org/Publications/Collections/Analyses/ST0215.pdf

[2]   Tyrrell, T. Anthropogenic modification of the oceans. *Philos. Trans. A Math. Phys. Eng. Sci.,* **2011**, *369*(1938), 887-908.
       [http://dx.doi.org/10.1098/rsta.2010.0334] [PMID: 21282152]

[3]   Levitus, S.; Antonov, J.; Boyer, T. Warming of the world ocean, 1955-2003. *Geophys. Res. Lett.,* **2005**, *32*(2), L02604.
       [http://dx.doi.org/10.1029/2004GL021592]

[4]   Levitus, S.; Antonov, J.; Boyer, T. Global ocean heat content 1955-2008 in light of recently revealed instrumentation problems. *Geophys. Res. Lett.,* **2009**, *36*(7), L07608.
       [http://dx.doi.org/10.1029/2008GL037155]

[5]   Gleckler, P.J.; Durack, P.J.; Stouffer, R.J. Industrial-era global ocean heat uptake doubles in recent decades. *Nat. Clim. Chang.,* **2016**, *6*, 394-398.
       [http://dx.doi.org/10.1038/nclimate2915]

[6]   BADC British Atmospheric Data Centre. http://badc.nerc.ac.uk/home/index.html**2006**.

[7]   Hausfather, Z.; Cowtan, K.; Clarke, D.C.; Jacobs, P.; Richardson, M.; Rohde, R. Assessing recent warming using instrumentally homogeneous sea surface temperature records. *Sci. Adv.,* **2017**, *3*(1), e1601207.
       [http://dx.doi.org/10.1126/sciadv.1601207] [PMID: 28070556]

[8]   Rahmstorf, S.; Box, J.E.; Feulner, G. Exceptional twentieth-century slowdown in Atlantic Ocean overturning circulation. *Nat. Clim. Chang.,* **2015**, *5*, 475-480.
       [http://dx.doi.org/10.1038/nclimate2554]

[9]   Yu, L.; Gao, Y.; Otterå, O.H. The sensitivity of the Atlantic meridional overturning circulation to enhanced freshwater discharge along the entire, eastern and western coast of Greenland. *Clim. Dyn.,* **2016**, *46*(5), 1351-1369.
       [http://dx.doi.org/10.1007/s00382-015-2651-9]

[10]   Perez, F.F.; Mercier, H.; Vázquez-Rodríguez, M. Atlantic Ocean CO_2 uptake reduced by weakening of the meridional overturning circulation. *Nat. Geosci.,* **2013**, *6*, 146-152.
       [http://dx.doi.org/10.1038/ngeo1680]

[11]   Haug, G.H.; Sigman, D.M.; Tiedemann, R. Onset of permanent stratification in the subarctic Pacific Ocean. *Nature,* **1999**, *40*, 779-782.
       [http://dx.doi.org/10.1038/44550]

[12]   Caldeira, K.; Wickett, M.E. Oceanography: anthropogenic carbon and ocean pH. *Nature,* **2003**, *425*(6956), 365.
       [http://dx.doi.org/10.1038/425365a] [PMID: 14508477]

[13]   Sabine, C.L.; Feely, R.A.; Gruber, N.; Key, R.M.; Lee, K.; Bullister, J.L.; Wanninkhof, R.; Wong, C.S.; Wallace, D.W.; Tilbrook, B.; Millero, F.J.; Peng, T.H.; Kozyr, A.; Ono, T.; Rios, A.F. The oceanic sink for anthropogenic CO_2. *Science,* **2004**, *305*(5682), 367-371.
       [http://dx.doi.org/10.1126/science.1097403] [PMID: 15256665]

[14]   Le Quéré, C.; Moriarty, R.; Andrew, R.M. Global Carbon Budget 2015. *Earth Syst. Sci. Data,* **2015**, *7*, 349-396.
       [http://dx.doi.org/10.5194/essd-7-349-2015]

[15]   IPCC. *Carbon Cycle Feedbacks to Changes in Atmospheric Carbon Dioxide,* **2007**. https://www.ipcc.ch/publications_and_data/ar4/wg1/en/ch7s7-3-4-2.html

[16]   Keeling, R.F. A.; Kortzinger, A.; Gruber, N. Ocean Deoxygenation in a Warming World. *Annu. Rev. Mar. Sci.,* **2010**, *2*, 199-229.
       [http://dx.doi.org/10.1146/annurev.marine.010908.163855]

[17] Long, M.C.; Deutsch, C.; Ito, T. Finding forced trends in oceanic oxygen. *Global Biogeochem. Cycles,* **2016,** *30*(2), 381-397.
       [http://dx.doi.org/10.1002/2015GB005310]

[18] Löscher, C.R.; Fischer, M.A.; Neulinger, S.C. Hidden biosphere in an oxygen-deficient Atlantic open ocean eddy: future implications of ocean deoxygenation on primary production in the eastern tropical North Atlantic. *Biogeosci. Discuss.,* **2015,** *12*, 1-39.
       [http://dx.doi.org/10.5194/bgd-12-14175-2015]

[19] Schmidtko, S.; Stramma, L.; Visbeck, M. Decline in global oceanic oxygen content during the past five decades. *Nature,* **2017,** *542*(7641), 335-339.
       [http://dx.doi.org/10.1038/nature21399] [PMID: 28202958]

[20] Müller, S.M. *Out Of Sight, Out Of Mind - the Politics and Culture of Waste*; Mauch, C., Ed.; RCC Perspectives, Transformations in Environment and Society: Munich, **2016,** Vol. 1, pp. 13-19.

[21] Goudie, A.S. *The Human Impact on the Natural Environment. Past, Present and Future,* 7th ed; Wiley-Blackwell: Oxford, **2013.**

[22] *Rapport d'activité Centre de documentation, de recherche et d'expérimentations sur les pollutions accidentelles des eaux,* **2014.** http://wwz.cedre.fr/en/content/download/8413/134016/file/ rapport-activite-2014web.pdf

[23] Amos, H.M.; Jacob, D.J.; Streets, D.G. Legacy impacts of all-time anthropogenic emissions on the global mercury cycle. *Global Biogeochem. Cycles,* **2013,** *27*, 410-421.
       [http://dx.doi.org/10.1002/gbc.20040]

[24] Selin, N.E.; Jacob, D.J.; Yantosca, R.M. Global 3-D land-ocean-atmosphere model for mercury: Present-day *versus* preindustrial cycles and anthropogenic enrichment factors for deposition. *Global Biogeochem. Cycles,* **2008,** *22*, GB2011.

[25] International Atomic Energy Agency. *Estimation of global inventories of radioactive waste and other radioactive materials,* **2009.** http://www-pub.iaea.org/MTCD/publications/PDF/te_1591_web.pdf

[26] Koo, Y-H.; Yang, Y-S.; Song, K-W. Radioactivity release from the Fukushima accident and its consequences: A review. *Prog. Nucl. Energy,* **2014,** *74*, 61-70.
       [http://dx.doi.org/10.1016/j.pnucene.2014.02.013]

[27] Behrens, E.; Schwarzkopf, F.U.; Lübbecke, J.F. Model simulations on the long-term dispersal of ^{137}Cs released into the Pacific Ocean off Fukushima. *Environ. Res. Lett.,* **2012,** *7*, 034004.
       [http://dx.doi.org/10.1088/1748-9326/7/3/034004]

[28] Deepwater Horizon Study Group. *Final Report on the Investigation of the Macondo Well Blowout,* **2011.** http://ccrm.berkeley.edu/pdfs_papers/bea_pdfs/dhsgfinalreport-march2011-tag.pdf

[29] *On Scene Coordinator Report Deepwater Horizon Oil Spill. Submitted to the National Response Team,* **2011.** http://www.uscg.mil/foia/docs/dwh/fosc_dwh_report.pdf

[30] Feistel, S.; Feistel, R.; Nehring, D. Hypoxic and anoxic regions in the Baltic Sea, 1969. *Meereswiss. Ber.,* **2015,** *2016*, 100. [Warnemünde]

[31] Gao, X.; Zhou, F.; Chen, C-T. Pollution status of the Bohai Sea: an overview of the environmental quality assessment related trace metals. *Environ. Int.,* **2014,** *62*, 12-30.
       [http://dx.doi.org/10.1016/j.envint.2013.09.019] [PMID: 24161379]

[32] Liu, W.X.; Chen, J.L.; Lin, X.M.; Tao, S. Distribution and characteristics of organic micropollutants in surface sediments from Bohai Sea. *Environ. Pollut.,* **2006,** *140*(1), 4-8.
       [http://dx.doi.org/10.1016/j.envpol.2005.08.074] [PMID: 16253407]

[33] The World Bank. *China to integrate water and environment management with GEF support,* **2016.** http://www.worldbank.org/en/news/press-release/2016/05/09/china-to-integrate-water-and-environment-management-with-gef-support

[34]   Rabotyagov, S.S.; Campbell, T.D.; White, M.; Arnold, J.G.; Atwood, J.; Norfleet, M.L.; Kling, C.L.; Gassman, P.W.; Valcu, A.; Richardson, J.; Turner, R.E.; Rabalais, N.N. Cost-effective targeting of conservation investments to reduce the northern Gulf of Mexico hypoxic zone. *Proc. Natl. Acad. Sci. USA,* **2014**, *111*(52), 18530-18535.
[http://dx.doi.org/10.1073/pnas.1405837111] [PMID: 25512489]

[35]   Bristow, L.A.; Callbeck, C.M.; Larsen, M. N production rates limited by nitrite availability in the Bay of Bengal oxygen minimum zone. *Nat. Geosci.,* **2017**, *10*, 24-29.
[http://dx.doi.org/10.1038/ngeo2847]

[36]   Jambeck, J.R.; Geyer, R.; Wilcox, C.; Siegler, T.R.; Perryman, M.; Andrady, A.; Narayan, R.; Law, K.L. Marine pollution. Plastic waste inputs from land into the ocean. *Science,* **2015**, *347*(6223), 768-771.
[http://dx.doi.org/10.1126/science.1260352] [PMID: 25678662]

[37]   Ebbesmeyer, C.C.; Ingraham, W.J., Jr; Royer, T.C. Tub toys orbit the Pacific Subarctic Gyre. *Eos (Wash. D.C.),* **2007**, *88*(1), 1-12.
[http://dx.doi.org/10.1029/2007EO010001]

[38]   Kershaw, P.; Katsuhiko, S.; Lee, S. *In: UNEP Year Book; 2011,* **2011**. http://www.unep.org/yearbook/2011

[39]   Eriksen, M.; Lebreton, L.C.; Carson, H.S.; Thiel, M.; Moore, C.J.; Borerro, J.C.; Galgani, F.; Ryan, P.G.; Reisser, J. Plastic pollution in the world's Oceans: More than 5 trillion plastic pieces weighing over 250,000 tons afloat at sea. *PLoS One,* **2014**, *9*(12), e111913.
[http://dx.doi.org/10.1371/journal.pone.0111913] [PMID: 25494041]

[40]   Cressey, D. Bottles, bags, ropes and toothbrushes: the struggle to track ocean plastics. *Nature,* **2016**, *536*(7616), 263-265.
[http://dx.doi.org/10.1038/536263a] [PMID: 27535517]

[41]   Cole, M.; Lindeque, P.; Halsband, C.; Galloway, T.S. Microplastics as contaminants in the marine environment: a review. *Mar. Pollut. Bull.,* **2011**, *62*(12), 2588-2597.
[http://dx.doi.org/10.1016/j.marpolbul.2011.09.025] [PMID: 22001295]

[42]   Bergmann, M.; Sandhop, N.; Schewe, I. Observations of floating anthropogenic litter in the Barents Sea and Fram Strait, Arctic. *Polar Biol.,* **2015**, *39*(3), 553-560.
[http://dx.doi.org/10.1007/s00300-015-1795-8]

[43]   De Torenhove. *The Ocean Cleanup,* http://www.theoceancleanup.com

[44]   Ellen McArthur Foundation. *The New Plastics Economy: Rethinking the future of plastics,* **2016**. http://www.ellenmacarthurfoundation.org/news/new-plastics-economy-report-offers-blueprint-to-design-a-circular-future-for-plastics

[45]   Tibbetts, J.H. Managing marine plastic pollution: policy initiatives to address wayward waste. *Environ. Health Perspect.,* **2015**, *123*(4), A90-A93.
[http://dx.doi.org/10.1289/ehp.123-A90] [PMID: 25830293]

[46]   Kirstein, I.V.; Kirmizi, S.; Wichels, A.; Garin-Fernandez, A.; Erler, R.; Löder, M.; Gerdts, G. Dangerous hitchhikers? Evidence for potentially pathogenic *Vibrio* spp. on microplastic particles. *Mar. Environ. Res.,* **2016**, *120*, 1-8.
[http://dx.doi.org/10.1016/j.marenvres.2016.07.004] [PMID: 27411093]

[47]   North Atlantic Treaty Organization. *Monitoring dumped munitions in the Baltic Sea,* **2016**. http://www.nato.int/cps/in/natohq/news_136380.htm?selectedLocale=en

[48]   Michel, J.; Gilbert, T.; Schmidt Etkin, D. *Int. Oil Spill Conf. Proc.,* **2005**, pp. 1-40.

[49]   *The Ocean Conference. Our oceans, our future: partnering for the implementation of Sustainable Development, UN, New York,* **2017**.

[50]   National Oceanic and Atmospheric Administration. *Ocean pollution,* **2011**. http://www.noaa.gov/resource-collections/ocean-pollution

[51]   Boudreau, B.P.; Middelburg, J.J.; Hofmann, A.F. Ongoing transients in carbonate compensation. *Global Biogeochem. Cycles,* **2010**, *24*, GB4010.
[http://dx.doi.org/10.1029/2009GB003654]

CHAPTER 13

Atmosphere and Lowermost Exosphere

Abstract: The anthropogenic modifications of the atmosphere's trace gases and their heat content occurred because of world population growth, agriculture, industrial economy, deforestation, land use change, mobility, and urbanisation. CO_2 concentration increased from ca. 280 ppm in 1780 to 402 ppm in 2016 and, according to the cumulative greenhouse effect, the near earth surface temperature rose on average by 0,9 °C from 1850 to 2014. An estimated 2 teraT of CO_2 have been emitted into the atmosphere since 1760. The emission histories of other greenhouse gases (CO, CH_4, N_2O, NO_x, CFC) and their cumulative radiative forcings are given (ca. $3W/m^2$ relative to 1750). NH_3-emissions increased from 6 MT in 1890 to 32.5 MT in 1995. They contribute to expansion of subtropical arid zones, northward shift of the intertropical convergence zone, and poleward retreat of mid-latitude low-pressure tracks. The eastern Mediterranean drought (1998-2012) is explained by climate warming. Frequency and magnitude of events during Australian wildfire season increased significantly from 1973-2010. Arctic amplification caused the long-lasting boreal summer weather extremes of the last decades. SO_2 pollution was reduced in industrialised countries due to implementation of filters in the 1970s. Sources and histories of particulate matters, as well as the development of their radiative forcings are given. Cumulative global Hg-emissions between 1850-2010 are as much as 350,000 T. Significant decreases in particulate matter in urban air have occured since the 1920s in developed countries. Radioactive releases due to nuclear havaries and above ground nuclear tests are given. The stratospheric fraction of the latter persists over decades. The lowermost exosphere is littered with astro garbage (ca. 23,000 objects), raising safety risks of operating spacecraft.

Keywords: Aerosol, Albedo, Arctic amplification, Astro garbage, Atmospheric circulation, Climate warming, Clouds, Emission histories, Greenhouse gases, Heat, Keeling curve, Moisture content, Ozone, Particulate matter, Pollution, Radiative forcing, Residence time, Stratosphere, Stratospheric jets, Trace gases, Troposphere.

Compared to the other spheres, the mobility of pollutants is greated in the atmosphere. Initiated by Claude Lorius in 1956, the development of atmosphere chemistry and the history of air pollution can be reconstructed by physico chemical analysis of air bubbles and particulate matter trapped in ice layers of polar and alpine regions.

This method is a precise instrument in analysing palaeoclimatic developments [1].

The amount and physico chemical states of atmospheric trace gases were modified in correlation with the growing of the world population and their manifold activities. The combustion of rising quantities of fossil hydrocarbons; increase of size and number of agricultural areas and industries with growing output; rising quantities of vehicles, ships, and airplanes with increasing power and weight; clearing forests; drainage of peatland; *etc.* have all caused emissions of greenhouse gases[1] like water vapour, CO_2, CO, CH_4, N_2O, NO_X, SF_6, CCl_4, chlorofluorocarbon, O_3, *etc.* [2]. From 1780 to Jan. 2016, the global atmospheric CO_2 concentration rose on average from 280 parts per million per volume (ppmv) to 402.59 ppmv [3, 4]. This occurred because the anthropogenic part of the carbon cycle is open[2]: more CO_2 was emitted by economic activity than absorbed by natural sinks, some of which have decreased in efficiency [5][3]. Current atmospheric CO_2 levels exceed the amounts at any time since the Pliocene Quaternary boundary 2.6 Ma ago [6].

Anthropogenic greenhouse gas emissions contributed substantially to the fact that, because of the first rule of thermodynamics, the average terrestrial near earth surface temperature increased by ca. 1.6 °C to 9.7 °C [7] and that from 1850 to 2014 the global average near land and ocean surface temperature rose by ca. 0.9 °C to 14.5 °C [8]. The globally averaged near surface air temperature in 2016 was ca. 1.1 °C higher than at the end of pre-industrial period; it was ca. 0.83 °C above the long-term average (14 °C) of the 1961-1990 reference period and ca. 0.07°C warmer than in 2015 [9]. The rising quantity of released waste heat[4] contributed to this development.

According to the thermodynamic rule of Clausius-Clapeyron[5], a larger fraction of a stable water-vapour-phase can exist in a warmer atmosphere, thus augmenting the greenhouse-effect. Natural contributions to the atmospheric temperature increase over the past half of the 20th century are considered to be low, because of the good fit between observed global and hemispheric mean temperatures and simulations, consisting of natural and anthropogenic components [10].

Industrial CO_2 emissions from fossil carbon started in 1760 with ca. 0.01 GT, by the year 1900 these had risen to 1.95 GT, and came in 2014 to ca. 35.9 GT; the latter is ca. 130 times more than the 0.26 GT[6] of annual background subaerial and submarine quiescent emissions of volcanogenic CO_2 exhalations [11 - 14]. Technical developments in decarbonisation, wind energy, as well as improvement of power and heat efficiency factors - *e.g.* thermal insulation of buildings, hydraulic balancing - between 2013 and 2015 caused global economic growth decoupling from rising greenhouse gas emissions [15]. However, the emissions began to rise again in 2017. The cumulative direct and indirect CO_2 emissions - combustion of fossil fuels, cement production, land use change, and other industrial sources - from 1760 to 2015 amounted to ca. 2.0 teraT of CO_2 [13]. According to latest estimations, ca. 2.075 ± 0.205 teraT CO_2[7] were emitted between 1870 and 2016 (75% from fossil fuel combustion and 25% to land use change) [16]. The amount of the latter - CO_2 emissions by land use change - is probably substantially underestimated, because tree harvesting and land clearing from shifting agriculture have not been considered thoroughly enough [17].

An estimated 842 GT of these emissions have accumulated in the atmosphere, 567 GT in the oceans, and ca. 586 GT in land sinks [14]. Between 1959 and 2010, ca. 350 GT of CO_2 were emitted by anthropogenic activity into the atmosphere and 192 GT of that amount were absorbed by terrestrial and marine sinks [18]. A recently identified, artificial CO_2 sink consists in manufactured, exposed cement: physico chemical modelling revealed that cement masses globally processed between 1930-2013 absorbed *via* chemical alteration - carbonation - ca. 10.5 GT C [19]. 2.73 teraT of CO_2 are expected to be emitted globally in future according to an assessment in 2013 concerning proven fossil carbon reserves[8] [20].

Anthropogenic CO emissions (*e.g.* from industry, incomplete combustion in stoves, gas firing motors) increased continually from 298.06 MT in 1890 to ca. 1.1 GT in 1995 [21].

Anthropogenic SO_2 emissions, originating from the combustion of fossil fuels, metal smelting, flaring of H_2S during exploitation of natural sour gas deposits, waste, grasslands and forest burning *etc.* rose from ca. 3×10^6 T in 1850 to ca. 1.3×10^8 T in 1974, and declined until 2005 to ca. 1.2×10^8 T [22]. Global anthropogenic SO_2 emissions in 2011 are estimated to be 1.0×10^8 T [23]. This is at least twice the amount of volcanic SO_2 emissions. Estimates concerning average annual volcanic SO_2 emissions, calculated for the time interval between 1970 and 2000, are $1.5-5.0 \times 10^7$ T [24, 25]. Cumulative anthropogenic SO_2 emissions between 1850-2011 are estimated to be ca. 2.042 GT [22, 23]. Emissions peaked in 1975 in Europe (42 MT) and the USA/Canada (36 MT), and in 1990 in former Soviet Union (20 MT). The other industrial regions are characterised by rising quantities, topped by China with 35 MT in 2005 [22].

SO_2 emissions are transformed in the atmosphere into sulphate aerosols, which reflect solar irradiation and act as condensation nuclei for clouds; both result in a net cooling effect of the atmosphere [26].

Deposition of sulphate aerosols causes damage to ecosystems because of degradation of plants and acidification of porewaters in soils and rock[9], as well as to building stones because of corrosion of mineral grain fabric [2, 27] and metal fittings; corrosion due to acid rain has caused irreparable damage to buildings, monuments, and edifices considered to be cultural heritage [28].

During the 20th century, mining, sulphide roasting, and smelting activities at *e.g.* Sudbury (Canada) caused the emission of more than 100 MT of SO_2 and tens of thousands of T of aerosols containing heavy metals. Due to amendments in environmental legislation and changes in production, SO_2 emissions declined from 2.6×10^6 T in 1960 to 2.1×10^5 T in 1994. An annual amount of 15,000 T of particulate matter containing Fe, Cu, Ni, Pb, and As were released into the atmosphere between 1973 and 1981 [29].

Global anthropogenic N_2O emissions (resulting *e.g.* from nitrification and denitrification processes in sewerages and agricultural soil) rose from 2.15 MT in 1890 to 9.7 MT in 1995, compared to natural N_2O emissions of ca. 11.5 MT/a. Global anthropogenic NO_x emissions (*e.g.* from automotive exhaust gases) developed from 25.09 MT in 1890 to 137 MT NO_x in 1995 [21]. Long-term remote sensing of tropospheric NO_2 between 1996 and 2004 revealed a substantial reduction of concentration above Europe and the USA, but accelerated rates of increase (ca. 50%) above regions with rapidly developing economies like China [30].

In 1994, overall atmospheric emission of further reactive N compounds (NO_x, N_2O, NH_3) from livestock, applications of fertilisers, deposited waste, use of solvents and products, industrial processes, high temperature combustion in automotives, households and small consumers, industrial furnaces, power plants, and central heating facilities in Germany was 1.3 MT of nitrogen [31]. Annual NH_3 emissions from European food animal manure diminished by 30% between 1990 and 2014 to 1.23×10^6 T [32]. Global ammonia emissions (*e.g.* from agriculture, industry, sewerages) have increased from 5.95 MT in 1890 to 32.41 MT in 1995 [21]. Since the inventions of artificial fertilisers in 1860 and catalytic ammonia synthesis in 1918, the global nitrogen cycle has been altered significantly. Since then, the amount of reactive nitrogen produced increased globally to 190 MT/a[10] in 2013 and exceeds the nitrogen quantity cycled in natural flows by 78 MT. N_2O, released from crops and livestock, contributes to radiative forcing 300 times more effectively than CO_2 and causes depletion of the stratospheric ozone layer [33]. Because of progress in filter, catalyser, and combustion techniques, total annual emissions of NO_x and SO_2 in Europe diminished between 1990 and 2014 by 54% to 8.0×10^6 T of NO_x and by 87% to 3.3×10^6 T of SO_2 [32].

Intensified land use, agriculture, and growth of the world population, and industry has also contributed to the accumulation of the trace gas methane in the atmosphere. Emissions increased from ca. 100 MT CH_4 in 1890 - accelerating from 1950 onwards - to ca. 350 MT in 1995 [21]. Analyses by means of high resolution laser spectrometry of atmospheric methane trapped in air bubbles in ice, recovered from a 400 m long core drilled in the Greenland ice shield revealed a concentration increase from ca. 730 ppb CH_4 in 1700 to more than 1600 ppb CH_4 in 2000, which is singular in the last 1800 years [34].

In 2011, industrial agriculture contributed 26% of global methane emissions [35]. Concerning the anthropogenic part, the amount in 2010 was the equivalent of 6.875 GT of CO_2, which includes agriculture (54%), waste water (9%), oil and gas (20%), landfills (11%), and coal mining (6%) [36][11]. Methane from manure and enteric fermentation (ruminating), emitted in the top five countries (Brazil, China, India, the EU and the USA), was the equivalent of ca. 1.045 GT CO_2 in 2011 [35].

Global surface dry air fraction of atmospheric CH_4 rose on average from 1550 ppb in 1980 to 1810 ppb in 2012. Estimated output of the anthropogenic part (60%) of global methane emissions, due to the main sources fossil fuel production and use, agriculture, waste, biomass and biomass burning, amounted to 335 MT of CH_4 in 2012 [37].

A CH_4 blowout, caused by a ruptured pipe connecting to an underground storage facility at Alison Canyon (Los Angeles) resulted in the release of 97,100 T of CH_4 into the atmosphere [38].

Anthropogenic emissions of volatile organic compounds (except CH_4) - *e.g.* from the chemical industry, oil drill sites, waste repositories, and gas-firing motors - rose from 26.87 MT in 1890 to 207.73 MT in 1995 [21].

Enhanced unconventional development of heavy oil and tar sand open pit mining, combined with the application of hot water steam and chemicals as a separation technique, raise the concern that evaporation and atmospheric photo oxidation of low-volatility organic compounds from processed oil sands lead to the production of aerosols, affecting air quality and climate. For example, exploitation of oil sands in Alberta, which started in 1967, caused the emission of at least 45 T of secondary organic aerosols per day, which contaminated ambient ecosystems - forests, surface waters - by immission of polycyclic aromatic compounds, methyl Hg, sulphuric acid, *etc.* [39].

Emissions of the long-lived greenhouse gases perfluorocarbons, sulphurhexafluoride, and hydrofluorocarbons - all used in industry and applied in industrial products - increased from 25.17 kT in 1970 to 124.13 kT in 1995 [21].

Industrially produced quantities of chlorofluorocarbons - detected in 1928 by T. Midgley - rose from 100 kT in 1945 to 760 kT in 1985 [21]. These and homologous halocarbons turned out to cause depletion of the stratospheric ozone layer, due to catalytic destruction of O_3 by *e.g.* chlorine atoms from chloro fluoro methane and photodissociation by ultraviolet irradiation[12]. Since the international halocarbon ban passed in the Montreal Resolution and the consequent production stop, atmospheric concentrations dropped significantly, but stratospheric ozone concentrations have not recovered to preindustrial values.

Industrially developed chemical alternatives to chlorofluorocarbons - hydrofluorocarbons, perfluorocarbons, SF_6 - do not decompose stratospheric ozone, but contribute considerably to the anthropogenic greenhouse effect. Emissions from implemented industrial devices and components (refrigerators, air conditioning units, polyurethane insulation foam, *etc.*) persist [40]; atmospheric concentrations of these gases have been observed to decline since ca. 2000 [41].

Continuous and careful atmospheric monitoring found that the earthquake- and tsunami-related large-scale damage on 03/13/2011 of infrastructure in northeastern Japan resulted in additional halocarbon emissions of ca. 6600 T (equivalent to 19.2 MT of CO_2) in that year by ca. 30% of Japan's annual total halocarbon emissions (21,540 T). Japan's SF_6 emissions rose by 300 T to 540 T in 2011 [42].

The effects of atmospheric brown clouds, which result from aerosol plumes accumulating above mega cities and large industrialised areas, have been recognised to exert trans continental and trans oceanic effects on the climate system. They cause surface cooling by dimming and - due to the presence of black carbon and organics in the aerosol - atmospheric warming. Additional effects are decrease in monsoon circulation and meteoric precipitation on continents [43].

Carbonaceous aerosols, emitted between 1870 and 2000, have been estimated indirectly by scaling to CH_4 and CO_2 emissions, to population and crop production rates, to wood and fossil fuel consumption, heating and cooking practices *etc.*[13], and measured directly by satellite data since 1979 [44]:

Estimated black carbon emissions from combustion of:

- Fossil fuels rose from 8000 T/a in 1870 continually to 2.5 MT/a in 2000;
- Biofuels (*e.g.* agricultural waste) remained steady from 1870 to 1960 at ca. 0.7 MT/a, and then rose continually to 1.6 MT/a in 2000;
- Open vegetation increased irregularly from 0.8 MT/a in 1870 to 1.3 MT/a in 1950 and then climbed steeply to 2.3 MT/a in 1970; the remaining 30 years were characterised by an irregular and wide scatter between 1.8-2.4 MT/a.

Estimated particulate organic matter emissions from burning of:

- Fossil fuels rose very slowly from 2000 T/a in 1870 to 3.3 MT/a in 2000;
- Biofuels remained steady at 4 MT/a until 1960 and increased slowly to 7.5 MT/a in 2000;
- Open vegetation rose irregularly from 7.5 MT/a in 1870 to 12 MT/a in 1950, rose steeply to 20 MT/a in 1970, then was irregular between 15.5-22 MT/a through 2000.

To sum up, the global black carbon emissions increased from 2.1 MT/a in 1870 to 8.2 MT/a in 2000 and global particulate organic matter emissions rose in the same time interval from 15 MT/a to 41 MT/a[14]. More than 80% of these carbonaceous aerosols - except those from open vegetation burning - were emitted in the northern hemisphere, where the bulk of industrialised areas developed[15]. Contributions of these aerosols to climate change occur due to their radiative forcings: black carbon aerosols absorb short wave radiation and cause warming, whereas particulate organic matter aerosols scatter radiation, similar to sulphate aerosols (see above), and cause cooling of the atmosphere [44]. The best fit total aerosol forcing pathway, including the individual contributions from sulphate, black carbon, organic carbon, and clouds, is estimated from -0.05 W/m^2 in 1850 to a minimum of -1.0 W/m^2 in 1975, then increasing to -0.75 W/m^2 in 2010 [45]. This points to a decreasing atmospheric cooling effect of aerosols emitted from industrialised areas since 1975 as consequence of technical progress in implementing more efficacious particle filters and higher efficiency motors.

A metastudy evaluating 70 papers published between 2006 and 2015, in which heavy metals and metalloids, contained in PM_{10} (particulate matter < 10 micron) and $PM_{2.5}$ (particulate matter < 2.5 micron) in ambient air in different parts of the world were measured, revealed that, in general, these immissions were higher in developing countries than in developed ones. Most studies identified the sources of heavy metals contained in ambient air as vehicular and industrial emissions. Vehicles generate particulate matter containing Cr, Pb, Cu, Zn, and Cd. Industry emitts PM containing As, Mn, Ni, Hg, Cd, and Zn. Coal and oil combustion generate PM containing As, Hg, Cr, and Co. Brake wear contributes to the rise of Cu, Fe, Zn, Cr, Mn, and Ni content in PM. Generally, owing to rapid urbanisation and the increase in the number of vehicles and plants, averaged concentrations of Cd, Cr, Zn, and Hg has risen in ambient air. However, due to the ban on tetraethyl lead in gasoline by 1986, atmospheric Pb concentrations have declined remarkably [46].

In general, air pollution levels in European cities have fallen since the beginning of its accurate measurement in 1922 from an average of 400 $\mu g/m^3$ down to 10

$\mu g/m^3$ at its best in 2006 [47]. This contrasts with the development of air pollution in developing countries like India or China [48]: monitoring of concentrations of $PM_{2.5}$ and PM_{10}, including elementary C, organic matter, sulphate, nitrate, and ammonium in the megacities of eastern China between 1999-2006 revealed that the annual averaged values exceeded all World Health Organization norms and amounted to 156 $\mu g/m^3$. Measurements in 559 Chinese cities in 2006 found that air pollution in only 4,3% of them did not exceed 40 $\mu g/m^3$, and that in 37% it exceeds 100 $\mu g/m^3$. The fact that air pollution is elevated there can be seen as a consequence of China's GDP multiplying by 57 times between 1978 and 2006, that its industrial development depends heavily on coal, and that its vehicle fleet increased annually by 20% from 1990 to 2006 to ca. 2,700,000 automotives [49].

More than 543 above ground nuclear explosions - most of them between 1945 and 1963 - as well as nuclear havaries - have ejected large amounts of radioactive aerosols - containing toxic heavy metal radionuclides like ^{137}Cs, $^{238\text{-}241}$Pu, and ^{241}Am -, as well as ^{3}H and ^{14}C into the atmosphere. Cumulatively 440-604 MT of nuclear yield has been detonated. The tropospheric fractions of aerosol are removed by dry and wet deposition within weeks or months, but significant aerosol fractions < 0.1 μm remained drifting in the stratosphere over several decades because of its thermally stratified air masses, the separative effect of the tropopause, and because of the higher residence time of mentioned fractions. However, sporadic import of coarser aerosols from volcanic ash ejecta (*e.g.* Eyjafjallajökull 2010) have caused agglomeration with parts of the artificial radioactive aerosols because of self-charging effects in the stratosphere. In this way, sporadic natural events triggered import and redistribution of particulate contaminants in the lower atmosphere [50 - 52].

The positive effects of the test ban treaty 1963 concerning decreasing intensity of global radioactive fallout can be seen in the continuous monitoring conducted by the Meteorological Research Institute in Tsukuba, showing concentrations of ^{137}Cs and ^{90}Sr deposition in Japan decreasing from ca. 10^4 mBq/m^2/month in 1960 to less than 10^1 mBq/m^2/month in 2010 [53].

During the Chernobyl havary on 04/26/1986, clouds containing radioactive material rose up to ca. 1500 m altitude. Depending on atmospheric circulation, fallout was detected on 05/03/1986 in Japan and on 05/05/1986 in the USA/Canada [54].

On 03/11/2011 at Fukushima-Daiichi nuclear power plant, three core meltdowns, several hydrogen explosions and the subsequent venting of gaseous and volatile fission products resulted in the release of up to 1.9×10^{19} Bq of ^{133}Xe, 5×10^{17} Bq of ^{131}J, and 5×10^{16} Bq of ^{137}Cs into the atmosphere, with a cumulative radioactivity of ca. 2×10^{19} Bq. Due to atmospheric circulation systems, particulate matter containing ^{137}Cs from the Fukushima plant arrived in California on 03/17/2011 and in Europe on 03/22/2011 [55].

Nota bene: Modeling of the history of aerosol radiative forcing and cloud formation of the time between 1850 and 1950 indicates that its atmospheric cooling effect has been overestimated [56]. In addition, experimental investigation of aerosol and cloud formation under simulated, pre industrial, low sulphuric acid atmospheric conditions found that oxidation and ionisation of low-volatility organic trace vapours - here α-pinene, the most abundant monoterpene produced by conifers - resulted in the generation of substantial amounts of organic aerosol particles (condensation nuclei), thus indicating that cloud formation by industrial SO_2 emissions has been overestimated [57, 58]. As a consequence, the changing volume, sizes, and vitalities of the global conifer forests commercialised by the timber industry also influenced cloud development during the time considered.

Air pollution by toxic metal species is described here concerning mercury emissions:

Cumulative anthropogenic Hg emissions between 1850 and 2010 amount to ca. 350,000 T. According to the results from 38 land based monitoring sites and 309 tropospheric campaigns, the global anthropogenic mercury emissions in 2010 amounted to 2000-2400 T (industrial: 1300 T Hg^0, 650 T Hg^{2+}, 100 T $Hg^{particulate}$; biomass burning: 300 T Hg^0) and originated - in descending intensity - from artisanal and small scale gold production, fossil fuel combustion, nonferrous metal production, cement production, disposal and incineration of Hg-containing products, contaminated sites, ferrous metal production, chlor alkali production and oil refinery. For comparison, naturally emitted Hg from volcanism and hydrothermal vents amounts to ca. 90 T/a on average and is assumed to be constant. The amount of the anthropogenic reservoir mass of atmospheric Hg is calculated to be 5300 T and corresponds to an early industrial (1840) enrichment factor of 2.6; the global Hg burden has increased by 3-5 times as a result of human activity. 87% of atmospheric Hg deposition occurs due to primary and legacy anthropogenic contributions. Hg is recognised as mobile and bioaccumulative toxic trace heavy metal, having adverse neurological and other health effects [59 - 63]. In 2008, global Hg emissions from coal power plants amounted to 1423 T. Improvement in air pollution abatement techniques may reduce exhaust Hg emissions by 27 µgHg/m³ to 3 µgHg/m³ [64]. In 2013, Hg emissions of European coal-fired power plants were estimated to be ca. 12.3 T [65].

The average amount of Hg immissions are influenced by altered atmospheric states, as can be seen in the following: Overall losses from severe thunderstorms in Europe have increased on average since 1980, not only because of increase in destructible assets, but also because the frequency and intensity of thunderstorms has changed. Long-term observations - from 1978 to 2009 - at weather observatories found that potential thunderstorm energy - *i.e.* the convective available potential energy - has significantly risen. The main reason for this development is recent anthropogenic warming, which causes more evaporation of water vapour and rising moisture content in a warmer atmosphere. Therefore, more heat energy is transferred during the rising vapour-laden air into the upper troposphere and during phase transitions between gas, liquid, and solid, thus convection is increased [66]. This confirms the simulation results derived from a global climate model ensemble, which indicated a robust increase of heavy thunderstorm events in the USA in response to global warming [67]. Because the concentration of water soluble atmospheric Hg^{2+} increases with altitude above ground, moisture droplets convectively transported upward in more intense, and eventually more frequent, thunderstorms *in statu nascendi* will absorb Hg^{2+} in the upper troposphere, causing deposition of more Hg^{2+} during precipitation of rain water or hailstones on the ground [68].

Rising emissions of aerosols (from industrial plants, automotives, locomotive engines, aeroplane jet engines, smelters, residential stoves and heating systems, forest fires, *etc.*), and formation of organic and inorganic secondary aerosols has altered the heat content of and cloud formation in the atmosphere [69, 70]. The main sources of particulate air pollutants are identified as emissions from agrarian land (aerosol consisting of liquid ammonia), traffic, power stations, stoves, and open fireplaces [71]. Two examples:

• Nine months monitoring aerosol concentrations at three representative atmospheric observation stations on the Baltic Sea coast revealed that emissions from intens ship activities contribute considerably to the increase of these particulate atmospheric pollutants [72].
• The total number of airplanes globally in 2015 was ca. 20,400. The number of passengers transported increased elevenfold from 0.31 billion in 1970 to 3.441 billion in 2015 [73]. Between 1970 and 2012, CO_2 emissions from global aviation rose from 340 MT to 760 MT [74]. In 2006, global combustion of an estimated 188 MT kerosene for commercial aviation resulted in the production of 594 MT of CO_2, of 232 MT of water vapour, of 0.8 MT NO_x, of 0.7 MT CO, and of 98,000 T non-methane hydrocarbons [75]. Calculated global annual mean increase in radiative forcing by air traffic (commercial, freight, military) has risen from ca. 0 mW/m^2 in 1950 to ca. 65 mW/m^2 in 2015. The aviation industry has contributed to air pollution and climate warming by kerosene combustion in

jet engines, which release heat, black carbon, and other aerosols, CO_2, water vapour (which forms condensation trails and probably influences the formation of cirrus clouds), VOC, CO, NO_x, and SO_2 in the troposphere and tropopause [76]. Tourism has grown to an industrial economic branch depending heavily on the combustion of fossil energy carriers. It contributes 3.9-6% of the global emissions of greenhouse gases [77].

Some effects of the anthropogenic modulation of the chemical and physical states of the atmosphere are as follows:

• Analyses of global patterns of atmospheric circulation, derived from satellite cloud records between the 1980s and 2000s, revealed similarity between observed data and simulated cloud change patterns, driven by greenhouse gas concentrations: expansion of subtropical arid zones, poleward retreat of mid-latitude low-pressure tracks, northward shift of intertropical convergence zone, and increasing height of the tops of cumuli at all latitudes [78].
• The increasing number of long-lasting boreal summer weather extremes - floods and droughts - in recent decades can be explained by synchronous processes in a modified atmospheric circulation scheme: Arctic warming has caused weakening of the zonal mean jet stream and persistent high-amplitude, quasi-stationary, resonance-amplified Rossby waves developed, which control mid-latitude persistent surface weather conditions on monthly time scale [79].
• Studies of Mediterranean palaeoclimate and drought variability between the years 1100 and 2012 characterises the recent 15 year drought in eastern Mediterranean (1998-2012) as exceptional relative to natural oscillations during the last centuries and confirms other studies, which found evidence of anthropogenic influence: climate warming by emission of greenhouse gases has caused drying in that region [80]. Recent atmospheric warming is supposed to be the cause of more frequent disruptions of the atmospheric quasi-biennial oscillation of jet streams in the stratosphere above the tropical zone [81].
• Monitoring of surface air temperature data at meteorological stations concerning the variability of summer temperature anomalies (relative to the 1951-1980 mean) on northern hemisphere land areas revealed that the frequency of occurrences of more positive anomalies - *i.e.* more hot summers - has shifted from an occurrence probability of 33% in 1951 to ca. 75% in 2011 and that anthropogenic climate warming has contributed to more frequent heat waves. Additionally, area affected by heat waves rose from 1% of Earth's terrestrial surface in the 1950s to 10% in the earliest 21st century [82]. Thus the occurrence probability of wildland fires and related emissions has risen:

According to a 40 year inventory of global wildland fire emissions, based on literature studies and satellite-data, the mean global emission of CO_2 increased steadily from ca. 1.7 GT/a in 1960 to ca. 2.6 GT/a in 2000, and of CO from ca. 250 MT/a in 1960 to 415 MT in 2000. A maximum of 3.14 GT of carbon was emitted in 1992. Wildland fire emissions contain trace gases, whose average annual amounts are estimated to be: 4.6 MT/a of N from NO_x, 3.9 MT/a of formaldehyde, 15.4 MT/a CH_4, 2.2 MT/a SO_2. The main contributions to the carbon emissions occurred in South America and Africa (ca. 75%, equivalent to an average of 1.54 GT/a) [83]. Monitoring of Australian fires from 1973 to 2010, conducted at 38 stations, revealed a significant rise of number and magnitude of events [84].

To sum up: The increase of direct radiative forcing of CO_2 since preindustrial times (1750) to 2015 has been calculated to be ca. 1.9 Wm^{-2}, increasing by 50% just since 1990. Increase of cumulative direct radiative forcing of major greenhouse gases CO_2, CH_4, N_2O, CFC-11, CFC-12, as well as a dozen minor long-lived halogenated gases from 1750 to 2015 amounts to 3.0 Wm^{-2}. Atmospheric concentrations of anthropogenic CO_2 und CO_2 equivalents rose from 275 ppm in 1750 to 485 ppm in 2015 [41]. Due to the long residence time of greenhouse gases, their radiative forcing effect would outlast a shutdown of anthropogenic greenhouse gas emissions from fossil carbon combustion by a minimum of 100 years [2].

A large portion of the anthropogenic part of the carbon cycle is still open and combustion of fossil carbon continues, so atmospheric CO_2 concentration will continue to increase. Higher greenhouse gas concentration will alter atmospheric heat and moisture content, as well as convective transport, causing rising adaptive stress to life [8].

Satellite and ocean heat monitoring between 2005 and 2010 carried out to determine the planetary radiation energy budget, found that the human made greenhouse effect contributed to the observed radiation imbalance of +0.58 ± 0.15 W/m^{2}, meaning that more heat is stored than backscattered to space [85].

Although several periods of air pollution - due to metallurgic activities - prior to the beginning of industrialisation have been identified globally in annually resolvable natural archives like ice layers [86], lacustrine deposits, peat bogs, deposition and accumulation of particulate matter containing toxic trace metals (*e.g.* Pb, As, Bi, Cd, Cr, Cu, Mo, Sb) and organic contaminants due to anthropogenic activities during the last 250 years are unprecedented over all of human history [87, 88].

Lowermost exosphere: The adverse consequences of 50 years of spaceflight are detailed by the U.S. Space Surveillance Network: an estimated 23,000 objects (astro garbage) > 7.5 cm in diameter, catalogued in the year 2012, litter useful orbits with derelict satellites, burnt out rocket stages, payload and collision debris, as well as discarded trash. Their presence and ongoing fragmentation by collisional cascading raises safety risks of operating active spacecrafts and of ca. 1000 active satellites, as first mentioned by Donald Kessler who warned of orbital space becoming impassable in 1978 [89].

To avoid collisions, astro garbage is monitored by high discrimination tracking and imaging radar (TIRA) in orbital space. The trajectories of orbiting objects are recorded and - in case of menacing collisions - timely evading manœuvres are made. Lines of flight of large de-orbiting objects were calculated to start timely protection precautions in the target areas.

The Clean Space Initiative was started 2012, which encouraged consideration of the entire life-cycle of space activities including removal of large pieces of orbiting space debris and design of non-debris creating missions [90].

NOTES

[1] Based on the ideas of Jean B. J. Fourier and John Tyndall, who both recognised in 1824 and 1862 the greenhouse gas properties of CO_2, water vapour, and ozone and that CO_2 has substantial effects on the waxing and waning of ice ages, S. Arrhenius created the first climate model in 1896. He concluded that the atmospheric CO_2 concentration during the Late Glacial Maximum was 30% less than that at the end of preindustrial time [1]. In 1938, Guy J. Callendar was the first to demonstrate anthropogenic global warming. He also suggested that production of CO_2 from combustion of fossil fuels was the reason for that change. Gilbert Plass, Charles D. Keeling, and Roger Revell pioneered work in identifying and quantifying the effects of anthropogenic greenhouse gases on climate change and in calculating the thermodynamic factors controlling the CO_2 fluxes between the atmosphere and oceans [91, 92]. The predicted anthropogenic signatures in recent climate change were evidenced by simulations of global climate effects of time-dependent variations of trace gas concentrations (CO_2, CH_4, N_2O, CFC) and stratospheric aerosol: the time-series (1958-2058) showed that - despite natural variability - annual mean global surface air temperature will reach and maintain significant and robust warming compared to the climate conditions in the 1950s, even if drastic reductions of greenhouse gases have occurred [93].

[2] Open C cycle: this means that the amount of emitted CO_2 cannot be fully absorbed by biological or physicochemical processes (plant photosynthesis, hydrolysis, silicate weathering) or by technical measures. Therefore, anthropogenic CO_2 accumulates in the atmosphere.

[3] However, the latter statement is doubted by other authors, who analysed in a mass balance calculation that land and ocean sinks absorbed annually ca. 50 MT more CO_2 between 1960 and 2010 [18]. 20 years of biomonitoring of net carbon uptake in temperate forest affirms a negative feedback to climate warming: altered timing of plant phenology caused increase of photosynthetic activity and carbon assimilation by the forest [94]. See also discussions in the chapters about oceans and plants.

[4] F. W. Ostwald realised in 1908 that, according to the theorems of N. L. S. Carnot 1824, the transfer of heat energy into mechanical work remains principally incomplete and that the resulting relict quantities of heat accumulating over time in the spheres colonised by faunas and floras will finally cause heat death of all forms of life on Earth.

[5] Clausius-Clapeyron equation (1834, 1850): it describes the temperature dependency of equilibrium pressure between solid, liquid, and vapour phases, as well as the phase transformation heat characterising one and multi component systems [95].

[6] This value corresponds to modelled anthropogenic CO_2 emissions of the year 1855, when the world population was ca. 1.3 billion [13].

[7] Excluded is additional CO_2 from oxidized anthropogenic CO and CH_4.

[8] They total 746 GT fossil organic carbon, consisting of 1668 gigabarrels of petroleum and LNG, 6558 trillion cubic feet of natural gas, and 892 GT of coal [20]. Medium-term utilisable resources are estimated to be 22.6 teraT of fossil organic carbon [96].

[9] Acid pore waters in soil and rock mobilise toxic chemical elements contained in mineral grains as major constituents like aluminium and also accessories like heavy metals, *e.g.* copper, zinc, cadmium, *etc.*

[10] 121 MT NH_3 is produced by the Haber-Bosch process, 40 MT by cultivation of leguminous crops and 25 MT by high temperature combustion of fossil fuels per annum [97]. Emissions from domestic waste deposits and from slash and burn land clearence contribute to the remaining part. Anthropogenic production of reactive nitrogen exceeded natural nitrogen fixation (112 MT/a) in 1983 [98].

[11] A large scale revision in atmospheric methane isotope source signatures revealed that anthropogenic fossil fuel methane emissions are 20-60% greater than previously estimated; the difference amounts to ca. 30 MT CH_4/a [99].

[12] Stratospheric ozone protects the lower atmosphere from ultraviolet irradiation with wavelengths between 320nm-600nm. The photocatalytic decomposition of ozone by CFC was discovered 1974 by physicochemist Mario Molina and chemist Frank Rowland [100]. In the same year atmospheric chemist Paul Crutzen set up a photochemical model of that destruction process of ozone, catalysed by chlorine atoms [101] and mentioned the stratospheric ozone destruction by N_2O, generated in agriculture, in the exhaust gases of supersonic jets and by above ground atomic tests [102]. In 1985 physical observations found the regular annual development of an ozone hole during September/October in the stratosphere above the Antarctic Pole. Monitoring revealed that O_3 concentrations have steadily decreased since 1960 by 200 units to 100 Dobson units (1 Dobson unit is equivalent to 10µm columnar hight of O_3 at normal conditions [0°C and 1013 hPa]) and that the size of the ozone hole increased to 24 million km^2, while at the same time industrial production of CFC rose from 0.18 MT/a to 0.9 MT/a. Special physical conditions like continuous sunlight during polar summer and the existence of a very cold, almost isolated circulation vortex, in which the products of photochemical reactions remained internally contained, effected the origination of the ozone hole [2]. Subsequent to the Montreal Protocol in 1987 - a CFC ban, signed by 160 states - CFC production decreased sharply until 1999 to ca. 0.045 MT. But Antarctic O_3 concentration persisted at comparably low values, due to the fact that halogens in photocatalytic processes are almost inert and that their residence times in stratosphere are up to 150 years [2].

[13] Measurements include direct primary emissions of particles < 2.5 µm, excluding secondary organic aerosols.

[14] Another source mentioned the following values: annual emission of black carbon / organic carbon in 1890: 2.13 MT / 7.36 MT; 1995: 13.43 MT / 44.05 MT [21].

[15] *e.g.* Germany: Primary total emissions of suspended matter from sources like processing bulk goods, industry, dwellings, agriculture, forestry, farming, traffic, brake wear, construction sites, power plants, *etc.* came up in 1999 to 1.95 MT and declined by 86% until 2003 to ca. 0.27 MT; PM_{10} emissions amounted in 2006 to 0.23 MT [103].

CONFLICT OF INTEREST

The author (editor) declares no conflict of interest, financial or otherwise.

ACKNOWLEDGEMENTS

Declare none.

REFERENCES

[1]   Jouzel, J. A brief history of ice core science over the last 50 yr. *Clim. Past,* **2013**, *9*, 2525-2547.
        [http://dx.doi.org/10.5194/cp-9-2525-2013]

[2]   Goudie, A.S. *The Human Impact on the Natural Environment. Past, Present and Future,* 7th ed; Wiley-Blackwell: Oxford, **2013**.

[3]   Scripps Institute of Oceanography. https://scripps.ucsd.edu/programs/keelingcurve/

[4]   Earth System Research Laboratory. *Trends in Atmospheric Carbon Dioxide, Recent Global CO$_2$,* http://esrl.noaa.gov/gmd/ccgg/trends/global.html#global

[5]   Canadell, J.G.; Le Quéré, C.; Raupach, M.R.; Field, C.B.; Buitenhuis, E.T.; Ciais, P.; Conway, T.J.; Gillett, N.P.; Houghton, R.A.; Marland, G. Contributions to accelerating atmospheric CO$_2$ growth from economic activity, carbon intensity, and efficiency of natural sinks. *Proc. Natl. Acad. Sci. USA,* **2007**, *104*(47), 18866-18870.
        [http://dx.doi.org/10.1073/pnas.0702737104] [PMID: 17962418]

[6]   Masson-Delmotte, V.; Schulz, M.; Abe-Ouchi, A. *Climate Change 2013: The Physical Science Basis. Contribution of Working Group I to the fifth Assessment Report of the Intergovernmental Panel on Climate Change*; Stocker, T., Ed.; Cambridge University Press, **2013**, pp. 383-464.

[7]   Rohde, R.; Muller, R.A.; Jacobsen, R. A new estimate of the average earth surface land temperature spanning 1753 to 2011. *Geoinfor. Geostat. An Overview,* **2013**, *1*, 1.

[8]   Field, B.; Barros, V.R.; Mastrandrea, M.D. *Climate Change 2014: Impacts, Adaptation, and Vulnerability. Summary for policymakers,* **2014**. https://ipcc-wg2.gov/AR5/images/uploads/IPCC_WG2AR5_SPM_Approved.pdf

[9]   WMO World Meteorological Organization. *WMO confirms 2016 as hottest year on record, about 1.1°C above pre-industrial era,* **2017**. https://public.wmo.int/en/media/press-release/wmo-confirs-s-2016-hottest-year-record-about-11%C2%B0c-above-pre-industrial-era

[10]   Mann, M.E.; Rahmstorf, S.; Steinman, B.A.; Tingley, M.; Miller, S.K. The likelihood of recent record warmth. *Sci. Rep.,* **2016**, *6*, 19831.
        [http://dx.doi.org/10.1038/srep19831] [PMID: 26806092]

[11]   Gerlach, T.M. Volcanic *versus* anthropogenic carbon dioxide. *Eos (Wash. D.C.),* **2011**, *92*(24), 201-208.
        [http://dx.doi.org/10.1029/2011EO240001]

[12]   USGS. *Volcanic Gases and Climate Change Overview,* https://volcanoes.usgs.gov:443/hazards/gas/climate.php

[13]   Saussay, A. *Interactive map: Historical emissions around the world,* **2015**. http://www.carbonbrief.org/interactive-map-historical-emissions-around-the-world

[14]   Global Carbon Project. *Emissions from fossil fuels and industry,* **2015**. http://www.globalcarbon project.org/carbonbudget/15/hl-full.htm

[15]   International Energy Agency. *Decoupling of global emissions and economic growth confirmed,* **2016**. http://www.iea.org/newsroomandevents/pressreleases/2016/march/decoupling-of-global-emissions-an

d-economic-growth-confirmed.html

[16]   Le Quéré, C.; Andrew, M. R.; Canadell, J. G. Global Carbon Budget 2016. *Earth Syst. Sci. Data,* **2016**, *8*, 605-649.
[http://dx.doi.org/10.5194/essd-8-605-2016]

[17]   Arneth, A.; Sitch, S.; Pongratz, J. Historical carbon dioxide emissions caused by land-use changes are possibly larger than assumed. *Nat. Geosci.,* **2017**, *10*, 79-84.
[http://dx.doi.org/10.1038/ngeo2882]

[18]   Ballantyne, A.P.; Alden, C.B.; Miller, J.B.; Tans, P.P.; White, J.W. Increase in observed net carbon dioxide uptake by land and oceans during the past 50 years. *Nature,* **2012**, *488*(7409), 70-72.
[http://dx.doi.org/10.1038/nature11299] [PMID: 22859203]

[19]   Xi, F.; Davis, S.J.; Ciais, P. Substantial global carbon uptake by cement carbonation. *Nat. Geosci.,* **2016**, *9*, 880-883.
[http://dx.doi.org/10.1038/ngeo2840]

[20]   Heede, R.; Oreskes, N. Potential emissions of CO_2 and methane from proved reserves of fossil fuels: An alternative analysis. *Glob. Environ. Change,* **2016**, *36*, 2-20.
[http://dx.doi.org/10.1016/j.gloenvcha.2015.10.005]

[21]   Asadoorian, M.O.; Sarofim, M.C.; Reilly, J.M. *Historical Anthropogenic Emissions Inventories for Greenhouse Gases and Major Criteria Pollutants,* **2006**. http://globalchange.mit.edu/files/ document/MITJPSPGC_TechNote8.pdf

[22]   Smith, S.J.; van Aardenne, J.; Klimont, Z. Anthropogenic sulfur dioxide emissions: 1850-2005. *Atmos. Chem. Phys.,* **2011**, *11*, 1101-1116.
[http://dx.doi.org/10.5194/acp-11-1101-2011]

[23]   Klimont, Z.; Smith, S.J.; Cofala, J. *The last decade of global anthropogenic sulfur dioxide: 2000-2011 emissions. Supplementary material,* http://iopscience.iop.org/1748-9326/8/1/014003/media/erl441620 suppdata.pdf

[24]   Textor, C.; Graf, H-F.; Timmreck, C. *Emissions of Chemical Compounds and Aerosols in the Atmosphere*; Granier, C.; Reeves, C.; Artaxo, C., Eds.; Kluwer: Dordrecht, **2003**.

[25]   Halmer, M.M.; Schmincke, H-U.; Graf, H-F. The annual volcanic gas input into the atmosphere, in particular into the stratosphere: a global data set for the past 100 years. *J. Volcanol. Geotherm. Res.,* **2002**, *115*, 511-528.
[http://dx.doi.org/10.1016/S0377-0273(01)00318-3]

[26]   Forster, P.; Ramaswamy, V.; Artaxo, P. *Changes in Atmospheric Constituents and in Radiative Forcing,* **2007**. https://www.ipcc.ch/publications_and_data/ar4/wg1/en/ch2s2-4.html

[27]   Prikryl, R.; Smith, B.J., Eds. *Building Stone Decay: from Diagnosis to Conservation*; Geol. Soc. Spec. Publ., **2007**, p. 271.

[28]   Saiz-Jimenez, C., Ed. Air Pollution and Cultural Heritage. In: *Proc. of the Internat. Workshop on Air Pollution and Cultural Heritage*; Taylor & Francis Group plc: London, UK, **2004**.
[http://dx.doi.org/10.1201/b17004-38]

[29]   Potvin, R.R.; Negusanti, J.J. Restoration and Recovery of an Industrial Region. In: *Progress in Restoring the Smelter-Damaged Landscape Near Sudbury, Canada*; Gunn, J.M., Ed.; Springer: New York, **1995**; pp. 51-65.

[30]   Richter, A.; Burrows, J.P.; Nüss, H.; Granier, C.; Niemeier, U. Increase in tropospheric nitrogen dioxide over China observed from space. *Nature,* **2005**, *437*(7055), 129-132.
[http://dx.doi.org/10.1038/nature04092] [PMID: 16136141]

[31]   WHO. *Chapter 1 - Origin of nitrate in drinking water,* **2001**. https://www.yumpu.com/ en/document/view/10554959/chapter-1-origin-of-nitrate-in-drinking-water-world-health-

[32]   EEA European Energy Agency. *Air pollutant emissions data viewer,* **2016**. http://www.eea.

europa.eu/data-and-maps/data/data-viewers/air-emissions-viewer-lrtap

[33]   UNEP Year Book 2014 emerging issues update. *Excess nitrogen in the environment,* **2014**. http://www.unep.org/yearbook/2014/PDF/chapt1.pdf

[34]   Rhodes, R.H.; Faïn, X.; Stowasser, C.I Continuous methane measurements from a late Holocene Greenland ice core: Atmospheric and *in-situ* signals. *Earth Planet. Sci. Lett.,* **2013**, *368*, 9-19.
[http://dx.doi.org/10.1016/j.epsl.2013.02.034]

[35]   Knapp, J.R.; Laur, G.L.; Vadas, P.A.; Weiss, W.P.; Tricarico, J.M. Invited review: Enteric methane in dairy cattle production: quantifying the opportunities and impact of reducing emissions. *J. Dairy Sci.,* **2014**, *97*(6), 3231-3261.
[http://dx.doi.org/10.3168/jds.2013-7234] [PMID: 24746124]

[36]   Global Methane Initiative. *Global methane emissions and mitigation opportunities,* http://www.globalmethane.org/documents/analysis_fs_en.pdf

[37]   Saunois, M.; Bousquet, P.; Poulter, B. The global methane budget 2000-2012. *Earth Syst. Sci. Data,* **2016**, *8*, 697-751.
[http://dx.doi.org/10.5194/essd-8-697-2016]

[38]   Conley, S.; Franco, G.; Faloona, I. *Methane emissions from the 2015 Aliso Canyon blowout in Los Angeles, CA,* **2016**.
[http://dx.doi.org/10.1126/science.aaf2348]

[39]   Liggio, J.; Li, S-M.; Hayden, K.; Taha, Y.M.; Stroud, C.; Darlington, A.; Drollette, B.D.; Gordon, M.; Lee, P.; Liu, P.; Leithead, A.; Moussa, S.G.; Wang, D.; O'Brien, J.; Mittermeier, R.L.; Brook, J.R.; Lu, G.; Staebler, R.M.; Han, Y.; Tokarek, T.W.; Osthoff, H.D.; Makar, P.A.; Zhang, J.; Plata, D.L.; Gentner, D.R. Oil sands operations as a large source of secondary organic aerosols. *Nature,* **2016**, *534*(7605), 91-94.
[http://dx.doi.org/10.1038/nature17646] [PMID: 27251281]

[40]   Japan Ministry of Economy, Trade and Industry. *Chemical substances control law, History of chlorofluorocarbons,* http://www.meti.go.jp/policy/chemical_management/ozone/files/pamplet/panel/08e_basic.pdf**2011**.

[41]   Butler, J.H.; Montzka, S.A. *The NOAA annual greenhouse gas index (AGGI). NOAA Earth System Research Laboratory, Global Monitoring Division,* **2016**. http://www.esrl.noaa.gov/gmd/aggi/aggi.html

[42]   Saito, T.; Fang, X.; Stohl, A. Extraordinary halocarbon emissions initiated by the 2011 Tohoku earthquake. *Geophys. Res. Lett.,* **2015**, *42*(7), 2500-2507.
[http://dx.doi.org/10.1002/2014GL062814]

[43]   Ramanathan, V.; Feng, Y. Air pollution, greenhouse gases and climate change: Global and regional perspectives. *Atmos. Environ.,* **2009**, *43*, 37-50.
[http://dx.doi.org/10.1016/j.atmosenv.2008.09.063]

[44]   Ito, A.; Penner, J.E. Historical emissions of carbonaceous aerosols from biomass and fossil fuel burning for the period 1870-2000. *Global Biogeochem. Cycles,* **2005**, *19*, GB2028.
[http://dx.doi.org/10.1029/2004GB002374]

[45]   Smith, S.J.; Bond, T.C. Two hundred fifty years of aerosols and climate: the end of the age of aerosols. *Atmos. Chem. Phys.,* **2014**, *14*, 537-549.
[http://dx.doi.org/10.5194/acp-14-537-2014]

[46]   Suvarapu, L.N.; Baek, S-O. Determination of heavy metals in the ambient atmosphere. *Toxicol. Ind. Health,* **2017**, *33*(1), 79-96.
[http://dx.doi.org/10.1177/0748233716654827] [PMID: 27340261]

[47]   Anderson, H.R. Air pollution and mortality: A history. *Atmos. Environ.,* **2009**, *43*, 142-152.
[http://dx.doi.org/10.1016/j.atmosenv.2008.09.026]

[48]   Marlier, M.E.; Jina, A.S.; Kinney, P.L. Extreme Air Pollution in Global Megacities. *Curr. Clim. Change Rep.,* **2016**, *2*(1), 15-27.
[http://dx.doi.org/10.1007/s40641-016-0032-z]

[49]   Fang, M.; Chan, C.K.; Yao, X. Managing air quality in a rapidly developing nation: China. *Atmos. Environ.,* **2009**, *43*, 79-86.
[http://dx.doi.org/10.1016/j.atmosenv.2008.09.064]

[50]   Alvarado, J.A.; Steinmann, P.; Estier, S.; Bochud, F.; Haldimann, M.; Froidevaux, P. Anthropogenic radionuclides in atmospheric air over Switzerland during the last few decades. *Nat. Commun.,* **2014**, *5*, 3030.
[http://dx.doi.org/10.1038/ncomms4030] [PMID: 24398434]

[51]   Griffith, C.; Rossenfeld, C. *Worldwide Effects of Nuclear War, Radioactive Fallout,* **2015**.
http://www.atomicarchive.com/Docs/Effects/wenw_chp2.shtml

[52]   Center for Disease Control and Prevention. *Feasibility Study of Weapons Test Fallout,* **2014**.
http://www.cdc.gov/nceh/radiation/fallout/feasibilitystudy/technical_vol_1_chapter_3.pdf

[53]   Geochemical Research Department and Atmospheric Environment and Applied Meteorology Research Department Meteorological Research Institute, Japan. *Artificial radionuclides in the environment,* **2013**. http://www.mri-jma.go.jp/Dep/ap/ap4lab/recent/ge_report/2013Artifi_Radio_report/2013Artifi_Radio_report.pdf

[54]   Deutsches Atomforum, V. *Der Reaktorunfall in Tschernobyl - Unfallursachen, Unfallfolgen und deren Bewältigung; Sicherung und Entsorgung des Kernkraftwerks Tschernobyl, Berlin,* **2011**.
http://www.kernenergie.de/kernenergie-wAssets/docs/service/025reaktorunfall_tschernobyl2011.pdf

[55]   Koo, Y-H.; Yang, Y-S.; Song, K-W. Radioactivity release from the Fukushima accident and its consequences: A review. *Prog. Nucl. Energy,* **2014**, *74*, 61-70.
[http://dx.doi.org/10.1016/j.pnucene.2014.02.013]

[56]   Stevens, B. Rethinking the lower bound on aerosol radiative forcing. *J. Clim.,* **2015**, *28*, 4794-4819.
[http://dx.doi.org/10.1175/JCLI-D-14-00656.1]

[57]   Kirkby, J.; Duplissy, J.; Sengupta, K.; Frege, C.; Gordon, H.; Williamson, C.; Heinritzi, M.; Simon, M.; Yan, C.; Almeida, J.; Tröstl, J.; Nieminen, T.; Ortega, I.K.; Wagner, R.; Adamov, A.; Amorim, A.; Bernhammer, A.K.; Bianchi, F.; Breitenlechner, M.; Brilke, S.; Chen, X.; Craven, J.; Dias, A.; Ehrhart, S.; Flagan, R.C.; Franchin, A.; Fuchs, C.; Guida, R.; Hakala, J.; Hoyle, C.R.; Jokinen, T.; Junninen, H.; Kangasluoma, J.; Kim, J.; Krapf, M.; Kürten, A.; Laaksonen, A.; Lehtipalo, K.; Makhmutov, V.; Mathot, S.; Molteni, U.; Onnela, A.; Peräkylä, O.; Piel, F.; Petäjä, T.; Praplan, A.P.; Pringle, K.; Rap, A.; Richards, N.A.; Riipinen, I.; Rissanen, M.P.; Rondo, L.; Sarnela, N.; Schobesberger, S.; Scott, C.E.; Seinfeld, J.H.; Sipilä, M.; Steiner, G.; Stozhkov, Y.; Stratmann, F.; Tomé, A.; Virtanen, A.; Vogel, A.L.; Wagner, A.C.; Wagner, P.E.; Weingartner, E.; Wimmer, D.; Winkler, P.M.; Ye, P.; Zhang, X.; Hansel, A.; Dommen, J.; Donahue, N.M.; Worsnop, D.R.; Baltensperger, U.; Kulmala, M.; Carslaw, K.S.; Curtius, J. Ion-induced nucleation of pure biogenic particles. *Nature,* **2016**, *533*(7604), 521-526.
[http://dx.doi.org/10.1038/nature17953] [PMID: 27225125]

[58]   Tröstl, J.; Chuang, W.K.; Gordon, H.; Heinritzi, M.; Yan, C.; Molteni, U.; Ahlm, L.; Frege, C.; Bianchi, F.; Wagner, R.; Simon, M.; Lehtipalo, K.; Williamson, C.; Craven, J.S.; Duplissy, J.; Adamov, A.; Almeida, J.; Bernhammer, A.K.; Breitenlechner, M.; Brilke, S.; Dias, A.; Ehrhart, S.; Flagan, R.C.; Franchin, A.; Fuchs, C.; Guida, R.; Gysel, M.; Hansel, A.; Hoyle, C.R.; Jokinen, T.; Junninen, H.; Kangasluoma, J.; Keskinen, H.; Kim, J.; Krapf, M.; Kürten, A.; Laaksonen, A.; Lawler, M.; Leiminger, M.; Mathot, S.; Möhler, O.; Nieminen, T.; Onnela, A.; Petäjä, T.; Piel, F.M.; Miettinen, P.; Rissanen, M.P.; Rondo, L.; Sarnela, N.; Schobesberger, S.; Sengupta, K.; Sipilä, M.; Smith, J.N.; Steiner, G.; Tomè, A.; Virtanen, A.; Wagner, A.C.; Weingartner, E.; Wimmer, D.; Winkler, P.M.; Ye, P.; Carslaw, K.S.; Curtius, J.; Dommen, J.; Kirkby, J.; Kulmala, M.; Riipinen, I.; Worsnop, D.R.; Donahue, N.M.; Baltensperger, U. The role of low-volatility organic compounds in initial particle growth in the atmosphere. *Nature,* **2016**, *533*(7604), 527-531.

[http://dx.doi.org/10.1038/nature18271] [PMID: 27225126]

[59] Selin, N.E.; Jacob, D.J.; Yantosca, R.M. Global 3-D land-ocean-atmosphere model for mercury: Present-day *versus* preindustrial cycles and anthropogenic enrichment factors for deposition. *Global Biogeochem. Cycles*, **2008**, *22*, GB2011.

[60] Global Mercury Observation System. *Mid term results,* http://www.gmos.eu/freedocuments/ 140615_brochure_midterm_low.pdf

[61] Holmes, C.D.; Jacob, D.J.; Corbitt, E.S. Global atmospheric model for mercury including oxidation by bromine atoms. *Atmos. Chem. Phys.*, **2010**, *10*, 12037-12057.
[http://dx.doi.org/10.5194/acp-10-12037-2010]

[62] Amos, H.M.; Jacob, D.J.; Streets, D.G. Legacy impacts of all-time anthropogenic emissions on the global mercury cycle. *Global Biogeochem. Cycles,* **2013**, *27*, 410-421.
[http://dx.doi.org/10.1002/gbc.20040]

[63] De Simone, F.; Gencarelli, C.N.; Hedgecock, I.M. Global atmospheric cycle of mercury: a model study on the impact of oxidation mechanisms. *Environ. Sci. Pollut. Res., Special issue: Heavy Metals in the Environment: Sources Interactions and Human Health,* **2014**, *21*(6), 4110-4123.

[64] Weem, A.P. *Reduction of mercury emissions from coal fired power plants.,* **2011**. http://www.unece.org/fileadmin/DAM/env/documents/2011/eb/wg5/WGSR48/Informal%20docs/Info. doc.3_Reduction_of_mercury_emissions_from_coal_fired_power_plants.pdf

[65] Jones, D.; Huscher, J.; Myllyvirta, L. *Europe's dark cloud - How coal-burning countries are making their neighbours sick, CAN Europe, HEAL, Sandbag, WWF,* **2016**. http://www.caneurope.org/ docman/position-papers-and-research/coal-2/ 2924-report-europe-s-dark-cloud-how-coal-burning-countries-are-making-their-neighbours-sick/file

[66] Faust, E. *Severe thunderstorms in Europe,* **2016**. https://www.munichre.com/en/reinsurance/magazine/ topics-online/2016/ topicsgeo2015/severe-thunderstorms-in -europe/index.html

[67] Diffenbaugh, N.S.; Scherer, M.; Trapp, R.J. Robust increases in severe thunderstorm environments in response to greenhouse forcing. *Proc. Natl. Acad. Sci. USA,* **2013**, *110*(41), 16361-16366.
[http://dx.doi.org/10.1073/pnas.1307758110] [PMID: 24062439]

[68] Holmes, C.D.; Krishnamurthy, N.P.; Caffrey, J.M.; Landing, W.M.; Edgerton, E.S.; Knapp, K.R.; Nair, U.S. Thunderstorms increase Mercury wet deposition. *Environ. Sci. Technol.,* **2016**, *50*(17), 9343-9350.
[http://dx.doi.org/10.1021/acs.est.6b02586] [PMID: 27464305]

[69] Jimenez, J.L.; Canagaratna, M.R.; Donahue, N.M.; Prevot, A.S.; Zhang, Q.; Kroll, J.H.; DeCarlo, P.F.; Allan, J.D.; Coe, H.; Ng, N.L.; Aiken, A.C.; Docherty, K.S.; Ulbrich, I.M.; Grieshop, A.P.; Robinson, A.L.; Duplissy, J.; Smith, J.D.; Wilson, K.R.; Lanz, V.A.; Hueglin, C.; Sun, Y.L.; Tian, J.; Laaksonen, A.; Raatikainen, T.; Rautiainen, J.; Vaattovaara, P.; Ehn, M.; Kulmala, M.; Tomlinson, J.M.; Collins, D.R.; Cubison, M.J.; Dunlea, E.J.; Huffman, J.A.; Onasch, T.B.; Alfarra, M.R.; Williams, P.I.; Bower, K.; Kondo, Y.; Schneider, J.; Drewnick, F.; Borrmann, S.; Weimer, S.; Demerjian, K.; Salcedo, D.; Cottrell, L.; Griffin, R.; Takami, A.; Miyoshi, T.; Hatakeyama, S.; Shimono, A.; Sun, J.Y.; Zhang, Y.M.; Dzepina, K.; Kimmel, J.R.; Sueper, D.; Jayne, J.T.; Herndon, S.C.; Trimborn, A.M.; Williams, L.R.; Wood, E.C.; Middlebrook, A.M.; Kolb, C.E.; Baltensperger, U.; Worsnop, D.R. Evolution of organic aerosols in the atmosphere. *Science,* **2009**, *326*(5959), 1525-1529.
[http://dx.doi.org/10.1126/science.1180353] [PMID: 20007897]

[70] Jokinen, T.; Berndt, T.; Makkonen, R.; Kerminen, V.M.; Junninen, H.; Paasonen, P.; Stratmann, F.; Herrmann, H.; Guenther, A.B.; Worsnop, D.R.; Kulmala, M.; Ehn, M.; Sipilä, M. Production of extremely low volatile organic compounds from biogenic emissions: Measured yields and atmospheric implications. *Proc. Natl. Acad. Sci. USA,* **2015**, *112*(23), 7123-7128.
[http://dx.doi.org/10.1073/pnas.1423977112] [PMID: 26015574]

[71] Lelieveld, J.; Evans, J.S.; Fnais, M.; Giannadaki, D.; Pozzer, A. The contribution of outdoor air pollution sources to premature mortality on a global scale. *Nature,* **2015**, *525*(7569), 367-371.

[http://dx.doi.org/10.1038/nature15371] [PMID: 26381985]

[72] Kecorius, S.; Kivekäs, N.; Kristensson, A. Significant increase of aerosol number concentrations in air masses crossing a densely trafficked sea area. *Oceanologia,* **2016**, *58*(1), 1-12.
[http://dx.doi.org/10.1016/j.oceano.2015.08.001]

[73] World Bank. *Air transport, passengers carried,* **2016**. http://data.worldbank.org/indicator/ IS.AIR.PSGR?name_desc=true

[74] Bows-Larkin, A.; Mander, S.L.; Traut, M.B. *Encyclopedia of Aerospace Engineering*; Richard Blockley, R.; Shyy, W., Eds.; John Wiley & Sons, Ltd.: UK, **2016**.

[75] Brasseur, G.P.; Gupta, M.; Anderson, B.E. Impact of aviation on climate: FAA's Aviation Climate Change Research Initiative (ACCRI) Phase II. *Bull. Am. Meteorol. Soc.,* **2016**, *97*(4), 561-584.
[http://dx.doi.org/10.1175/BAMS-D-13-00089.1]

[76] Frömmig, C.; Ponater, M.; Dahlmann, K. Aviation-induced radiative forcing and surface temperature change in dependency of the emission altitude. *J. Geophys. Res.,* **2012**, *117*, D19104.

[77] *Klimafakten, Klimawandel - Was er für den Tourismus bedeutet - Kernergebnisse aus dem fünften Sachstandsbericht des IPCC,* **2016**. https://www.klimafakten.de/meldung/der-tourismus-leidet-un-er-dem-klimawandel-und-treibt-ihn

[78] Norris, J.R.; Allen, R.J.; Evan, A.T.; Zelinka, M.D.; O'Dell, C.W.; Klein, S.A. Evidence for climate change in the satellite cloud record. *Nature,* **2016**, *536*(7614), 72-75.
[http://dx.doi.org/10.1038/nature18273] [PMID: 27398619]

[79] Coumou, D.; Petoukhov, V.; Rahmstorf, S.; Petri, S.; Schellnhuber, H.J. Quasi-resonant circulation regimes and hemispheric synchronization of extreme weather in boreal summer. *Proc. Natl. Acad. Sci. USA,* **2014**, *111*(34), 12331-12336.
[http://dx.doi.org/10.1073/pnas.1412797111] [PMID: 25114245]

[80] Cook, B.I.; Anchukaitis, K.J.; Touchan, R. Spatiotemporal drought variability in the Mediterranean over the last 900 years. *J. Geophys. Res.,* **2016**, *121*, 2060-2074.

[81] Osprey, S.M.; Butchart, N.; Knight, J.R.; Scaife, A.A.; Hamilton, K.; Anstey, J.A.; Schenzinger, V.; Zhang, C. An unexpected disruption of the atmospheric quasi-biennial oscillation. *Science,* **2016**, *353*(6306), 1424-1427.
[http://dx.doi.org/10.1126/science.aah4156] [PMID: 27608666]

[82] Hansen, J.; Sato, M.; Ruedy, R. Perception of climate change. *Proc. Natl. Acad. Sci. USA,* **2012**, *109*(37), E2415-E2423.
[http://dx.doi.org/10.1073/pnas.1205276109] [PMID: 22869707]

[83] Schultz, M.G.; Heil, A.; Hoelzemann, J.J. Global wildland fire emissions from 1960 to 2000. *Global Biogeochem. Cycles,* **2008**, *22*, GB2002.
[http://dx.doi.org/10.1029/2007GB003031]

[84] Clarke, H.; Lukas, C.; Smith, P. Changes in Australian fire weather between 1973 and 2010. *Int. J. Climatol.,* **2013**, *33*(4), 931-944.
[http://dx.doi.org/10.1002/joc.3480]

[85] Hansen, J.; Sato, M.; Kharecha, P. Earth's energy imbalance and implications. *Atmos. Chem. Phys.,* **2011**, *11*, 13421-13449.
[http://dx.doi.org/10.5194/acp-11-13421-2011]

[86] More, A.F.; Spaulding, N.E.; Bohleber, P. *Next generation ice core technology reveals true minimum natural levels of lead (Pb) in the atmosphere: insights from the Black Death. GeoHealth,* **2017**.
http://onlinelibrary.wiley.com/doi/10.1002/2017GH000064/pdf

[87] Uglietti, C.; Gabrielli, P.; Cooke, C.A.; Vallelonga, P.; Thompson, L.G. Widespread pollution of the South American atmosphere predates the industrial revolution by 240 y. *Proc. Natl. Acad. Sci. USA,* **2015**, *112*(8), 2349-2354.

[http://dx.doi.org/10.1073/pnas.1421119112] [PMID: 25675506]

[88]   Gabrielli, P.; Vallelonga, P. *Developments in Paleoenvironmental Research. Environmental Contaminants - Using natural archives to track sources and long-term trends of pollution*; Blais, J.M.; Rosen, M.; Smol, J.P., Eds.; Springer: Dordrecht, **2015**, Vol. 18, pp. 393-430.

[89]   Kessler, D.J. Collisional cascading: The limits of population growth in low earth orbit. *Adv. Space Res., 1991, 11*(12), 63-66.
        [http://dx.doi.org/10.1016/0273-1177(91)90543-S]

[90]   ESA. *Clean Space,* **2016**. http://www.esa.int/Our_Activities/Space_Engineering_Technology/ Clean_Space/The_ Challenge

[91]   Revelle, R.; Suess, H.E. Carbon Dioxide Exchange Between Atmosphere and Ocean and the Question of an Increase of Atmospheric CO_2 during the Past Decades. *Tellus,* **1957**, *9*(1), 18-27.
        [http://dx.doi.org/10.3402/tellusa.v9i1.9075]

[92]   Hawkins, E.; Jones, P.D. On increasing global temperatures: 75 years after Callendar. *Q. J. R. Meteorolog. Soc. Part B,* **2013**, *139*(677), 1961-1963.

[93]   Hansen, J.; Fung, I.; Lacis, A. Global climate changes as forecast by Goddard Institute for Space Studies three-dimensional model. *J. Geophys. Res.,* **1988**, *93*(D8), 9341-9364.
        [http://dx.doi.org/10.1029/JD093iD08p09341]

[94]   Keenan, T.F.; Gray, J.; Friedl, M.A. Net carbon uptake has increased through warming-induced changes in temperate forest phenology. *Nat. Clim. Chang.,* **2014**, *4*, 598-604.
        [http://dx.doi.org/10.1038/nclimate2253]

[95]   Greulich, W. *Lexikon der Physik*; Spektrum Akademischer Verlag GmbH: Heidelberg, **1999**, pp. 1-5.

[96]   Bertau, M. *Energie und Rohstoffe - Gestaltung unserer nachhaltigen Zukunft*; Kausch, P.; Bertau, M.; Gutzmer, J., Eds.; Springer Spektrum Akademischer Verlag: Heidelberg, **2011**, pp. 135-149.

[97]   *EU Science for Environment Policy In-depth report - Nitrogen Pollution and the European Environment - Implications for Air Quality Policy,* **2013**. http://ec.europa.eu/environment/integration/ research/newsalert/pdf/IR6_en.pdf

[98]   Fields, S. Global nitrogen: cycling out of control. *Environ. Health Perspect.,* **2004**, *112*(10), A556-A563.
        [http://dx.doi.org/10.1289/ehp.112-a556] [PMID: 15238298]

[99]   Schwietzke, S.; Sherwood, O.A.; Bruhwiler, L.M.; Miller, J.B.; Etiope, G.; Dlugokencky, E.J.; Michel, S.E.; Arling, V.A.; Vaughn, B.H.; White, J.W.; Tans, P.P. Upward revision of global fossil fuel methane emissions based on isotope database. *Nature,* **2016**, *538*(7623), 88-91.
        [http://dx.doi.org/10.1038/nature19797] [PMID: 27708291]

[100]   Molina, M.J.; Rowland, F.S. Stratospheric sink for chlorofluoromethanes: chlorine atom-catalysed destruction of ozone. *Nature,* **1974**, *249*, 810-812.
        [http://dx.doi.org/10.1038/249810a0]

[101]   Crutzen, P.J. Estimates of possible future ozone reductions from continued use of fluoro-chlor--methanes (CF_2Cl_2, $CFCl_3$). *Geophys. Res. Lett.,* **1974**, *1*(5), 205-208.
        [http://dx.doi.org/10.1029/GL001i005p00205]

[102]   Crutzen, P.J. Estimates of possible variations in total ozone due to natural causes and human activities. *Ambio,* **1974**, *3*(6), 201-210.

[103]   Umweltbundesamt, U.B. *Feinstaubbelastung in Deutschland,* **2009**. https://www.umweltbundesamt. de/sites/ default/files/medien/publikation/ long/3565.pdf

CHAPTER 14

Biosphere

Abstract: The human species has, due to its creativity and innate behavioural properties, always been driver of major extinctions. Modern species losses exceed, by far, the natural background trends of defaunation and homogenisation. The main reason is the violent expansion of human beings into the biosphere, its displacement, utilisation, and transformation. This development was accelerated at the beginning of industrialisation by motors driven by combustion of fossil energy carriers. The biosphere and its natural cycles were degraded because huge artificial material cycles exceeded the turnovers of natural cycles and immense amounts of waste were disposed of in nature. Awareness about biodiversity loss and the fact that 4/5 of the estimated 10 million species are still unknown, gave rise to the Census of Marine Life. The main threats to biodiversity are the primary and subsequent effects of industrialisation and overpopulation. The impacts on species include: physiology, demography, evolution, species interactions, range dynamics, and responses to environmental change. These negative impacts must be minimised, because conservation of biodiversity and functioning of ecosystem services are essential for human well being.

Keywords: Adaptation, Biodiversity, Biosphere transformation, Colonisation, Conservation, Defaunation, Displacement, Ecosystem services, Expansion pressure, Extinction, Homogenisation, Secondary habitat, Species census, Species response, Technosphere.

Human beings have often been major drivers of extinctions, causing the decimation of a number of megafauna species and individuals of many other species [1 - 6]. Modern species losses are supposed to exceed, by far, the natural background [7]. Since the beginning of industrialisation, the synergetic combination of human ingenuity, creative spirit, discoveries, inventions, availability of huge amounts of natural resources, economic growth, and expansion pressure caused the biosphere to be exposed to more and more artificial substances and processes, to new kinds of applied chemical compounds (Al, Si, nitroglycerine, penicillin, polyvinylchloride, *etc.*), chemical elements and isotopes, to huge amounts of disposed aggregated matter (waste), and intense fluxes of chemical compounds (CO_2, CH_4, NH_3, phosphate, *etc.*) from point and diffuse sources, to the effects of new kinds of artificial physical processes in technical apparatuses (*e.g.* gas-firing motors, turbines, jet engines) that generate artificial spectra of electromagnetic waves (ranging from infrared light, radar, radio waves to roentgen rays, and gamma rays, polarised light waves, laser, maser, terahertz radiation), emissions like artificial heat, particulate matter and exhaust gases as well as artificial sonic frequency spectra ranging into the infra and ultrasonic parts.

It is obvious that there exist only a few plant and animal species in the wild, that have since the beginning of industrialisation, profited from machine and energy carrier-supported expansion of human beings into the diverse biospheres and from their subsequent transformations. Among many other causes, species extinctions result from mining. Pioneer faunal and floral elements have been endangered because of the prevention of natural dynamics like erosion, flooding, and fire. In a few cases, high species diversity was temporarily achieved in cultivated landscapes, during early successional stages in secondary habitats (*e.g.* the Lusatian opencast mine sites provided very large, undivided, nutrient-poor, and structurally diverse secondary biotopes). However, this was observed to disappear after mine closure, reclamation of land, and approach to terminal successional stage [8].

Although habitat dissection and destruction by anthropogenic activity prevail by far, new habitat areas for flora and fauna have originated at the fringes of glaciers and ice masses, which retreat because of climate warming.

The fact that roughly 4/5 of the ca. 10 million species on Earth are estimated to be still unknown to science inspired the global research program on marine biodiversity: the Census of Marine Life 2000-2010. Research activities of 2700 scientists and 535 field expeditions into ca. 2/5 of global marine regions resulted in the identification of 230,000 species of an estimated 1.4 million marine species present in those areas. Regional estimations of undescribed species are 40-60% in Antarctica, 38% in South Africa, 70% in Japan, 75% in the Mediterranean deep sea, and 80% in Australia. The study included regional species richness and number of endemic and alien species. The program also addressed the threats to biodiversity, which can be seen as resulting from primary and subsequent effects of industrial activities and overpopulation: overfishing, bycatch, dynamite fishing, pollution, intrusion of xenospecies, acidification, habitat loss (due to coastal urbanisation, seabed mining, *etc.*), climate change-induced warming, transfer of infectious diseases from aquacultures, maritime traffic, sediment runoff from land, eutrophication and hypoxia due to land derived nutrients consisting of sewage and fertilisers, sea level rise, fishery bottom trawling and dredging, aggregate dredging and extraction, oil pollution, discharge of garbage, plastics, toxic metals and persistent organic pollutants, toxic antifouling agents like tributyltin, *etc.* [9].

In general, species respond to anthropogenic impacts on the environment by means of six key biological mechanisms:

- Physiology: adaptation of energy and mass balance;
- Demography: alteration of life history, phenology, reproduction, and mortality (population dynamics);
- Evolution: genetic adaptive processes;
- Species interactions: changes in predatory and symbiotic behaviours;
- Range dynamics: alterations in dispersal and colonisation;
- Responses of species to environmental change [10].

Conservation of biodiversity, achievement of sustainable use of resources and maintenance of quality, goods, services, and functions of ecosystems - all together essentials for human well-being and livelihood - were recognised globally as urgent and important. So these topics made their ways into the programs of the UN, IPCC and many other national political and NGO organisations. Examples: The Biodiversity Network Bonn (Germany) consists of high-profile experts dealing with: basic research (monitoring, assessment, prediction), biodiversity and human development, biodiversity for food, agriculture and bioeconomy, ethical aspects and societal impacts of biodiversity, and public awareness about biodiversity [11]. The Sustainability Development Solutions Network Germany, founded by leading knowledge centres of this country, recognised sustainability as universal guiding principle that must be integrated into economic and environmental developments. Its most important points are: clean production of green energy carriers (decarbonisation), restricting global warming, responsible consumption of resources (dematerialisation) and protection of aquatic and terrestrial biospheres [12].

In the following, some examples are given concerning the impacts resulting from anthropogenic activities - industry, technology, economy, growth, consumption, concentration, mobility, pollution, *etc.* - on the historic development and state of global flora and fauna, as well as on human beings themselves.

CONFLICT OF INTEREST

The author (editor) declares no conflict of interest, financial or otherwise.

ACKNOWLEDGEMENTS

Declare none.

REFERENCES

[1]   Faurby, S. S.; Svenning, J.-C. Historic and prehistoric human-driven extinctions have reshaped global mammal diversity patterns. *Divers. Distrib.,* **2015**, *21*(10), 1155-1166.

[http://dx.doi.org/10.1111/ddi.12369]

[2]   Kolbert, E. *The Sixth Extinction: An Unnatural History*; Henry Holt & Co.: New York, **2014**.

[3]   Sandom, C.; Faurby, S.; Sandel, B. Global late Quaternary megafauna extinctions linked to humans, not climate change. *P. Roy. Soc. Lond. B Biol.,* **1787**, *2014*(281), 20133254.

[4]   Douglas, M.S.; Smol, J.P.; Savelle, J.M.; Blais, J.M. Prehistoric Inuit whalers affected Arctic freshwater ecosystems. *Proc. Natl. Acad. Sci. USA,* **2004**, *101*(6), 1613-1617. [http://dx.doi.org/10.1073/pnas.0307570100] [PMID: 14745043]

[5]   Bartlett, L.J.; Williams, D.R.; Prescott, G.W. Robustness despite uncertainty: regional climate data reveal the dominant role of humans in explaining global extinctions of Late Quaternary megafauna. *Ecography,* **2016**, *39*(2), 152-161. [http://dx.doi.org/10.1111/ecog.01566]

[6]   Saltré, F.; Rodríguez-Rey, M.; Brook, B.W.; Johnson, C.N.; Turney, C.S.; Alroy, J.; Cooper, A.; Beeton, N.; Bird, M.I.; Fordham, D.A.; Gillespie, R.; Herrando-Pérez, S.; Jacobs, Z.; Miller, G.H.; Nogués-Bravo, D.; Prideaux, G.J.; Roberts, R.G.; Bradshaw, C.J. Climate change not to blame for late Quaternary megafauna extinctions in Australia. *Nat. Commun.,* **2016**, *7*, 10511. [http://dx.doi.org/10.1038/ncomms10511] [PMID: 26821754]

[7]   Ceballos, G.; Ehrlich, P.R.; Barnosky, A.D.; García, A.; Pringle, R.M.; Palmer, T.M. Accelerated modern human-induced species losses: Entering the sixth mass extinction. *Sci. Adv.,* **2015**, *1*(5), e1400253. [http://dx.doi.org/10.1126/sciadv.1400253] [PMID: 26601195]

[8]   Xylander, W.E.; Bender, J. *Post-Mining Landscapes. Reclamation, Ecology, Nature Preservation and Socio-economy in Practice*; Xylander, W.E.R., Ed.; State Museum of Natural History: Görlitz, **2004**.

[9]   Costello, M.J.; Coll, M.; Danovaro, R.; Halpin, P.; Ojaveer, H.; Miloslavich, P. A census of marine biodiversity knowledge, resources, and future challenges. *PLoS One,* **2010**, *5*(8), e12110. [http://dx.doi.org/10.1371/journal.pone.0012110] [PMID: 20689850]

[10]   Urban, M.C.; Bocedi, G.; Hendry, A.P.; Mihoub, J.B.; Pe'er, G.; Singer, A.; Bridle, J.R.; Crozier, L.G.; De Meester, L.; Godsoe, W.; Gonzalez, A.; Hellmann, J.J.; Holt, R.D.; Huth, A.; Johst, K.; Krug, C.B.; Leadley, P.W.; Palmer, S.C.; Pantel, J.H.; Schmitz, A.; Zollner, P.A.; Travis, J.M. Improving the forecast for biodiversity under climate change. *Science,* **2016**, *353*(6304), aad8466. [http://dx.doi.org/10.1126/science.aad8466] [PMID: 27609898]

[11]   Biodiversity Network Bonn. *About us - working groups,* **2014**. http://www.bion-bonn.org/en/ueber-uns

[12]   German Development Institute. *Research - 2030 Agenda for Sustainable Development,* **2016**. https://www.die-gdi.de/en/2030-agenda/

CHAPTER 15

Flora

Abstract: Humans have converted 75% of global terrestrial biomes into anthomes. 30.1 million km^2 wilderness remain. Global tropical forests declined between 1990-2015 by 10% to 1,760,000 ha. The baseline value of the Samples Red Averaged List Index for global plants is 0.86; their main threats are biological resource use, immissions of reactive N, agri- and aquaculture, natural system modifications, residential and commercial development, invasive species and genes, mining and smelting, transportation and service corridors, human intrusions and disturbance, climate change, and pollution. In 58% of land areas, the planetary boundaries of functional and genetic biodiversity are transgressed. Warming has caused 5.7% of vegetated land to shift to warmer and dryer climate zones. 10% of vegetated land is sensitive to warming and 21% of vascular plants are threatened with extinction. Because gymnosperm species diversified, adapted, and radiated during the CO_2 decline in the Cretaceous, recent CO_2 increase will have profound implications, although the capacity of terrestrial flora to sequester carbon has increased in the 20th century. In 2014, global tree cover loss amounted to 18 million ha and decreased its carbon storage capacity. Global flora is threatened by extended wildfire seasons, degradation or loss of habitat zones, modified annual growth phases, and altered heat, moisture, and trace gas concentrations. Attempts have been made to attain genetically engineered crop plants resilient against heat and drought. A short history of protection measures for global flora is given. Due to warming and fertilisation with CO_2 and reactive N, boreal and austral floras has begun to sequester more carbon and to colonise deglaciated areas. Aquatic vegetation was impaired by pollutants, heat, harvesting, and intrusion of xenospecies. Experiments have revealed adaptive limits of primary producers to sea water acidification.

Keywords: Adaptive limits, Agriculture, Arctic greening, Biodiversity decline, Carbon sequestration, Climate classes, Extinction, Fertilisation, Forests, Genetic engineering, Habitat degradation, Homogenisation, Immissions, Land take, Nature parks, Planetary boundaries, Primary producers, Technogenic deserts, Warming, Xenospecies.

Land use on a global scale to provide food, fibre, water, energy carriers, mineral raw materials *etc*. has caused considerable reduction of biodiversity and ecosystem capacities [1]. Due to residential development (creation of villages, cities, metropolises, settlements, plants, lines of interconnection), biological resource use (logging, gathering terrestrial plants) agriculture (crop farming, pastures, wood and pulp plantation), setting up of dumping and mining sites, extraction of peat and unconsolidated rock, creation of dams, recreational activities, tourism, fire, displacement by invasives and import of pathogens, very large parts of the original terrestrial biospheres - biomes - have been transformed into cultivated land: anthropogenic biomes (anthomes) [2 - 4]. One third of that transformation was accomplished in the 20th century [2]. According to least conservative measures of land-use models, based on the History Database of the Global Environment [5] of biosphere transformations, ca. 50% were converted by 1750 and 75% were converted by 2000 into anthomes [6].

Since the onset of agriculture ca. 12,000 years ago, the number of trees has decreased by 46% to ca. 3 trillion [7], so that their total capacity as a CO_2 sink was considerably diminished. At present, ca. 15 billion trees are logged or combusted annually [8]. Landsat data in 2000 and 2012 of 768 ecological regions revealed substantial changes in forest cover, which can be summarised as a net loss of 3.76 million km^2 of forest interior area. This points to a shift to a more fragmented state of the global forests and to underestimated ecological risks [9]. In 2014 tree cover loss reached a total area of 18 million ha [8].

Humans utilised European forests prior to industrialisation in a very complex and diverse manner[1]. Concerning the UK, it was shown that, despite the switch from firewood-based to coal- and oil-based economic growth, new industrial applications of wood (demand as stabilisers in the workings of mines, railroad ties, furniture, for paper production, *etc.*), and growth in GDP and in population caused an increase of home-grown and imported wood from ca. 3 million m^3 in 1850 to 30 million m^3 in 1939 [10]. A very low level of forest cover level is reported at the beginning of the 20th century, which developed by 2004 to a level of only slightly more than 30% -, compared to an original cover of ca. 90% of intact original forest. It consists of highly fragmented areas, which grade northwards into a more contiguously forested region [11]. European forest management for the past 250 years, characterised by large-scale tree species conversion from mixed woodland to commercial timberland (mostly conifer forest), has resulted in a transformation of European forests from carbon sinks to sources because of industrial wood extraction and carbon release from litter, dead wood, and soil. The wood economy has effectively removed 3.1 GT carbon from the forests compared to 1750. In addition, climate warming was caused by diminished albedo and increased evapotranspiration of these transformed forests [12]. Carbon density monitoring revealed that tropical forests switched from carbon sinks to carbon sources, due to deforestation and land degradation [13].

Global cropland spread, at the expense of forest, from 2.65 million km^2 in 1700 to 15 million km^2 in 1980 [14]. Global pasture land increased from 2.83 million km^2

(2% of ice free land) in 1700 to 34 million km² (24%) in 2000 [5]. Due to direct and indirect anthropogenic impacts (industrial land use, urbanisation, climate change), ca. 9.6% (3.3 million km²) of the remaining terrestrial wilderness areas (30.1 million km²) have been lost since the early 1990s, causing deterioration of interlinked hydrological systems and habitat conditions of indigenous faunas. This has occurred despite intense nature conservation measures and biodiversity protections. Wilderness blocks, defined as globally significant contiguous areas > 10,000 km² hosting intact ecological communities lost, in the time considered, 37 (10%) of its elements. In three of 14 biome types² - tropical/subtropical coniferous forests, mangroves, and tropical/subtropical dry broadleaf forests - globally significant wilderness areas have disappeared. This development has impaired the carbon sequestration capacity of terrestrial ecosystems, which store ca. 360 GT more carbon than that contained in fossilised subterranean ecosystems (gas, oil, tar, and coal deposits: 1000 GT) plus in the atmosphere (CO_2, CH_4: 598 GT) [15].

Since 1700, ca. 25% of the global tropical forests were converted to nonforests or agricultural areas. Concerning the world's extant tropical forests, a recent estimate suggests that 24% are intact, 46% are fragmented, and 30% are otherwise degraded. 1.2-1.5 billion people directly rely on tropical forests [16]. In general, impacts of agriculture, deforestation, and climate change-induced droughts diminish the carbon storage capacity of tropical rainforests [17]. This trend has been confirmed by a 30 year analysis of biomass dynamics using a distributed network of 321 plots in a tropical rainforest, which resulted in the finding of a long-term decrease in C storage capacity because of climate change-induced increased mortality and reduced tree longevity due to atmospheric CO_2 fertilisation [18]. 17 months of CO_2 flux monitoring in a tropical woodland impinged by logging, slashing, burning, and by agriculture proved its reduced function as carbon sink [19]. According to prognoses derived from time series maps and based on accurate remote sensing data from above ground carbon stocks (265.57×10^6 T C) present on 3.68 million ha of tropical forest, business as usual land use and forest policy will result in the emission of 747.61×10^6 T CO_2 and the loss of 75% of forest by 2030. Forest conservation, however, will result in a loss of ca. 11% [20]. Between 2000 and 2010, slashing and burning of tropical forests decreased in a few developing countries and net reforestation occurred there [21].

Significant anthropogenic forcings between 1950 and 2010 caused changes in the regional boundaries between the five major Köppen climate classes³. They were designed to systematically explain the global geographic distribution pattern of biomes in 1936. Since then, ca. 5.7% of the global terrestrial area has shifted into warmer and drier climate types [22].

The globally extended wildfire season, the gradual shifts and diminution of habitat zones, as well as the modified annual growth phases caused by the climate change-induced fluctuations of atmospheric heat, moisture and trace gas content have caused adaptive stress for terrestrial vegetation [23].

Warming due to ongoing rise of anthropogenic CO_2 emissions is supposed to have profound implications for global plant biodiversity and composition of ecosystems because gymnosperms diversified, adapted, and radiated during the Cretaceous CO_2 decline [24]. Another point is the availability of reactive nitrogen, necessary to stimulate net primary productivity [25]. But according to proxy data from carbonylsulphide bubbles trapped in Antarctic firn and ice cores, the overall feedback of global net primary productivity by photosynthesising terrestrial plants to anthropogenic CO_2 emissions has increased by 31% in the 20th century [26]. Increase of atmospheric CO_2, deposition of reactive N, climate warming, land cover change, and regionally increased meteoric precipitation caused considerable spread and densification of global vegetation, as observed in spectroscopic data sets (leaf area indexes), recorded between 1982 and 2009 [27]. According to global carbon budget estimations, global vegetation models, as well as ground, atmosphere and satellite observations, the capacity of terrestrial ecosystems to absorb the large amounts of anthropogenic CO_2 emissions has increased between 2002 and 2012, although CO_2 emissions have continued and resulted CO_2 concentrations measured at Mauna Loa Observatory staying above 400 ppm for the entire year 2016 [28, 29].

Because anthropogenic climate change has accelerated shifts and altered sizes of climate zones, species that cannot keep pace with migrating zone boundaries or are fixed to stationary climate zones diminishing in size are endangered [30]. According to the extinction risk calculated from 7000 species randomly sampled from five major plant groups (bryophytes, pteridophytes, gymnosperms, monocotyledons, and legumes), the overall IUCN baseline value of the Samples Red Averaged List Index[4] for global plants is 0.86, with the species present in the tropical rain forests at higher risk. The threats are, in declining severity: biological resource use, agriculture and aquaculture, natural system modification, residential and commercial development, invasive species and genes, energy production and mining, transportation and service corridors, human intrusions and disturbance, climate change and pollution [31].

Highly profitable crop plants - developed for agroindustry and generated since ca. 1945 by cultivation and later by genetic engineering - replacing genetically diverse species developed over millennia, as well as application of fertilisers and biocides, endanger the conservation of the reservoir of biological diversity [2]. The expected decline in crop yield in tropical and subtropical regions caused by climate warming [32] might be met with genetically engineered phytochromes in the photoreceptors of plants, which function as their thermosensors and regulate their thermomorphogenesis [33].

Degradation of the quality of vegetation has occurred due to its exposure to atmospheric pollutants and to ingression of contaminants into bottom waters of the rhizosphere and its microorganisms [2]. Acidic meteoric precipitation caused forest damage [34]: this debilitated *e.g.* the function of protective forests to shelter built infrastructure against gravitative mass transport events like snow avalanches, mud flows, land slides, *etc.*

Chronic nitrogen immissions turned out to be strongly negatively correlated to plant species diversity: detailed analysis - determination of Ellenberg indicators, soil pH, leaf N, acid preference index, *etc.* - of 2790 vegetation plots drawn from heathland and acidic, calcareous, and mesotrophic grassland of the cool/temperate climate zone revealed that NO_2 and NH_3 related acidification and eutrophication contributed to loss of species richness [35]. If nitrogen deposition on calcareous grassland exceeds 15-25 kgN/ha/a, a decrease of species diversity and evenness (Shannon diversity index), a decline in the abundance of characteristic species, and a diminution of numbers of rare species - *e.g. Epipactis atrorubens, Trinia glauca, Pulsatilla vulgaris* - occurs. These results came from a comparison between the data from two surveys, conducted in 1990-1993 and 2006-2009 at 106 plots in nature reserves and calcareous grassland present in a cool/temperate climate zone [36].

Excess loads of immissions have resulted in the origination of technogenic deserts [37] (at Sudbury, Norilsk, Chuquicamata, Fushun, *etc.*). Acid rain from toxic immissions from smelters and associated industrial activities has heavily impacted ambient and downwind flora and fauna. At Sudbury for example, 150 years of wood industry and toxic immissions - from smelting and open roasting of sulphides generating aerosols containing sulphuric acid, aluminium, nickel, copper, *etc.* - has given rise to the origination of ca. 170 km^2 of barren land and 720 km^2 of semibarren, transitional land, on which the extinct natural plant community (originally a hemlock/white pine/northern hardwood forest) was transformed - due to revegetation programs (*e.g.* liming) - to a woodland containing birch, aspen, maples, willows, as well as metal acid-tolerating plant species like tickle grass (*Agrostis scabra*), tufted hairgrass (*Deschampsia caespitosa*), sorrel (*Rumex acetosella*), and the moss species *Pohlia* sp [38 - 41].

According to a synthesis of existing botanic information and new findings about the current state of global vascular plants:

• 391,000 species are known to science;

- 31,128 species have documented use in medicine, pharmacy, nutrition, gene sources, production of fuel, poison, or environmental application;
- 1771 important plant areas, predominantly without protection statuses, were identified globally;
- More than 10% of vegetated area is highly sensitive to climate change;
- 21% of all plant species are currently threatened with extinction (see above);
- Global tropical forest areas declined between 1990 and 2015 by 210,000 ha to 1,760,000 ha;
- A considerable part of biodiversity loss was caused by import: 13,168 species have become naturalised outside their native range due to international trade. Xenospecies imported into existing ecosystems have displaced native species or modified relationships between them [42, 43].

In a detailed and quantitative biodiversity assessment based on evaluation of 2.38 million records of about 40,000 terrestrial species present at nearly 19,000 locations, the response of species to land use and associated stress has been modeled. The result is that, across 65% of the terrestrial surface, biotic intactness has declined to less than 10%. On 58.1% of land area that is densely populated by 5.36 billion humans, the planetary boundaries of functional and genetic biodiversity[5] - biosphere integrity - have been transgressed. Grassland biomes, *e.g.* steppes, prairies, savannahs, shrub lands, and some wilderness areas are affected most of all [44].

Protective actions include emission controls (the application of machines with higher degrees of efficiency) and implementation of gas filters, *e.g.* for flue gas and for waste gases of vehicles. Other measures involve land use restrictions, beginning with the establishment of the world's first nature reserve by Charles Waterton in 1821 at Walton Hall, West Yorkshire, UK [45]. By 2014, ca. 209,000 landscapes and seascapes - 32.6 million km^2 - had protected statuses [46]. A few philanthropists and visionaries are engaged in preserving forests, wetlands, savannas, *etc.* [47]. More than 1/5 of the continent of Europe has achieved the status of protected area: designated as regional nature parks, national parks, Unesco Biosphere Reserves and World Heritage Sites [48].

Red List candidates classified by the International Union for the Conservation of Nature as "endangered", "rare", or "extinct in the wild" receive shelter in biotopes created in botanic gardens. New criteria to scrutinise the threat levels of species, *e.g.* distribution, ecological, and life-history data, are useful to calculate precise extinction risks and conservation priorities [49]. In the meantime, reforestation, as well as tree cover conservation, have been recognised locally as important and necessary measures.

The consequences of high latitude warming concerning plants are as follows:

High northern latitude terrestrial ecosystems take up progressively more CO_2 during annual growth phases, because the annual period of net carbon uptake has become more intense [50, 51]. Because soil organic matter is the principle source of plant-available nutrients (*e.g.* P, N) in arctic tundra, climate warming enhances their transfer to the plants and contributes to CO_2 assimilation, as well as C accumulation in these ecosystems [52].

Geomorphological analyses of ca. 2800 tundra ponds present in 22 drained thaw lake basins at Barrow Peninsula (Alaska) revealed that its number and surface areas decreased by 17.1% / 30.3% since 1948. These losses were caused by climate warming and enhanced growth of macrophytes like *Carex aquatilis* and *Arctophila fulva* encroaching on open waters of the ponds [53].

Age-corrected stemwood growth data, acquired between 1956 and 2001 from 3432 Douglas firs present in coastal and interior sites in boreal forests revealed that climate warming predominantly, but also CO_2 and N fertilisation, contributed to a significant and widespread rise in net primary production [54].

Long-standing biomonitoring experiments, during which several tree species of the boreal and temperate zone were exposed to 3.4 °C above ambient temperature, revealed an increase in plant respiration of only 5% on average due to leaf thermal acclimation. Less consumption of photosynthates in plant metabolism causes the increase of atmospheric CO_2 due to stronger plant respiration per °C to be less than expected and may not raise the atmospheric CO_2 concentrations as much as expected[6] [55].

Deglaciating areas of maritime west Antarctica transform into carbon sinks because of rapid development of soil formation and carbon sequestration, driven by physical, chemical, and biological weathering processes. The latter is characterised by coevolution of soils and a community of pioneer organisms and plant successions: cyanobacteria, eukaryotic algae, fungi, spores, lichens, mosses, liverworts, as well as the Antarctic hairgrass *Deschampsia antarctica* (E. Desv.) and Antarctic pearlwort *Colobanthus quitensis* (Kunth Bartl), restricted to ornithic soils [56].

Aquatic vegetation - in lakes, rivers, oceans and the seas - has deteriorated due to its exposure to rising quantities and types of pollutants, as well as transferred heat [2].

The responses of marine primary producers to anthropogenic global change concerning acidification, solar irradiation, and warming are complex. Principally, warming has caused shallowing of the surface mixed layer and its enhanced stratification. As a consequence, less nutrients are available because of reduced vertical flux from below and the primary producers are exposed to more UV irradiation. In addition, acidification has caused an increase of C:N and/or C:P ratios in phytoplankton species. Alteration of species' competitive behaviour, biogeochemical cycles (food chain), and shifts of ecological niches have been the consequences [57]. Here some examples:

Acidified sea water impairs chemotactic reactions and modulates infochemicals, *e.g.* in seagrass-associated communities [58]. Diversity and community structure of kelp and seaweeds, which represent important littoral to innershelf primary producers, function as carbon sinks and absorbers of nutrients (resulting in the production of turf) as well as ecosystem engineers (*e.g.* formating complex habitats), appear to have been impaired by anthropogenic global and local stressors. Invasion of alien seaweed species resulted in space monopolisation and shading, thus displacing native species. Sea water warming has caused range contractions of several seaweed species in the Mediterranean and along the Norwegian and Normandy coasts. Sea water acidification has caused adverse effects to some calcifying species like encrusting corallines, as well as red, green, and brown seaweeds, and has resulted in species composition shifts. Kelp and seaweeds worldwide are also affected by sea level rise and an elevated level of UVB radiation. Local disturbances of seaweed and kelp forests include trampling; exploitation for alginite fertilisers, cosmetic, and for health food industries; and overgrazing by sea urchins and fishes, whose predators were decimated due to overfishing. Eutrophication, which causes opportunistic phytoplankton blooms (most often the species *Ulva prolifera*), is also a problem and the resulting floating mats and green tides. These can persist for weeks and months, and reduce penetration depth of light into the water column, thus causing decline in the number of seaweed species [59, 60].

A four year long adaptive evolution experiment, including 2100 generations from a single clone of the species *Emiliana huxleyi*, the world's most abundant unicellular calcifying microalga, revealed complex phenotypic responses to heightened levels of pCO_2. Initial positive responses were later reversed, so that adaptation potential to compensate acidification was less, as expected due to results from earlier experiments. Fitness and growth rates increased only slightly, calcification was lower more than in non-adapted control groups and cell size decreased. Optimisation of the processes controlling calcification amplifies the negative effects of ocean acidification, because less inorganic carbon is sequestered by the photosynthesis of *E. huxleyi* [61].

NOTES

[1] Between the Neolithic Age and the onset of industrialisation, European forests were significantly impacted many times by activities resulting from human cultural development: creating agrarian land; producing fuel, bricks, glass, and charcoal to reduce ores; baking bread; building houses, civilian ships, and war fleets; boiling rock salt; *etc.* Reduced anthropogenic pressure in times of war or plagues and by administrative decrees of forest conservation effected temporary recovery [62]. In 1490 AD, central European forest cover amounted only ca. 20%. It is evident that past civilisations tended to overexploit forests and that this process contributed to the decline of cultures and prosperity. The Saxonian chief mining administrator Hans Carl von Carlowitz responded to a severe timber supply crisis by recommending a novel concept of sustainability - here concerning the yield of forestry - in his treatise "Sylvicultura oeconomica, oder haußwirthliche Nachricht und Naturmäßige Anweisung zur wilden Baum-Zucht" (1713). His ideas were partly realised in the 20th century in several ways: selection of strictly protected areas, forest plantations, and extensive management of sustainable utilisation [14].

[2] Terrestrial biome types: Deserts and xeric shrublands, tropical and subtropical grasslands and savannas, tropical and subtropical moist broadleaf forests, boreal forests/taiga, temperate broadleaf and mixed forests, temperate grasslands/savannas and shrublands, tundra, montane graslands and shrublands, temperate coniferous forests, Mediterranean forests/woodlands and scrub, tropical and subtropical dry broadleaf forests, flooded grasslands and savannas, tropical and subtropical coniferous forests, mangroves [15].

[3] Meteorologist Wladimir Peter Köppen defined the five major climate zones in 1936, which are as follows: arid, tropical, temperate, continental and polar. They are linked to diverse vegetation belts characterised by distinct moisture, temperature, and insolation parameters [63].

[4] Samples Red List Index is scaled between 1 (Least Concern) and 0 (Extinct).

[5] The threshold value of planetary boundary of biodiversity is defined as when more than 10% of the abundance of organisms and 20% of species are lost.

[6] The annual amount of emitted carbon due to plant respiration comes up to 60 GT, ca. six times more than emitted anthropogenic carbon [55].

CONFLICT OF INTEREST

The author (editor) declares no conflict of interest, financial or otherwise.

ACKNOWLEDGEMENTS

Declare none.

REFERENCES

[1]   Foley, J.A.; Defries, R.; Asner, G.P.; Barford, C.; Bonan, G.; Carpenter, S.R.; Chapin, F.S.; Coe, M.T.; Daily, G.C.; Gibbs, H.K.; Helkowski, J.H.; Holloway, T.; Howard, E.A.; Kucharik, C.J.; Monfreda, C.; Patz, J.A.; Prentice, I.C.; Ramankutty, N.; Snyder, P.K. Global consequences of land use. *Science,* **2005**, *309*(5734), 570-574.
[http://dx.doi.org/10.1126/science.1111772] [PMID: 16040698]

[2]   Goudie, A.S. *The Human Impact on the Natural Environment. Past, Present and Future,* 7th ed; Wiley-Blackwell: Oxford, **2013**.

[3]   Williams, M. *Deforesting the Earth: From Prehistory to Global Crisis*; University of Chicago Press, **2003**.

[4]   Willis, K.J.; Bachmann, S. *Kew Report, State of the World's Plants Annual cutting-edge horizon scan of global plant status, extinction risk,* **2016**. https://stateoftheworldsplants.com/extinction-risk

[5]   Klein Goldewijk, K.; Beusen, A.; van Drecht, G. The HYDE 3.1 spatially explicit database of human induced global land use change over the past 12000 years. *Glob. Ecol. Biogeogr.,* **2011**, *20*(1), 73-86.
[http://dx.doi.org/10.1111/j.1466-8238.2010.00587.x]

[6]   Ellis, E.C. Anthropogenic transformation of the terrestrial biosphere. *Philos Trans A Math Phys Eng Sci,* **2011**, *369*(1938), 1010-1035.
[http://dx.doi.org/10.1098/rsta.2010.0331] [PMID: 21282158]

[7]   Crowther, T.W.; Glick, H.B.; Covey, K.R.; Bettigole, C.; Maynard, D.S.; Thomas, S.M.; Smith, J.R.; Hintler, G.; Duguid, M.C.; Amatulli, G.; Tuanmu, M.N.; Jetz, W.; Salas, C.; Stam, C.; Piotto, D.; Tavani, R.; Green, S.; Bruce, G.; Williams, S.J.; Wiser, S.K.; Huber, M.O.; Hengeveld, G.M.; Nabuurs, G.J.; Tikhonova, E.; Borchardt, P.; Li, C.F.; Powrie, L.W.; Fischer, M.; Hemp, A.; Homeier, J.; Cho, P.; Vibrans, A.C.; Umunay, P.M.; Piao, S.L.; Rowe, C.W.; Ashton, M.S.; Crane, P.R.; Bradford, M.A. Mapping tree density at a global scale. *Nature,* **2015**, *525*(7568), 201-205.
[http://dx.doi.org/10.1038/nature14967] [PMID: 26331545]

[8]   Petersen, R.; Sizer, N.; Hansen, M. *Satellites uncover 5 surprising hotspots for tree cover loss,* **2015**.
http://www.wri.org/blog/2015/09/satellites-uncover-5-surprising-hotspots-tree-cover-loss

[9]   Riitters, K.; Wickham, J.; Costanza, J.K. A global evaluation of forest interior area dynamics using tree cover data from 2000 to 2012. *Landsc. Ecol.,* **2016**, *31*(1), 137-148.
[http://dx.doi.org/10.1007/s10980-015-0270-9]

[10]   Iriarte-Goñi, I.; Ayuda, M.-I. Not only subterranean forests: Wood consumption and economic development in Britain (1850-1938). *Ecol. Econ.,* **2012**, *77*, 176-184.
[http://dx.doi.org/10.1016/j.ecolecon.2012.02.029]

[11]   Mikusinski, G.; Angelstam, P. Occurrence of mammals and birds with different ecological characteristics in relation to forest cover in Europe - do macroecological data make sense? *Ecol. Bull.,* **2004**, *51*, 265-275.

[12]   Naudts, K.; Chen, Y.; McGrath, M.J.; Ryder, J.; Valade, A.; Otto, J.; Luyssaert, S. Europe's forest management did not mitigate climate warming. *Science,* **2016**, *351*(6273), 597-600.
[http://dx.doi.org/10.1126/science.aad7270] [PMID: 26912701]

[13] Baccini, A.; Walker, W.; Carvalho, L. Tropical forests are a net carbon source based on aboveground measurements of gain and loss. *Science,* **2017**. eaam5962, in press

[14] McNeely, J. A. Lessons from the past: forests and biodiversity. *Biodivers. Conserv.,* **1994**, *3*, 3-20.

[15] Watson, J.E.; Shanahan, D.F.; Di Marco, M.; Allan, J.; Laurance, W.F.; Sanderson, E.W.; Mackey, B.; Venter, O. Catastrophic declines in wilderness areas undermine global environment targets. *Curr. Biol.,* **2016**, *26*(21), 2929-2934.
[http://dx.doi.org/10.1016/j.cub.2016.08.049] [PMID: 27618267]

[16] Lewis, S.L.; Edwards, D.P.; Galbraith, D. Increasing human dominance of tropical forests. *Science,* 2015, *349*(6250), 827-832.
[http://dx.doi.org/10.1126/science.aaa9932] [PMID: 26293955]

[17] Davidson, E.A.; de Araújo, A.C.; Artaxo, P.; Balch, J.K.; Brown, I.F.; C Bustamante, M.M.; Coe, M.T.; DeFries, R.S.; Keller, M.; Longo, M.; Munger, J.W.; Schroeder, W.; Soares-Filho, B.S.; Souza, C.M., Jr; Wofsy, S.C. The Amazon basin in transition. *Nature,* **2012**, *481*(7381), 321-328.
[http://dx.doi.org/10.1038/nature10717] [PMID: 22258611]

[18] Brienen, R.J.; Phillips, O.L.; Feldpausch, T.R.; Gloor, E.; Baker, T.R.; Lloyd, J.; Lopez-Gonzalez, G.; Monteagudo-Mendoza, A.; Malhi, Y.; Lewis, S.L.; Vásquez Martinez, R.; Alexiades, M.; Álvarez Dávila, E.; Alvarez-Loayza, P.; Andrade, A.; Aragão, L.E.; Araujo-Murakami, A.; Arets, E.J.; Arroyo, L.; Aymard C, G.A.; Bánki, O.S.; Baraloto, C.; Barroso, J.; Bonal, D.; Boot, R.G.; Camargo, J.L.; Castilho, C.V.; Chama, V.; Chao, K.J.; Chave, J.; Comiskey, J.A.; Cornejo Valverde, F.; da Costa, L.; de Oliveira, E.A.; Di Fiore, A.; Erwin, T.L.; Fauset, S.; Forsthofer, M.; Galbraith, D.R.; Grahame, E.S.; Groot, N.; Hérault, B.; Higuchi, N.; Honorio Coronado, E.N.; Keeling, H.; Killeen, T.J.; Laurance, W.F.; Laurance, S.; Licona, J.; Magnussen, W.E.; Marimon, B.S.; Marimon-Junior, B.H.; Mendoza, C.; Neill, D.A.; Nogueira, E.M.; Núñez, P.; Pallqui Camacho, N.C.; Parada, A.; Pardo-Molina, G.; Peacock, J.; Peña-Claros, M.; Pickavance, G.C.; Pitman, N.C.; Poorter, L.; Prieto, A.; Quesada, C.A.; Ramírez, F.; Ramírez-Angulo, H.; Restrepo, Z.; Roopsind, A.; Rudas, A.; Salomão, R.P.; Schwarz, M.; Silva, N.; Silva-Espejo, J.E.; Silveira, M.; Stropp, J.; Talbot, J.; ter Steege, H.; Teran-Aguilar, J.; Terborgh, J.; Thomas-Caesar, R.; Toledo, M.; Torello-Raventos, M.; Umetsu, R.K.; van der Heijden, G.M.; van der Hout, P.; Guimarães Vieira, I.C.; Vieira, S.A.; Vilanova, E.; Vos, V.A.; Zagt, R.J. Long-term decline of the Amazon carbon sink. *Nature,* **2015**, *519*(7543), 344-348.
[http://dx.doi.org/10.1038/nature14283] [PMID: 25788097]

[19] Ago, E.E.; Serça, D.; Agbossou, E.K.; Galle, S.; Aubinet, M. Carbon dioxide fluxes from a degraded woodland in West Africa and their responses to main environmental factors. *Carbon Balance Manag.,* **2015**, *10*, 22.
[http://dx.doi.org/10.1186/s13021-015-0033-6] [PMID: 26413151]

[20] Thapa, R.B.; Motohka, T.; Watanabe, M.; Shimada, M. Time-series maps of aboveground carbon stocks in the forests of central Sumatra. *Carbon Balance Manag.,* **2015**, *10*, 23.
[http://dx.doi.org/10.1186/s13021-015-0034-5] [PMID: 26413152]

[21] Meyfroidt, P.; Lambin, E.F. Global forest transition: Prospects for an end to deforestation. *Annu. Rev. Environ. Resour.,* **2011**, *36*, 343-371.
[http://dx.doi.org/10.1146/annurev-environ-090710-143732]

[22] Chan, D.; Wu, Q. Significant anthropogenic-induced changes of climate classes since 1950. *Sci. Rep.,* 2015, *5*, 13487.
[http://dx.doi.org/10.1038/srep13487] [PMID: 26316255]

[23] Jolly, W.M.; Cochrane, M.A.; Freeborn, P.H.; Holden, Z.A.; Brown, T.J.; Williamson, G.J.; Bowman, D.M. Climate-induced variations in global wildfire danger from 1979 to 2013. *Nat. Commun.,* **2015**, *6*, 7537.
[http://dx.doi.org/10.1038/ncomms8537] [PMID: 26172867]

[24] McElwain, J.C.; Willis, K.J.; Lupia, R. A history of atmospheric CO_2 and its effects on plants, animals and ecosystems In: *Ecological Studies*; Ehleringer, J.R.; Cerling, T.E.; Dearing, M.D. Springer: New

York, **2005**; 177, pp. 133-165.

[25]   Norby, R.J.; Warren, J.M.; Iversen, C.M.; Medlyn, B.E.; McMurtrie, R.E. CO_2 enhancement of forest productivity constrained by limited nitrogen availability. *Proc. Natl. Acad. Sci. USA,* **2010**, *107*(45), 19368-19373.
       [http://dx.doi.org/10.1073/pnas.1006463107] [PMID: 20974944]

[26]   Campbell, J.E.; Berry, J.A.; Seibt, U.; Smith, S.J.; Montzka, S.A.; Launois, T.; Belviso, S.; Bopp, L.; Laine, M. Large historical growth in global terrestrial gross primary production. *Nature,* **2017**, *544*(7648), 84-87.
       [http://dx.doi.org/10.1038/nature22030] [PMID: 28382993]

[27]   Zhu, Z.; Piao, S.; Myneni, R.B. Greening of the Earth and its drivers. *Nat. Clim. Chang.,* **2016**, *6*, 791-795.
       [http://dx.doi.org/10.1038/nclimate3004]

[28]   Keenan, T.F.; Prentice, I.C.; Canadell, J.G.; Williams, C.A.; Wang, H.; Raupach, M.; Collatz, G.J. Recent pause in the growth rate of atmospheric CO_2 due to enhanced terrestrial carbon uptake. *Nat. Commun.,* **2016**, *7*, 13428.
       [http://dx.doi.org/10.1038/ncomms13428] [PMID: 27824333]

[29]   Scripps Institute of Oceanography. *The Keeling Curve,* https://scripps.ucsd.edu/programs/keeling curve/

[30]   Aldhous, P. Climate maps offer wildlife hope of sanctuary. *New Sci.,* **2008**, *199*(2668), 8-9.
       [http://dx.doi.org/10.1016/S0262-4079(08)61964-8]

[31]   Brummitt, N.A.; Bachman, S.P.; Griffiths-Lee, J.; Lutz, M.; Moat, J.F.; Farjon, A.; Donaldson, J.S.; Hilton-Taylor, C.; Meagher, T.R.; Albuquerque, S.; Aletrari, E.; Andrews, A.K.; Atchison, G.; Baloch, E.; Barlozzini, B.; Brunazzi, A.; Carretero, J.; Celesti, M.; Chadburn, H.; Cianfoni, E.; Cockel, C.; Coldwell, V.; Concetti, B.; Contu, S.; Crook, V.; Dyson, P.; Gardiner, L.; Ghanim, N.; Greene, H.; Groom, A.; Harker, R.; Hopkins, D.; Khela, S.; Lakeman-Fraser, P.; Lindon, H.; Lockwood, H.; Loftus, C.; Lombrici, D.; Lopez-Poveda, L.; Lyon, J.; Malcolm-Tompkins, P.; McGregor, K.; Moreno, L.; Murray, L.; Nazar, K.; Power, E.; Quiton Tuijtelaars, M.; Salter, R.; Segrott, R.; Thacker, H.; Thomas, L.J.; Tingvoll, S.; Watkinson, G.; Wojtaszekova, K.; Nic Lughadha, E.M. Green plants in the red: A baseline global assessment for the IUCN Sampled Red List Index for plants. *PLoS One,* **2015**, *10*(8), e0135152.
       [http://dx.doi.org/10.1371/journal.pone.0135152] [PMID: 26252495]

[32]   Battisti, D.S.; Naylor, R.L. Historical warnings of future food insecurity with unprecedented seasonal heat. *Science,* **2009**, *323*(5911), 240-244.
       [http://dx.doi.org/10.1126/science.1164363] [PMID: 19131626]

[33]   Jung, J.-H.; Domijan, M.; Klose, C.; Biswas, S.; Ezer, D.; Gao, M.; Khattak, A.K.; Box, M.S.; Charoensawan, V.; Cortijo, S.; Kumar, M.; Grant, A.; Locke, J.C.; Schäfer, E.; Jaeger, K.E.; Wigge, P.A. Phytochromes function as thermosensors in Arabidopsis. *Science,* **2016**, *354*(6314), 886-889.
       [http://dx.doi.org/10.1126/science.aaf6005] [PMID: 27789797]

[34]   Bresser, A.H.; Salomons, W., Eds. *Acidic Precipitation - International Overview and Assessment*; Springer: Heidelberg, Berlin, **1990**.
       [http://dx.doi.org/10.1007/978-1-4613-8941-5]

[35]   Maskell, L.C.; Smart, S.M.; Bullock, J.M. Nitrogen deposition causes widespread loss of species richness in British habitats. *Glob. Change Biol.,* **2010**, *16*, 671-679.
       [http://dx.doi.org/10.1111/j.1365-2486.2009.02022.x]

[36]   Van den Berg, L.J.; Vergeer, P.; Rich, T.C. Direct and indirect effects of nitrogen deposition on species composition change in calcareous grasslands. *Glob. Change Biol.,* **2011**, *17*(5), 1871-1883.
       [http://dx.doi.org/10.1111/j.1365-2486.2010.02345.x]

[37]   Bernhardt, A. *World's Worst Pollution Problems,* **2015**. http://www.worstpolluted.org/docs/WWPP_2015_Final.pdf

[38]   Winterhalder, K. *Restoration and Recovery of an Industrial Region - Progress in Restoring the Smelter-Damaged Landscape Near Sudbury, Canada*; Gunn, J., Ed.; Springer: New York, **1995**, pp. 93-102.
[http://dx.doi.org/10.1007/978-1-4612-2520-1_7]

[39]   Winterhalder, K. Environmental degradation and rehabilitation of the landscape around Sudbury, a major mining and smelting area. *Environ. Rev.,* **1996**, *4*(3), 185-224.
[http://dx.doi.org/10.1139/a96-011]

[40]   Courtin, G.M. The last 150 years: a history of environmental degradation in Sudbury. *Sci. Total Environ.,* **1994**, *148*(2-3), 99-102.
[http://dx.doi.org/10.1016/0048-9697(94)90388-3]

[41]   Courtin, G.M. *Restoration and Recovery of an Industrial Region - Progress in Restoring the Smelter-Damaged Landscape Near Sudbury, Canada*; Gunn, J., Ed.; Springer: New York, **1995**, pp. 233-245.
[http://dx.doi.org/10.1007/978-1-4612-2520-1_18]

[42]   Willis, K.J.; Bachmann, S. *Kew Report, State of the World's Plants,* **2016**.
https://stateoftheworldsplants.com/report/sotwp_2016.pdf

[43]   Davis, M.A. Invasion biology, 1. ed.; University Press: Oxford, **2009**.

[44]   Newbold, T.; Hudson, L.N.; Arnell, A.P.; Contu, S.; De Palma, A.; Ferrier, S.; Hill, S.L.; Hoskins, A.J.; Lysenko, I.; Phillips, H.R.; Burton, V.J.; Chng, C.W.; Emerson, S.; Gao, D.; Pask-Hale, G.; Hutton, J.; Jung, M.; Sanchez-Ortiz, K.; Simmons, B.I.; Whitmee, S.; Zhang, H.; Scharlemann, J.P.; Purvis, A. Has land use pushed terrestrial biodiversity beyond the planetary boundary? A global assessment. *Science,* **2016**, *353*(6296), 288-291.
[http://dx.doi.org/10.1126/science.aaf2201] [PMID: 27418509]

[45]   Edginton, B.W. *Charles Waterton: A Biography*; Lutterworth Press: Cambridge, **1996**.

[46]   Juffe-Bignoli, D.; Burgess, N.D.; Bingham, H. *Protected Planet Report,* **2014**.
https://portals.iucn.org/library/sites/library/files/documents/2014-043.pdf

[47]   Humes, E. *Eco Barons: The Dreamers, Schemers, and Millionaires Who Are Saving Our Planet, Ecco*; HarperCollins Publishers: New York, **2010**.

[48]   Hammer, T.; Mose, I.; Siegrist, D. *Parks of the Future - Protected areas in Europe challenging regional and global change, oekom verlag: Munich,* **2016**.
https://www.oekom.de/nc/buecher/gesamtprogramm/buch/parks-of-the-future.html

[49]   Tomović, L.; Urošević, A.; Vukov, T. Threatening levels and extinction risks based on distributional, ecological and life-history datasets (DELH) *versus* IUCN criteria: example of Serbian reptiles. *Biodivers. Conserv.,* **2015**, *24*(12), 2913-2934.
[http://dx.doi.org/10.1007/s10531-015-0984-7]

[50]   Keeling, C.D.; Chin, J.F.; Whorf, T.P. Increased activity of northern vegetation inferred from atmospheric CO_2 measurements. *Nature,* **1996**, *382*, 146-149.
[http://dx.doi.org/10.1038/382146a0]

[51]   Barlow, J.M.; Palmer, P.I.; Bruhwiler, L.M. Analysis of CO_2 mole fraction data: first evidence of large-scale changes in CO_2 uptake at high northern latitudes. *Atmos. Chem. Phys.,* **2015**, *15*, 13739-13758.
[http://dx.doi.org/10.5194/acp-15-13739-2015]

[52]   Jiang, Y.; Rocha, A.V.; Rastetter, E.B. C–N–P interactions control climate driven changes in regional patterns of C storage on the North Slope of Alaska. *Landsc. Ecol.,* **2016**, *31*(1), 195-213.
[http://dx.doi.org/10.1007/s10980-015-0266-5]

[53]   Andresen, C.G.; Lougheed, V.L. Disappearing Arctic tundra ponds: Fine-scale analysis of surface hydrology in drained thaw lake basins over a 65 year period (1948–2013). *J. Geophys. Res.,* **2015**, *120*(3), 466-479.

[http://dx.doi.org/10.1002/2014JG002778]

[54]   Wu, C.; Hember, R.A.; Chen, J.M.; Kurz, W.A.; Price, D.T.; Boisvenue, C.; Gonsamo, A.; Ju, W. Accelerating forest growth enhancement due to climate and atmospheric changes in British Colombia, Canada over 1956-2001. *Sci. Rep.,* **2014**, *4*, 4461.
[http://dx.doi.org/10.1038/srep04461] [PMID: 24844560]

[55]   Reich, P.B.; Sendall, K.M.; Stefanski, A.; Wei, X.; Rich, R.L.; Montgomery, R.A. Boreal and temperate trees show strong acclimation of respiration to warming. *Nature,* **2016**, *531*(7596), 633-636.
[http://dx.doi.org/10.1038/nature17142] [PMID: 26982730]

[56]   Boy, J.; Godoy, R.; Shibistova, O. Successional patterns along soil development gradients formed by glacier retreat in the Maritime Antarctic, King George Island. *Rev. Chil. Hist. Nat.,* **2016**, *89*, 6.
[http://dx.doi.org/10.1186/s40693-016-0056-8]

[57]   Gao, K.; Helbling, E.W.; Häder, D.-P. Responses of marine primary producers to interactions between ocean acidification, solar radiation, and warming. *Mar. Ecol. Prog. Ser.,* **2012**, *470*, 167-189.
[http://dx.doi.org/10.3354/meps10043]

[58]   Zupo, V.; Maibam, C.; Buia, M.C.; Gambi, M.C.; Patti, F.P.; Scipione, M.B.; Lorenti, M.; Fink, P. Chemoreception of the seagrass *Posidonia Oceanica* by benthic invertebrates is altered by seawater acidification. *J. Chem. Ecol.,* **2015**, *41*(8), 766-779.
[http://dx.doi.org/10.1007/s10886-015-0610-x] [PMID: 26318440]

[59]   Mineur, F.; Arenas, F.; Assis, J. European seaweeds under pressure: Consequences for communities and ecosystem functioning. *J. Sea Res.,* **2015**, *98*, 91-108.
[http://dx.doi.org/10.1016/j.seares.2014.11.004]

[60]   Hepburn, C.D.; Pritchard, D.W.; Cornwall, C.E. Diversity of carbon use strategies in a kelp forest community: implications for a high CO_2 ocean. *Glob. Change Biol.,* **2011**, *17*(7), 2488-2497.
[http://dx.doi.org/10.1111/j.1365-2486.2011.02411.x]

[61]   Schlüter, L.; Lohbeck, K.T.; Gröger, J.P.; Riebesell, U.; Reusch, T.B. Long-term dynamics of adaptive evolution in a globally important phytoplankton species to ocean acidification. *Sci. Adv.,* **2016**, *2*(7), e1501660.
[http://dx.doi.org/10.1126/sciadv.1501660] [PMID: 27419227]

[62]   Thirgood, J.V. Man's impact on the forests of Europe. *J. World Forest Resource Manag.,* **1989**, *4*, 127-167.

[63]   Köppen, W. *Handbuch der Klimatologie*; Köppen, W.; Geiger, G., Eds.; Borntraeger: Stuttgart, **1936**, 1, pp. 1-44.

Fauna

Abstract: Humans have intruded and degraded aquatic and terrestrial biospheres by hunting, overexploitation, land take, habitat dissection and destruction, agriculture, mining, damming, water management, pollution, and import of xenospecies and of pathogens. This caused the trend of anthropogenic defaunation and homogenisation. Littoral faunas are most endangered, because of the interfering impacts from both land and ocean. Fishery-induced evolutionary pressure resulted in smaller sized species. Global fish catch quota declined since 1996. Climate change-induced evolutionary and ecological responses in many equatorial marine species revealed the insufficiency of genetic shifts and adaptations to prevent range contractions and extinctions. But species more poleward have enlarged their ranges. Sea water warming, acidification, and deoxygenation are, except on *e.g.* cephalopoda, the main stressors. Calcifying species are impaired in the growth of their shells and skeletons. 75% of global coral reefs are threatened. Sea water warming has caused the displacement of native species by intruding thermophilic species and temporal mismatches between life-cycle phases and available food. Sea channel constructions have caused the intrusion of xenospecies. Marine faunas have been impaired by oil spills and bioaccumulating toxins. Homogenisation of land has resulted in habitat loss and biodiversity decline of *e.g.* avian species. Applied biocides have impaired the fitness of bee species and immune functions of wildlife populations. Immissions of reactive N have caused a decrease in plant diversity and, consequently, of specialised insects. Billions of animal fatalities have occurred annually by roadkill, other collisions, and nocturnal illumination. Warming has caused the average body mass of Alpine chamois and Arctic reindeer to decline. Episodes of zoonotic diseases were caused by industrial large scale livestock farming. Organic contaminants and heavy metals have been observed to accumulate in the food chain. Thawing permafrost has resulted in reactivation of buried pathogens. Soil microbiota have been impaired by acid and toxic immissions. Specialised microfaunas have developed in acid mine drainages and in the deep biosphere impacted by fracking fluids. The importance of biophilia is expressed.

Keywords: Acidification, Adaptation, Adaptive limits, Agriculture, Bioaccumulation, Biodiversity decline, Biophilia, Contamination, Defaunation, Deoxygenation, Habitat degradation, Homogenisation, Hunting, Immissions, Land take, Livestock, Pollution, Range, Warming, Xenospecies, Zoonotic diseases.

Hubert Engelbrecht

Due to massive expansion into natural biospheres, activities detrimental to the flora and the application of technologically more and more advanced hunting devices, marine and terrestrial faunas in the wild have suffered from a severe decimation in their numbers of species and populations sizes. Drivers like overexploitation, invasive species, contamination, acidification, eutrophication, warming, human introduced pathogens, utilisation, habitat dissection, fragmentation, ruptures, and losses [1] cause feedbacks that amplify anthropogenic defaunation [2].

Concerning the oceans and seas, modelling of human impacts threatening deep to shallow marine ecosystems found that these multiple drivers[1] correlate with a large fraction of ecosystems (41%) being strongly affected and that the highest cumulative impacts were at the continental shelves and slopes because these regions contain multiple ecosystems and receive impacts from both land and oceans: *e.g.* North Sea, Norwegian Shelf, Bering Sea, Persian Gulf, South and East China Sea, Gulf of Bengal, Eastern Caribbean, northwestern Atlantic Shelf. Ocean acidification appears to have had the maximum impact [3]. This prognosis has been confirmed in many parts by a detailed synopsis of numerous ecological studies on the state of marine fauna, which are developing from a complex ecosystem towards a more simplified one consisting of *e.g.* jellyfish, cephalopods and metabolically flexible microbes and algae [4]. Similar correlations and developments occur in terrestrial ecosystems (see below).

The responses of coastal marine ecosystems to anthropogenically induced physical and chemical transformations occur because of rising temperature, rising sea level, acidity of sea water, UV radiation, altered nutrient supply, higher CO_2 content, storminess, and altered circulation patterns. The general consequences are horizontal and vertical distributional shifts, changes in bioproductivity and biodiversity, and microevolutionary change [5].

Recent climate change-caused declines in the suitability of habitats and hunting grounds present in ecological and climatic niches have provoked many species to respond *via* shifts in time (phenology), space (range), and self (physiology) [6].

The results of a five year assessment concerning number of taxa, distribution, habitats, and population trends of the world's wild terrestrial and marine mammal species revealed that 25% are threatened with extinction, that the population of one species in two is declining, that 188 species are critically endangered, and that the main reasons are habitat loss, harvesting, accidental mortality, and pollution. Land mammals are highly adversely affected in Asia due to hunting and habitat loss. Marine species are seriously impaired by increasingly intensive utilisation of the oceans [7].

Ecological and evolutionary responses to anthropogenic climate change consist in compensating for the temporal advance of spring events and increasing trophic asynchrony in interacting predator/prey and insect/plant systems. However, mostly negative consequences result from compensations like timing of life cycles, poleward range shifts and range extension of species in terrestrial and marine areas, and range contractions of range-restricted species present in polar and alpine regions. In addition, most negative impacts on tropical coral reef communities and amphibians indicate that observed adaptations and genetic shifts are insufficient to prevent range contractions and predicted extinctions [8].

These general statements on impairments of and adaptive stresses imposed on faunas are detailed in the following:

Fishery: World capture of fisheries in lakes and rivers rose continually from 2 MT in 1950 to 11.6 MT in 2012. Concerning marine waters, the increase during the same time interval was from 20 MT, peaking in 1995 at 85 MT, and diminishing slowly to 79.7 MT in 2012 [9]. Overfishing has caused declines in species diversity, as well as number of individuals, *e.g.* the cod population present in the western North Atlantic nearly vanished after severe exploitation between 1958-1972 [1]. Industrial, artisanal, and subsistence fisheries decimated the number and abundance of species of large parts of the oceans (especially the Exclusive Industrial Zones and the High Seas). According to more precise reconstructions, there was a general and steady decline of catch quotas subsequent to a maximum of catch in 1996 (130 MT) [10]. Literature research and monitoring of *Arapaima* spp. populations in Amazon floodplains found that fishing pressure continues even in depleted populations, which can result in extinctions ("fishing down"). This contradicts the bioeconomic theory of sustainable behaviour: fishing cannot cause extinction of species, because - according to the primrose path - fishing effort moves away from depleted areas [11]. Fishing has pushed all sturgeon species towards the state of endangerment [12]. Fishery-induced selection has caused evolution towards slow growth, early maturation at smaller sizes, higher reproductive investment, and reduction of functional genetic variations [13].

Despite bans on whaling, several species did not recover; in 2012, 37% of marine mammals were estimated to be at risk of extinction [14].

The population size of Antarctic krill (*Euphausia superba*) has declined since 1965 due to industrial fishing [15].

Hunting and poaching, combined with habitat degradation and other subordinate reasons: A standardised survey of the development of the population of the African savannah elephant (*Loxodonta africana*) between 2007 and 2014 found that its overall number decreased by ca. 144,000 to 378,000. Overall, the premature mortality rate between 2010 and 2014 has been estimated at 8% due to habitat loss and poaching [16]. The population of the forest elephant (*Loxodonta cyclotis*) present in remote and sheltered areas of Gabon (Central Africa) declined between 2004 and 2014 by ca. 80% because of poaching and cross border poaching [17].

The African chimpanzee species *Pan troglodytes* was exposed to a significant population reduction in the past 30 years. Its estimated population size in 2003 was only 172,700 - 299,700. The reasons for the decline were growth of the human population in Sub-Saharan Africa and related transfer of infectious diseases, poaching for bushmeat, and habitat degradation and loss because of industrial agriculture and clearcutting of forests. The latter amounted to ca. 80% in western Africa during the early 2000s [18]. In addition, contentions and misunderstandings arouse about the implementation of protected areas [19].

Pan-African populations of vulture species like *Necrosyrtes monachus*, *Gyps africanus* and *G. rueppellii* declined between 1961 and 2014 by 62% on average because of habitat loss, poisoning, killing for bushmeat, trade in traditional medicine, and collisions with technical infrastructures. All African vulture species are assigned the status of critically endangered. This crisis implies the risk of transmission of diseases like rabies by feral dogs [20].

Investigations about the population decline of amphibians revealed, among other reasons (see below), their commercial use: From 1987 to 1997 a minimum 47,000 T of frogs were harvested globally. Between 1998 and 2002 the estimates of frogs caught in the wild reached up to 52,000 T, corresponding to 15 million individuals. As a consequence, transfer of pathogens occurred, *e.g.* 28 million amphibians, imported into the US-markets for bait trade between 2000 and 2005 contained many infected individuals [21].

Artificial noise and light: Ship noise and certain sonar frequencies affect marine life; the auditory sensitivity of several marine species to artificial noise has experimentally been proven [22]. Sonar-induced disruption of deep feeding and stress, probably causing mass-stranding, has been observed in tagged blue whales (*Balaenoptera musculus*) [23].

Widespread artificial nocturnal illumination causes navigational disorientation in *e.g.* marine turtles and migrant birds, as well as disruption of natural foraging and mating cycles of aquatic animal species [24]. Artificial nocturnal illumination causes the premature death of billions of flying insects annually.

Collisions with infrastructure and fast moving means of transportation: Ca. 1 million mammals, birds, reptiles, amphibians, and invertebrates were roadkilled daily in the USA in 2005. Globally, billions of vertebrates are killed annually by collision with automotives, trains, ships and airplanes, because of expanding modern transportation and infrastructure [25]. Direct human-caused mortality of avian species killed by collision with windows, wires, and fast-moving cars, predation by cats, poisoning, environmental pollution, and habitat loss exceeds one billion of fatalities annually [26, 27]. Global monitoring of abundance changes in more than 800 avian species between urbanised and non-urbanised surroundings revealed that species loss in the former case is due to the lack of developing appropriate adaptations like finding food and habitat or avoiding risks [28]. The number of fatalities concerning insects are a thousand times higher.

Pollution (accidentally and deliberately), in combination with other reasons: Sea water contamination by hydrocarbons and the resulting damage to ecosystems has occurred during pipeline leakages, offshore blowouts, and oil tanker havaries[2]. Despite increases in seaborne oil trade, the number of accidental spills has decreased. Since 1970, ca. 10,000 incidents have been reported, the vast majority of which fall into the small category of < 7 T. In sum, ca. 5.75 MT of oil has poured into seawater since 1970 [29]. Sea water contamination also occurred due to the Deepwater Horizon havary on 04/20/2010 and resulted in the impairment of marine fauna and flora of the Gulf of Mexico, known as biodiversity hotspot. According to a species inventory - completed in 2009 - the Gulf of Mexico hosted 8332 species of plants and animals (including 1461 molluscs, 604 polychaetes, 1503 crustaceans, 1270 fishes, four sea turtles, 218 birds, and 29 marine mammals) in the sector of the spill [30, 31]. Deep water coral communities ca. 11 km distant from the oil spill were found in Nov. 2010 covered with petroleum residues - flocculent material and droplets - and showed signs of stress like tissue loss, sclerite enlargement, excess mucous production, and bleached symbiotic ophiuroids. Hopanoid biomarkers of sampled hydrocarbon residues provided evidence that they contained degraded material from the Deepwater Horizon oil spill [32]. This oil spill deteriorated the life conditions especially of marine mammals (*e.g.* dolphins), sea turtles, migratory birds (*e.g.* pelicans), endangered species (*e.g.* Gulf sturgeon, green sea turtle, wood stork, finback whale, Alabama beach mouse), but also these of cetaceans, pinnipeds, manatees and sea otters. Many of them were debilitated from oiling and/or habitat degradation by oil washed ashore [33]. Biomonitoring of the larvae of large, pelagic, predatory fish species (bluefin and yellowfin tuna, amberjack) exposed to realistic concentrations (1-15 µg/L total PAH) of Deepwater Horizon oil samples revealed in all species the formation of defects in cardiac functions and heart development, causing acute or delayed mortality [34].

Industrial coal development put further ecological stress on the Great Barrier Reef World Heritage Site, which has had its coral cover decline by 50% during the past

27 years [35].

More than 10 years of biomonitoring of the abundance of the solitary cold water coral species *Desmophyllum dianthus* (Esper, 1794) in a Patagonian fjord region found that, very probably, a synergistic effect, resulting from several coeval occurrences - ambient submarine effusion of CH_4- and sulphide-containing cold fluids, hypoxic conditions due to intensified industrial aquafarming, and frequent algae blooms - caused a mass die-off of *D. dianthus* species [36].

Concentration of microplastic in aquatic milieus has become of ecological concern because it is ingested by a wide variety of aquatic organisms (zooplancton, polychaetes, echinoderms, decapods, fishes, *etc.*) and accumulate in the food web; but the indisputable levels of microplastic uptake is unclear [37]. The chemical, physical, and biological effects of ingested microplastic particles on marine organisms are in an early stage of investigation [38].

A review of literature about the taxonomy, ecology, distribution, threats, and population trends of crayfish revealed that, of 590 species globally, 32% are threatened with extinction and that the reasons for this development are pollution, urban development, damming, water management, climate change, harvesting, agriculture, and invasive species, thus confirming the current freshwater biodiversity crisis [39].

Severe salinisation of the river Werra (Germany), caused by discharge of brines from industrial potassium mining, resulted in a significant decline of the original fluvial fauna and flora, recurrent blooms of halophile diatome species *Thalassiosira fluviatilis* HUSTEDT, growth of the kelp species *Ulva intestinalis* (Linnaeus) Link 1820, and immigration of the marine alga *Ectocarpus confervoides* (ROTH) KJELLMAN. The lowered freezing point of strongly salinised waters of the Werra caused hypothermia of fish species; many eels died during winter of 1962/63. Less potassium production and decommissioning of several potassium mining sites in the Werra valley has resulted in a slight improvement of the hydro chemical and ecological state, as well as biodiversity of the Werra since 1990 [40].

Extremophiles have developed globally in acid mine drainage of mining regions. For example in the fluvial ecosystem of Rio Tinto, draining the mined Iberian pyrite belt, have evolved chemolithotrophic, sulphur- and iron-oxidizing, acidophile microbial communities: bacteria, proteobacteria, and archaea (*e.g.* *Leptospirillum ferrooxidans, Ferrimicrobium acidiphilum, Ferroplasma acidophilum, Acidithiobacillus ferrooxidans, Acidiphilum multivorum*) [41]. But also acid and radiation-tolerant eukaryota can exist in extreme habitats. In the abandoned underground uranium mine Königstein (Germany), where leaching techniques with sulphuric acid were also applied, workings 250 m below the surface were found with biofilms containing species of ciliates, flagellates, amoebae, heterolobosea, fungi, apicomplexa, stramenopiles, rotifers, and arthropoda [42].

Geo-engineering of the deep biosphere: 328 days of biomonitoring of microbial diversity and its metabolic products was carried out in fracking fluids (consisting of injected fluids and hydrocarbons) recovered from 2,5 km depth in the Marcellus and Utica shales (USA). Hydraulic fracturing provided microorganismal and chemical inputs for persistent colonisation in the deep terrestrial subsurface. Despite the presence of biocides, co-injected low-abundance microorganisms, imported from surface- and groundwater, as well as from fracking additives, adapted to 65 °C, 25 MPa of hydrostatic pressure and to increasing salinity (caused by halite dissolution from shale), and evolved to thermo- and halotolerant bacteria, archaea, and opportunistic viruses. Adaptation to the rising salinity of the fluids occurred *via* production of intracellular osmoprotectants (*e.g.* glycine betaine) in *e.g. Methanohalophilus*. The new ecosystem established and stabilised itself energetically in the deep biosphere by metabolising injected organic fluids and *in situ* hydrocarbons. Interconnected metabolic processes catalysed by microorganisms are as follows: Decomposition of *in situ* hydrocarbons and of imported glucose, sucrose, cellulose, acetate, *etc.* (fracking additives) is accomplished *via* aerobic oxidation and nitrate respiration in *Marinobacter* and Halomonadaceae. Fermentation of metabolic glycine betaine - released into the fluids by viral lysis of bacteria and archaea - is processed by *Halanaerobium* and *Frackibacter.* Fermentation of injected methanol and methylamines then occurs in *Methanohalophilus* and *Methanolobus* [43].

Habitat degradation in combination with other reasons: Monitoring of southeastern German butterfly and burnet moth communities (Rhopalocera, Zygaenidae) on calcareous oligotrophic grassland from 1840 to 2013 resulted in the observation that the numbers of species declined by 46 to 71 and a proportional decrease of habitat specialists, *e.g.* those with restricted dispersal behaviour and those adapted to nutrient poor soils. This transformation, the gradual shift from a species assemblage hosting many habitat specialists to a community with few generalists, was caused by climate change, habitat fragmentation, excess atmospheric nitrogen deposition, and related species community changes in larval host plants [44]. Lack of nectar availability, habitat fragmentation, and degradation of the overwintering colonies, as well as reduced abundance of milkweed host plants for larvae, contributed to the population decline of the monarch butterfly (*Danaus plexippus*) in the USA [45]. Evolutionary limits to climate change-driven range extension are obtained when key interacting species are critically limited and new biotic interactions cannot be developed rapidly enough. Significant loss of adaptive variation has been observed in the case of the recent poleward extension of the UK brown argus butterfly, affecting its ability to adapt to new main host plants [46]. Combined environmental pressures like climate warming and habitat restriction by high-intensity land use has fostered decline of *e.g.* cold-adapted bird and butterfly species in the UK [47].

European hamsters (*Cricetus cricetus*) present in cereal monocultures of farmlands are endangered because a monotonous maize-based diet causes vitamin B3 deficiency. This effects a disease comparable to *pellagra*. One of its symptoms is manifested in impaired social behaviour like maternal infanticide. Simultaneously occurring other detrimental factors, like pesticide toxicity and agricultural ploughing, caused considerable decline in abundance of this species [48]. Homogenisation of industrial agrarian land use has caused biodiversity decline and habitat loss [49].

Commercial interests have created methods to industrially prospect marine regions in order to mine and process ores containing technologically indispensable elements like manganese, cobalt, copper, *etc.*, present at hydrothermal vents, cold seeps, and deep sea regions. However, these areas are situated in extreme marine environments. The ecological impacts of exploiting *e.g.* polymetallic massive sulphide nodules were experimentally studied in the Peru Basin. Simulation of commercial extraction by sea floor ploughing at 4150 m water depth was carried out in 1989. During return visits to the impacted region (ca. 6 km^2) it was investigated and documented how and to what extent recolonisation of fauna occurred and whether the ecosystem had recovered. The 2015 revision concluded that the mechanical scars of the plough tracks have remained nearly unchanged as steep morphological features over 26 years and that a near total loss of the sessile fauna (*e.g.* crinoids, sponges, corals, polychaetes) and its symbionts in ploughed areas and their reduced abundance adjacent to the tracks has resulted [50][3]. Proposed seabed mining on less deep, technically better accessible continental shelf areas [51] will probably result in more ecological damage, because these regions are denser colonised by fauna and flora than the extreme habitats of deep sea regions.

Sea water warming, acidification, eutrophication, deoxygenation, and sea level rise: According to a metastudy consisting of 208 published investigations resulting from biomonitoring spanning > 19 years, including phenology, distribution, abundance, community composition, demography, and calcification - in which 1735 biological responses of 857 marine species and assemblages are described, 96% of observations identified climate change-induced temperature stress as primary driver and the remainder related biological change to acidification, sea ice cover, sea level rise, and brief climate variations. 81-83% of all observations were consistent with expected impacts - *e.g.* shifts in range, phenology, *etc.* - of recent climate change [52]. Sea water warming, acidification, and deoxygenation will affect population dynamics of fish stocks, will influence their spatial distribution and range, and might cause regime shifts in the food web [53].

The earlier reproduction phase of the bivalve *Macoma baltica* in boreal coastal waters fails to coincide with the phytoplankton bloom, causing a larval production and essential food supply temporal mismatch [5]. A similar decoupling problem occurs for shrimps (*Pandalus borealis*) present in North Atlantic shelf waters [54].

Long-term monitoring (1972-2014) of trends of total abundance and species composition of crustacean and estuarine fish species in the German Wadden Sea revealed abundance increases in *Carcinus maenas*, *Liocarcinus holsatus*, and *Pomatoschistus* spp. in the late 1980s, while juvenile species *Pleuronectes platessa*, and *Limanda limanda* decreased. *Solea solea* has increased significantly since the early 2000s. A correlation analysis indicates that these shifts are influenced by climate warming effects [55].

Warmer sea water has increased the abundance of the predatory sea star *Pisaster ochraceus*, which progressively eliminates mid-intertidal mussel beds and thereby displaces hundreds of symbiotic species [5].

The combined effects of warming and deoxygenation of sea water has begun to tighten a metabolic constraint on marine fauna concerning the energy transfer of individual species in their original habitats. According to their thermal and hypoxic tolerances, species - *e.g.* cod - shifted their ranges to deeper regions and poleward to maintain the critical energetic requirements for organismal activity [56].

Sea water acidification causes chemical corrosion of the aragonitic shell of the "sea butterfly": the snail *Limacina helicina* [57, 58].

Monitoring of abundance of 35 cephalopod species between 1953 and 2013 present in demersal, benthopelagic, and pelagic areas revealed that they have responded to anthropogenic influences with an increase in the number of individuals, because their thermal range is not exceeded and elevated temperatures accelerate their life cycles. In addition, predating and competing fish stocks were diminished by harvesting [59].

The endemic species *Melomys rubicola*, an exotic and non-target murid rodent from only one small, low-lying, isolated tropical reef island in the Torres Strait, went extinct in the wild during the last two decades because of more frequent ocean inundations, increased frequency and intensity of storms, decline of floristic diversity, and severe diminution of extent of vegetated areas [60].

Sea water acidification and warming diminishes growth rates and reduced predatory behaviour in the Port Jackson shark (*Heterodontus portusjacksoni*) because its olfactory ability to locate food gets restricted [61].

Continuous warming of sea water causes differing reactions among organisms due to their thermosensitivities and corresponding vulnerabilities. Many reef-building tropical corals are characterised by limited temperature tolerance, resulting in more frequent mass coral-bleaching events (expulsing of intracellular endosymbiotic algae) and reef degradation [5]. These global-scale events have been reported since 1982 [62] and are linked with unusually high sea surface temperatures [63]. Sessile (*e.g.* algae, anemones, bivalves, cnidarians) and motile (*e.g.* gobies, Alpheus shrimps, butterflyfish, damselfish, and velvetfish) symbiotic reef organisms were observed to be negatively affected by mass coral bleaching, which caused shelter and habitat loss, as well as diminution of food availability [64, 65].

The rise in sea water temperatures - according to monitoring between 1878 and 2012 - by average +0.8 °C at Florida Keys coral reef habitats caused declines in *Acropora palmata* stands, coral bleaching events beginning in 1973, shifts in species distributions and reduction of reef accretion [66]. Most of the tropical corals exist at their upper thermal limits (28 °C - 30 °C), because they are unable to acclimate or adapt fast enough to keep pace with the present rapid rate of warming of tropical oceans [67][4]. This thermal tolerance limit has been experimentally verified by exposure of *Acropora aspera* and *Dipsastreae* sp. to temperatures only slightly above the maximum monthly mean and for only a few days, which appeared to cause bleaching [68]. Ca. 75% of the world's tropical coral reefs are evaluated as threatened, if local sources of pressure are combined with temperature stress. Only 27% (67,350 km^2) of the entire coral reef areas (250,000 km^2) are provided protected status [69]. The continual and significant decline of coral reefs in Bahrain has been observed since 1970. Observed species shifts and biodiversity losses are attributed to anthropogenic coastal development[5] and elevated sea surface temperature events [70].

Assessment of extinction risk of all zooxanthellate reef-building coral species (845) known to science revealed that 32.8% have an elevated risk of extinction and that this fraction has increased quickly in recent decades. Reasons for this are global warming, sea water acidification, and local conditions like coastal development, turbid water due to inappropriate land use or watershed management, sewage input, eutrophication resulting from washed out agrochemicals, overfishing, and coral mining [71].

The presumption that anthropogenic ocean acidification is one of several essential reasons for calcification decline in tropical coral reefs has been proven by *in situ* measurement of biotic responses to an experimental reversal of ocean acidification. In sea water manipulated to pre industrial chemistry, a fraction of added alkalinity was taken up by the reef, due to an increase in net calcification [72]. A combination of outdoor tank and *in situ* field experiments revealed that further increase of acidification of sea water favours competitive advances of several common seaweed species over the coral *Acropora intermedia*, confining the already observed beginning global shift to macroalgal dominated tropical reefs [73].

Simulation of repeated bleaching events, to which *Porites divaricata, P. astreoidea,* and *Orbitella faveolata* were exposed, revealed that P. *astreoides* had a diminished recovery capacity. This indicates that repeated bleaching events result in disproportional impairment of long-term recovery and in change of coral community composition, diversity, as well as reef function [74].

According to experimental evidence, the widespread and biogeochemically important cyanobacteria species *Trichodesmium* sp. has been irreversibly adapted to higher CO_2 levels, is able to fix more nitrogen, and has elevated reproduction kinetics. These microbial evolutionary responses, which persist even after the original conditions return to the environment, will probably significantly modify the nutrient content of the sea water. The degree of complexity of interacting modifying processes - sea water warming, expanding anoxia, intensified UV-irradiation, rising acidification, and eutrophication - and adaptive responses prevents a solid prediction of biodiversity development [75].

The continuance of the Antarctic krill (*Euphausia superba)* as an important species in the marine food web is at risk because of the adaptive limits of this species to the warming of the circum-Antarctic waters, reduction and delayed advance of sea ice cover, less availability of chlorophyll a, and biota colonising the contact surface between sea ice and sea water [15].

Thinning and earlier melting of Arctic winter sea ice has caused downward shift of the bioluminescence compensation depth of plankton emitting chemically-generated light like *Thysanoessa inermis, Metridia longa,* and *Mertensia ovum* and has resulted in disturbance of the high Arctic food chain [76].

The biodiversity trend in marine species, influenced by human-made modifications to sea water properties as described above in several examples, is discussed. Based on the ecological data of 13,000 species and global climate models, it can be expected that in the 21st century biodiversity in equatorial areas will shrink; in poleward regions, where range expansions prevail, biodiversity will increase [77]. But homogenisation effects and the poleward expansion of opportunistic generalists will significantly perturb existing species communities [78].

Terrestrial warming: Biomonitoring (1994-2010) of successive cohorts of wild reindeer revealed an average decline in mature skeletal size and in body mass by 12% (55 kg to 48 kg). These effects are explained by climate warming, which causes warmer summers, resulting in greater primary productivity and population growth, but in winter more frequent resource limitations and food unavailability because of icing. And smaller phenotypes are disadvantaged in winter, because of elevated metabolic rate and less on-board reserves [79].

Warming-caused physiological and physical detriments have been found in the species *Rupicapra rupicapra* (Alpine chamois). Study of long-term (1979-2010) body mass development of 9388 juvenile chamois, derived from hunting protocols prepared in three neighbouring areas, revealed a striking body-mass decline of up to 20% over the time considered, which is interpreted as a direct consequence of behavioural change resulting from increasing population densities and temperatures 3-5 °C higher in spring and summer. Warming implies higher thermoregulatory costs and is seen, therefore, as factor limiting the ability of individuals to acquire resources, as directly observed in decrease of time spent foraging and in a modest upslope range shift [80, 81].

Solar irradiation: Due to increased catalytic decomposition of stratospheric ozone by discharge of CFCs since ca. 1930, terrestrial and marine life depending on solar irradiance is exposed to an additional dosage of UV-B [1].

Altered ocean surface currents: According to ocean current models and biomonitoring *via* satellite tracking, a relationship was found between ocean surface currents and long-distance migration behaviour of adult sea turtle species. They learned to find migration destinations through their past experiences as juveniles, when they were dispersed by passive drift with ocean currents [82]. This raises a question about the effects of ocean currents being altered by anthropogenic influences and whether these turtles will find areas appropriate for foraging.

Bioaccumulation of toxins (biocides, persistent organic pollutants, heavy metals, radionuclides) and habitat degradation: Acidification and accumulation of toxic metal species in soils near smelters has caused degradation of soil biota (viruses, bacteria, fungi, algae, *etc.*) and disruption of their activities so that soil formation, nitrogen fixation, nutrient cycling and production of organic matter was impaired [83].

Organic contaminants of emerging concern (polybrominated diphenyl ethers, thiabendazole, diethyl-meta-toluamide, bisphenol A, *etc.*), have bioaccumulated in the tissues of bats (*e.g. Myotis lucifugus, Eptesicus fuscus),* affecting their physiological functions necessary for hibernation and immune functions [84].

Pesticides with broad-spectrum toxicity (*e.g.* fipronil) have been found to exert directly and indirectly (*e.g.* in the food chain) adverse effects (geno and cytotoxicity, impaired immune function, reduced growth and reproductive success) on terrestrial and aquatic vertebrate wildlife populations [85].

Monitoring (1984-2014) of 144 European avian species revealed overall loss concerning biomass (ca. 420 million individuals less) and abundance of more common species. This unprecedented bulk loss has been caused predominantly by intensification of industrial agriculture [86].

Neonicotinoid pesticides turned out to be detrimental to the colony growth and fitness of bees [87, 88].

Rarefaction analyses, based on comparison of taxa richnesses[6] present in several fluvial ecosystems contaminated at different levels, revealed that pesticides caused significant negative effects on both species and family richness of stream macroinvertebrates, resulting in losses in taxa up to 42% of the recorded taxonomic pools [89].

Biochemical analysis of blubber from ca. 1000 marine mammals from European seas revealed that many of them were impaired by high concentrations - up to 200 mg/kg - of bioaccumulated polychlorinated biphenyls[7] (toxicity threshhold: 9 mg/kg) despite being banned in the EU since 1980. This degree of contamination probably explains the population decline and missing population recovery of dolphins and orcas [90].

Transfers of pathogens and toxins: Transfer of infectious diseases from cultivated species in industrial aquafarms to free living species has occurred (*e.g.* salmon anaemia, sea louse) [91, 92].

The thawing of permafrost caused reactivation of relict bacteria, archaea, algae, fungi, viruses, and protozoa, trapped for even millions of years in cryogenic sediment [93, 94]. Viable microorganisms have been redispersed and reintegrated into the life cycles of the Arctic environment. Pathogenic microbial species from that permafrost pool have been supposed to be very likely the source of recent anthrax infections in Siberia; more than 6500 burial grounds of carcasses of cattle and reindeer, infected during past outbreaks of anthrax, have been located in permafrost soil of Siberia [95].

Application of pesticides has contributed directly and indirectly to the decline of many avian species: predatory species - vultures, ospreys, eagles, owls, hawks - were indirectly poisoned by having consumed bodies that pesticides had bioaccumulated in, and insect- and grain-eating species - *e.g.* partridges, grouse, pheasants - decreased because of diminished insect population and of contaminated grain from agricultural land [96].

Bioaccumulation of persistent organic pollutants - DDE (major metabolite of DDT), PCBs, PBDE (polybrominated diphenyl ether) - and Hg in the bodies of already critically endangered coastal condors (*Gymnogyps californianus*) appeared to be significantly higher in those scavenging dead stranded marine mammals than in noncoastal condors. Thus, coastal condors suffer from the transfer of marine contaminants, which cause elevated risk of reproductive impairment [97].

Xenospecies: Construction and extension of the Suez Canal has, in combination with recent sea water warming, caused more than 1000 alien species to enter the eastern Mediterranean Sea. One of the invaders, the thermophilic lionfish *Pterois miles*, colonised the entire south eastern coast of Cyprus within one year. Because of the high reproductive rate of this species, that has only very few predators, and it being a generalist carnivore, it threatens the 17,000 species present in the Mediterranean Sea, recognised as biodiversity hotspot [98]. A few interoceanic species colonisations have occurred *via* the Panama Canal [99].

The introduction of predatory Nile Perch in the Lake Victoria in the 1950s resulted in the near-total extinction of 400 endemic species of haplochromites (*tilapia* sp.) and the replacement of diatoms by blue-green algae as primary producers [100].

Xenospecies caused hybridisation or displacement of native fauna by transfer of diseases, competition, or predation [1].

Damming: The industrial implementation of numerous large hydropower dams in the Mekong River caused blocking or reduction of long-distance migration of several fish species to their spawning sites [101].

Dam constructions and large-scale diversion of water of the Amu-Darya (draining the Aral Sea Basin) since the 1960s, performed to enable irrigation of adjacent agrarian land in semiarid climate, altered the ecosystem of the lower part of the river. Endemic fish species like *Pseudoscaphirhynchus fedtschenkoi, P. hermani, P. kaufmanni, Salmo trutta aralensis, Aspiolucius esocinus, Acipenser nudiventris* and *Barbus brachycephalus* went extinct, annual fish catch declined from 8800 T in 1972 to 1500 T in 1992, agrochemicals applied in the Aral Sea Basin like DDT and hexachlorane bioaccumulated in fish species, and the salinity of reservoirs rose to a level unsuitable to introduced species [102].

Multiple reasons: The diminution of arctic sea ice coverage aggravates food access for the polar bear (*Ursus maritimus*). Some of the southern subpopulations of this species have already declined because of poorer body condition of adults and lower survival rate of the cubs [103]. This effect is probably enhanced by the fact that this species is at elevated (above threshold) ecotoxicological risk, because it is exposed to easily transported, legacy, and emerging persistent organic pollutants (*e.g.* chlordanes, polychlorinated biphenyl, perfluorooctane sulphonate), which bioaccumulate in the Arctic food chain and cause deleterious health effects to the polar bear (*e.g.* alteration of liver and endocrine system, neurological damage) [104]. Northward range expansion of the brown bear (*Ursus arctos*) and range contraction of the polar bear provoked origination of a hybrid subspecies by interbreeding and introgression [105].

Several assessments of the health statuses of global amphibian populations found that decline in species and individuals began in the 1970s. At present the > 6000 amphibian species known to science are threatened to a higher degree than avians or mammals due to habitat loss, overutilisation and enigmatic processes (*e.g.* the fungal disease chytridiomycosis: see below). 32.5% are globally threatened by species decline, predominantly neotropical, montane, stream-associated species, and 43.2% have experienced population decrease. Since 1980, nine species became "formally extinct" and further 122 species are listed as "possibly extinct" [106]. The reasons, which caused degradation:

- Introduction of xenospecies, land use change, and habitat loss.
- Contamination with chemicals (*e.g.* higher concentrated levels of pesticide atrazine causes feminisation of the male leopard frog) and pollution with mineralised nitrogen.
- Climate change, by which new habitats are generated, as well as existing ones destroyed.
- Stimulation of pathogen emergences.
- Infectious diseases: iridoviruses (*e.g. Ambystoma tigrinum*) and the chytrid fungus *Batrachochytrium dendrobatidis,* the spread of which, propagated by

climate change-induced warming, has probably reached pandemic dimension [21].

The International Union for Conservation of Nature (IUCN) identified globally 1437 mammal species and 4263 avian species, subjected to defaunation, with the highest impact in tropical regions. 177 mammal species have lost more than 50% of their range [2]. A compiled index of all invertebrate populations globally showed an overall 45% decline between 1970 and 2008 [2]. Studies quantifying the number of species, whose populations were already impacted by climate change, resulted in the facts that 47% of non-volant mammals (out of 873 threatened species) and 23.4% of avians (out of 1372 threatened species) have been impaired in at least part of their range [107].

To mitigate the danger of loss of biodiversity and ecosystem services, Red Lists were established by IUCN to protect extant species. Australia, which has lost at least 130 species since European settlement 200 years ago due to the fatal Columbian Exchange, industrialisation and climate change, started an environmental protection program to reduce threats of extinction of its unique fauna and flora [108]. The Intergovernmental Platform of Biodiversity and Ecosystem Services was founded in 2012 and it makes science policy interfaces work better, provides important information about species and proposes measures to prioritise support for biodiversity, and that biodiversity concerns are taken into consideration in political and economic decisions [109].

Establishment of menageries, zoological parks and aquaria in the 19th century informed the public about biodiversity, demonstrated by the exotic fauna and their habitats present in foreign countries, allowed them to express biophilia, and later to conserve threatened species, study animal behaviour, develop caring, and to motivate urban people to live a more harmonious and sustainable relationship with nature [110].

Animal experiments for scientific purposes and mass utilisation of animals for industrial/economic purposes, *e.g.* animal testing in the pharmaceutical industry and factory farming of animals in very large quantities, are moral issues. In addition, the factory farming is detrimental to human health and environmental quality: large-scale livestock farming has caused dangerous infectious zoonotic diseases [111]. Ca. 65 billion food animals - buffalo: 0.024 bn, cattle: 0.3 bn, pig: 1.4 bn, goat: 0.43 bn, sheep: 0.5 bn, turkey: 0.65 bn, goose: 0.65 bn, duck: 2.81 bn, chicken: 58.1 bn - were slaughtered 2011 globally [112].

NOTES

[1] *e.g.*: commercial shipping, ocean-based pollution, species invasion, climate

change (UV-radiation, acidification, warming), fishing (catch and bycatch), pollution, and nutrient input.

[2] A few examples of larger oil spills by tanker accidents: 1955, English Channel (Johannishus), 20,000 T; 1960, south Atlantic (Petrolore), 60,000 T; 1971, North Sea (Texaco), 100,000 T; 1977, Honolulu (Hawaian Patriot), 95,000 T; 1978, coast of Brittany (BP), 223,000 T; 1988, coast of Canada (Odyssey), 132,000 T; 1996, Milford Haven (Liberia), 72,000 T; 2000, eastern Mediterranean, 35,000 T; 2002, western Atlantic (Prestige), 63,000 T; 2010, Dalian (China), 60,000 T [29].

[3] It is a reasonable decision of the Conference in Transparency and best Practices for Deep Seabed Mining, which consists of representatives from industrial, academic, and civil societies, national governments, and international organisations, to designate areas as protected from deep seabed mining and to mitigate impacts on species and ecosystems [113].

[4] However, there are exceptions in the reefs of the Persian/Arabian Gulf, where *e.g.* the species *Porites* spp. contains symbionts tolerating sea surface temperatures up to 36 °C [114].

[5] Nevertheless, a few coral species revealed unique abilities to adapt: *e.g.*: reef systems in the Arabian Gulf, contaminated by crude oil seeps, contain corals (*Acropora clathrata, Porites compressa),* which harbour oil-degrading bacteria (*Gammaproteobacteria, Actinobacteria, etc.*), capable of diminishing the degree of contamination and self-purifying the entire reef [115].

[6] Taxa richnesses are reliably determined by taxa accumulation or rarefaction graphs, which measure the relationship between the number of individuals *versus* number of taxa.

[7] Thermally and chemically stable, organic liquids, used in devices as electric isolators, hydraulic liquids, and cooling agents. The carcinogenic substances are lipidophilic and bioaccumulating in the food chain. Applications are limited to closed technical systems.

CONFLICT OF INTEREST

The author (editor) declares no conflict of interest, financial or otherwise.

ACKNOWLEDGEMENTS

Declare none.

REFERENCES

[1] Goudie, A.S. *The Human Impact on the Natural Environment. Past, Present and Future,* 7th ed; Wiley-Blackwell: Oxford, **2013**.

[2] Dirzo, R.; Young, H.S.; Galetti, M.; Ceballos, G.; Isaac, N.J.; Collen, B. Defaunation in the Anthropocene. *Science,* **2014**, *345*(6195), 401-406.
[http://dx.doi.org/10.1126/science.1251817] [PMID: 25061202]

[3] Halpern, B.S.; Walbridge, S.; Selkoe, K.A.; Kappel, C.V.; Micheli, F.; D'Agrosa, C.; Bruno, J.F.; Casey, K.S.; Ebert, C.; Fox, H.E.; Fujita, R.; Heinemann, D.; Lenihan, H.S.; Madin, E.M.; Perry, M.T.; Selig, E.R.; Spalding, M.; Steneck, R.; Watson, R. A global map of human impact on marine ecosystems. *Science,* **2008**, *319*(5865), 948-952.
[http://dx.doi.org/10.1126/science.1149345] [PMID: 18276889]

[4] Jackson, J.B. Colloquium paper: ecological extinction and evolution in the brave new ocean. *Proc. Natl. Acad. Sci. USA,* **2008**, *105*(1) Suppl. 1, 11458-11465.
[http://dx.doi.org/10.1073/pnas.0802812105] [PMID: 18695220]

[5] Harley, C.D.; Randall Hughes, A.; Hultgren, K.M.; Miner, B.G.; Sorte, C.J.; Thornber, C.S.; Rodriguez, L.F.; Tomanek, L.; Williams, S.L. The impacts of climate change in coastal marine systems. *Ecol. Lett.,* **2006**, *9*(2), 228-241.
[http://dx.doi.org/10.1111/j.1461-0248.2005.00871.x] [PMID: 16958887]

[6] Bellard, C.; Bertelsmeier, C.; Leadley, P.; Thuiller, W.; Courchamp, F. Impacts of climate change on the future of biodiversity. *Ecol. Lett.,* **2012**, *15*(4), 365-377.
[http://dx.doi.org/10.1111/j.1461-0248.2011.01736.x] [PMID: 22257223]

[7] Schipper, J.; Chanson, J.S.; Chiozza, F.; Cox, N.A.; Hoffmann, M.; Katariya, V.; Lamoreux, J.; Rodrigues, A.S.; Stuart, S.N.; Temple, H.J.; Baillie, J.; Boitani, L.; Lacher, T.E., Jr; Mittermeier, R.A.; Smith, A.T.; Absolon, D.; Aguiar, J.M.; Amori, G.; Bakkour, N.; Baldi, R.; Berridge, R.J.; Bielby, J.; Black, P.A.; Blanc, J.J.; Brooks, T.M.; Burton, J.A.; Butynski, T.M.; Catullo, G.; Chapman, R.; Cokeliss, Z.; Collen, B.; Conroy, J.; Cooke, J.G.; da Fonseca, G.A.; Derocher, A.E.; Dublin, H.T.; Duckworth, J.W.; Emmons, L.; Emslie, R.H.; Festa-Bianchet, M.; Foster, M.; Foster, S.; Garshelis, D.L.; Gates, C.; Gimenez-Dixon, M.; Gonzalez, S.; Gonzalez-Maya, J.F.; Good, T.C.; Hammerson, G.; Hammond, P.S.; Happold, D.; Happold, M.; Hare, J.; Harris, R.B.; Hawkins, C.E.; Haywood, M.; Heaney, L.R.; Hedges, S.; Helgen, K.M.; Hilton-Taylor, C.; Hussain, S.A.; Ishii, N.; Jefferson, T.A.; Jenkins, R.K.; Johnston, C.H.; Keith, M.; Kingdon, J.; Knox, D.H.; Kovacs, K.M.; Langhammer, P.; Leus, K.; Lewison, R.; Lichtenstein, G.; Lowry, L.F.; Macavoy, Z.; Mace, G.M.; Mallon, D.P.; Masi, M.; McKnight, M.W.; Medellín, R.A.; Medici, P.; Mills, G.; Moehlman, P.D.; Molur, S.; Mora, A.; Nowell, K.; Oates, J.F.; Olech, W.; Oliver, W.R.; Oprea, M.; Patterson, B.D.; Perrin, W.F.; Polidoro, B.A.; Pollock, C.; Powel, A.; Protas, Y.; Racey, P.; Ragle, J.; Ramani, P.; Rathbun, G.; Reeves, R.R.; Reilly, S.B.; Reynolds, J.E., III; Rondinini, C.; Rosell-Ambal, R.G.; Rulli, M.; Rylands, A.B.; Savini, S.; Schank, C.J.; Sechrest, W.; Self-Sullivan, C.; Shoemaker, A.; Sillero-Zubiri, C.; De Silva, N.; Smith, D.E.; Srinivasulu, C.; Stephenson, P.J.; van Strien, N.; Talukdar, B.K.; Taylor, B.L.; Timmins, R.; Tirira, D.G.; Tognelli, M.F.; Tsytsulina, K.; Veiga, L.M.; Vié, J.C.; Williamson, E.A.; Wyatt, S.A.; Xie, Y.; Young, B.E. The status of the world's land and marine mammals: diversity, threat, and knowledge. *Science,* **2008**, *322*(5899), 225-230.
[http://dx.doi.org/10.1126/science.1165115] [PMID: 18845749]

[8] Parmesan, C. Ecological and evolutionary responses to recent climate change. *Annu. Rev. Ecol. Evol. Syst.,* **2006**, *37*, 637-669.
[http://dx.doi.org/10.1146/annurev.ecolsys.37.091305.110100]

[9] FAO. *The state of world fishery and aquaculture,* **2012**. http://www.fao.org/docrep/

016/i2727e/i2727e.pdf

[10] Pauly, D.; Zeller, D. Catch reconstructions reveal that global marine fisheries catches are higher than reported and declining. *Nat. Commun.,* **2016,** *7*(10244), 10244.
[http://dx.doi.org/10.1038/ncomms10244] [PMID: 26784963]

[11] Castello, L.; Chaves Arantes, C.; Gibbs Mcgrath, D. Understanding fishing-induced extinctions in the Amazon. *Aquat. Conserv.,* **2015,** *25*(5), 587-598.
[http://dx.doi.org/10.1002/aqc.2491]

[12] Aldhous, P. Snobbish attitude could spell the end of sturgeon. *New Sci.,* **2008,** *199*(2665), 15.
[http://dx.doi.org/10.1016/S0262-4079(08)61774-1]

[13] Marty, L.; Dieckmann, U.; Ernande, B. Fisheries-induced neutral and adaptive evolution in exploited fish populations and consequences for their adaptive potential. *Evol. Appl.,* **2015,** *8*(1), 47-63.
[http://dx.doi.org/10.1111/eva.12220] [PMID: 25667602]

[14] Davidson, A.D.; Boyer, A.G.; Kim, H.; Pompa-Mansilla, S.; Hamilton, M.J.; Costa, D.P.; Ceballos, G.; Brown, J.H. Drivers and hotspots of extinction risk in marine mammals. *Proc. Natl. Acad. Sci. USA,* **2012,** *109*(9), 3395-3400.
[http://dx.doi.org/10.1073/pnas.1121469109] [PMID: 22308490]

[15] Piñones, A.; Fedorow, A.V. Projected changes of Antarctic krill habitat by the end of the 21st century. *Geophys. Res. Lett.,* **2016,** *43*(16), 8580-8589.
[http://dx.doi.org/10.1002/2016GL069656]

[16] Chase, M.J.; Schlossberg, S.; Griffin, C.R.; Bouché, P.J.; Djene, S.W.; Elkan, P.W.; Ferreira, S.; Grossman, F.; Kohi, E.M.; Landen, K.; Omondi, P.; Peltier, A.; Selier, S.A.; Sutcliffe, R. Continent-wide survey reveals massive decline in African savannah elephants. *PeerJ,* **2016,** *4*, e2354.
[http://dx.doi.org/10.7717/peerj.2354] [PMID: 27635327]

[17] Poulsen, J.R.; Koerner, S.E.; Moore, S.; Medjibe, V.P.; Blake, S.; Clark, C.J.; Akou, M.E.; Fay, M.; Meier, A.; Okouyi, J.; Rosin, C.; White, L.J. Poaching empties critical Central African wilderness of forest elephants. *Curr. Biol.,* **2017,** *27*(4), R134-R135.
[http://dx.doi.org/10.1016/j.cub.2017.01.023] [PMID: 28222286]

[18] IUCN. *Red List of Threatened Species, Pan troglodytes,* **2016.** http://www.iucnredlist.org/details/15933/0

[19] Leblan, V. Territorial and land-use rights perspectives on human-chimpanzee-elephant coexistence in West Africa (Guinea, Guinea-Bissau, Senegal, nineteenth to twenty-first centuries). *Primates,* **2016,** *57*(3), 359-366.
[http://dx.doi.org/10.1007/s10329-016-0532-4] [PMID: 27038218]

[20] Ogada, D.; Shaw, P.; Beyers, R.L. Another continental vulture crisis: Africa's vultures collapsing toward extinction. *Conserv. Lett.,* **2016,** *9*(2), 89-9.
[http://dx.doi.org/10.1111/conl.12182]

[21] Collins, J.P. Amphibian decline and extinction: what we know and what we need to learn. *Dis. Aquat. Organ.,* **2010,** *92*(2-3), 93-99.
[http://dx.doi.org/10.3354/dao02307] [PMID: 21268970]

[22] Popper, A.N. The Effects of Noise on Aquatic Life. In: *Advances in Experimental Medicine and Biology*; **2012,** Vol. 730.

[23] Goldbogen, J.A.; Southall, B.L.; DeRuiter, S.L.; Calambokidis, J.; Friedlaender, A.S.; Hazen, E.L.; Falcone, E.A.; Schorr, G.S.; Douglas, A.; Moretti, D.J.; Kyburg, C.; McKenna, M.F.; Tyack, P.L. Blue whales respond to simulated mid-frequency military sonar. *Proc. Biol. Sci.,* **2013,** *280*(1765), 20130657.
[http://dx.doi.org/10.1098/rspb.2013.0657] [PMID: 23825206]

[24] Davies, T.W.; Bennie, J.; Inger, R. Artificial light alters natural regimes of night-time sky brightness. *Nat. Sci. Rep.,* **2013,** *3*, 1722.

[http://dx.doi.org/10.1038/srep01722]

[25]   Seiler, A.; Helldin, J.O. *The Ecology of Transportation: Managing Mobility for the Environment*; Davenport, J.L., Ed.; Springer: New York, **2006**, pp. 165-189.
[http://dx.doi.org/10.1007/1-4020-4504-2_8]

[26]   Loss, S.R.; Will, T.; Marra, P.P. Direct human-caused mortality of birds: improving quantification of magnitude and assessment of population impact. *Front. Ecol. Environ.,* **2012**, *10*(7), 357-364.
[http://dx.doi.org/10.1890/110251]

[27]   Klem, D., jr. Collisions between birds and windows: Mortality and prevention. *J. Field Ornithol.,* **1990**, *61*(1), 120-128.

[28]   Sol, D.; González-Lagos, C.; Moreira, D.; Maspons, J.; Lapiedra, O. Urbanisation tolerance and the loss of avian diversity. *Ecol. Lett.,* **2014**, *17*(8), 942-950.
[http://dx.doi.org/10.1111/ele.12297] [PMID: 24835452]

[29]   The International Tanker Owners Pollution Federation Limited. *Oil Tanker Spill Statistics 2015,* **2015**. http://www.itopf.com/knowledge-resources/data-statistics/statistics/

[30]   Moretzsohn, F.; Brenner, J.; Michaud, P. *Biodiversity of the Gulf of Mexico Database (BioGoMx). Version 1.0. Harte Research Institute for Gulf of Mexico Studies (HRI), Texas A and M University-Corpus Christi (TAMUCC), Texas,* **2016**. http://www.gulfbase.org/biogomx/about.php

[31]   Felder, D.L.; Camp, D.K. *Gulf of Mexico-Origins, Waters, and Biota,* **2009**. http://www.tamupress.com/product/Gulf-of-Mexico-Origin-Waters-and-Biota,5338.aspx

[32]   White, H.K.; Hsing, P.-Y.; Cho, W.; Shank, T.M.; Cordes, E.E.; Quattrini, A.M.; Nelson, R.K.; Camilli, R.; Demopoulos, A.W.; German, C.R.; Brooks, J.M.; Roberts, H.H.; Shedd, W.; Reddy, C.M.; Fisher, C.R. Impact of the *Deepwater Horizon* oil spill on a deep-water coral community in the Gulf of Mexico. *Proc. Natl. Acad. Sci. USA,* **2012**, *109*(50), 20303-20308.
[http://dx.doi.org/10.1073/pnas.1118029109] [PMID: 22454495]

[33]   *On Scene Coordinator Report Deepwater Horizon Oil Spill, Submitted to the National Response Team,* **2011**.http://www.uscg.mil/foia/docs/dwh/fosc_dwh_report.pdf

[34]   Incardona, J.P.; Gardner, L.D.; Linbo, T.L.; Brown, T.L.; Esbaugh, A.J.; Mager, E.M.; Stieglitz, J.D.; French, B.L.; Labenia, J.S.; Laetz, C.A.; Tagal, M.; Sloan, C.A.; Elizur, A.; Benetti, D.D.; Grosell, M.; Block, B.A.; Scholz, N.L. Deepwater Horizon crude oil impacts the developing hearts of large predatory pelagic fish. *Proc. Natl. Acad. Sci. USA,* **2014**, *111*(15), E1510-E1518.
[http://dx.doi.org/10.1073/pnas.1320950111] [PMID: 24706825]

[35]   Brodie, J.; McCulloch, M.; Coles, R. *Declaration by concerned scientists on industrial development of the Great Barrier Reef coast,* **2013**. http://www.earth2ocean.com/endorsing/Final%20 science%20statement_ 030613.pdf

[36]   Försterra, G.; Häussermann, V.; Laudien, J. Mass die-off of the cold-water coral Desmophyllum dianthus in the Chilean Patagonian fjord region. *Bull. Mar. Sci.,* **2014**, *90*(3), 895-899.
[http://dx.doi.org/10.5343/bms.2013.1064]

[37]   Vandermeersch, G.; Van Cauwenberghe, L.; Janssen, C. R. A critical view on microplastic quantification in aquatic organisms. *Environ. Res.,* **2015**, *143*(B), 46-55.
[http://dx.doi.org/10.1016/j.envres.2015.07.016]

[38]   Rummel, C. *Occurrence and potential effects of plastic ingestion by pelagic and demersal fish from the North Sea and Baltic Sea,* **2014**. http://epic.awi.de/35956/

[39]   Richman, N.I.; Böhm, M.; Adams, S.B.; Alvarez, F.; Bergey, E.A.; Bunn, J.J.; Burnham, Q.; Cordeiro, J.; Coughran, J.; Crandall, K.A.; Dawkins, K.L.; DiStefano, R.J.; Doran, N.E.; Edsman, L.; Eversole, A.G.; Füreder, L.; Furse, J.M.; Gherardi, F.; Hamr, P.; Holdich, D.M.; Horwitz, P.; Johnston, K.; Jones, C.M.; Jones, J.P.; Jones, R.L.; Jones, T.G.; Kawai, T.; Lawler, S.; López-Mejía, M.; Miller, R.M.; Pedraza-Lara, C.; Reynolds, J.D.; Richardson, A.M.; Schultz, M.B.; Schuster, G.A.; Sibley, P.J.; Souty-Grosset, C.; Taylor, C.A.; Thoma, R.F.; Walls, J.; Walsh, T.S.; Collen, B. Multiple drivers of

decline in the global status of freshwater crayfish (Decapoda: Astacidea). *Philos. Trans. R. Soc. Lond. B Biol. Sci.,* **2015**, *370*(1662), 20140060.
[http://dx.doi.org/10.1098/rstb.2014.0060] [PMID: 25561679]

[40]   Hübner, G. *Ökologisch-faunistische Fließgewässerbewertung am Beispiel der salzbelasteten unteren Werra und ausgewählter Zuflüsse,* **2007**. http://www.uni-kassel.de/upress/online/ frei/978-3-899-8-295-6.volltext.frei.pdf

[41]   González-Toril, E.; Llobet-Brossa, E.; Casamayor, E.O.; Amann, R.; Amils, R. Microbial ecology of an extreme acidic environment, the Tinto River. *Appl. Environ. Microbiol.,* **2003**, *69*(8), 4853-4865.
[http://dx.doi.org/10.1128/AEM.69.8.4853-4865.2003] [PMID: 12902280]

[42]   Zirnstein, I.; Arnold, T.; Krawczyk-Bärsch, E.; Jenk, U.; Bernhard, G.; Röske, I. Eukaryotic life in biofilms formed in a uranium mine. *Microbiol. Open,* **2012**, *1*(2), 83-94.
[http://dx.doi.org/10.1002/mbo3.17] [PMID: 22950016]

[43]   Daly, R.A.; Borton, M.A.; Wilkins, M.J.; Hoyt, D.W.; Kountz, D.J.; Wolfe, R.A.; Welch, S.A.; Marcus, D.N.; Trexler, R.V.; MacRae, J.D.; Krzycki, J.A.; Cole, D.R.; Mouser, P.J.; Wrighton, K.C. Microbial metabolisms in a 2.5-km-deep ecosystem created by hydraulic fracturing in shales. *Nat. Microbiol.,* **2016**, *1*, 16146.
[http://dx.doi.org/10.1038/nmicrobiol.2016.146] [PMID: 27595198]

[44]   Habel, J.C.; Segerer, A.; Ulrich, W.; Torchyk, O.; Weisser, W.W.; Schmitt, T. Butterfly community shifts over two centuries. *Conserv. Biol.,* **2016**, *30*(4), 754-762.
[http://dx.doi.org/10.1111/cobi.12656] [PMID: 26743786]

[45]   Inamine, H.; Ellner, S.P.; Springer, J.P. Linking the continental migratory cycle of the monarch butterfly to understand its population decline. *Oikos,* **2016**, *125*, 1081-1091.
[http://dx.doi.org/10.1111/oik.03196]

[46]   Buckley, J.; Bridle, J.R. Loss of adaptive variation during evolutionary responses to climate change. *Ecol. Lett.,* **2014**, *17*(10), 1316-1325.
[http://dx.doi.org/10.1111/ele.12340] [PMID: 25104062]

[47]   Oliver, T.H.; Gillings, S.; Pearce-Higgins, J.W. Large extents of intensive land use limit community reorganization during climate warming. *Glob. Change Biol.,* in press

[48]   Tissier, M.L.; Handrich, Y.; Dallongeville, O. Diets derived from maize monoculture cause maternal infanticides in the endangered European hamster due to a vitamin B3 deficiency. *P. Roy. Soc. Lond. B Biol.,* **2017**, *284*, 1847.

[49]   White, E.V.; Roy, D.P. A contemporary decennial examination of changing agricultural field sizes using Landsat time series data. *Geo,* **2015**, *2*(1), 33-54.
[http://dx.doi.org/10.1002/geo2.4] [PMID: 27669424]

[50]   Purser, A.; Marcon, Y.; Boetius, A. Return to DISCOL: Megafauna distribution 26 years after simulated nodule mining. *Newsletter: Managing Impacts of Deep Sea Resource Exploitation,* **2015**, *6*, 1-3.

[51]   Hannington, M.; Petersen, S.; Krätschell, A. Subsea mining moves closer to shore. *Nat. Geosci.,* in press

[52]   Poloczanska, E.S.; Brown, C.J.; Sydeman, W.J. Global imprint of climate change on marine life. *Nat. Clim. Chang.,* **2013**, *3*, 919-925.
[http://dx.doi.org/10.1038/nclimate1958]

[53]   Koenigstein, S.; Mark, F.C.; Gößling-Reisemann, S. Modelling climate change impacts on marine fish populations: process-based integration of ocean warming, acidification and other environmental drivers. *Fish Fish.,* **2016**, *17*(4), 972-1004.
[http://dx.doi.org/10.1111/faf.12155]

[54]   Koeller, P.; Fuentes-Yaco, C.; Platt, T.; Sathyendranath, S.; Richards, A.; Ouellet, P.; Orr, D.; Skúladóttir, U.; Wieland, K.; Savard, L.; Aschan, M. Basin-scale coherence in phenology of shrimps

and phytoplankton in the North Atlantic Ocean. *Science,* **2009**, *324*(5928), 791-793.
[http://dx.doi.org/10.1126/science.1170987] [PMID: 19423827]

[55]   Meyer, J.; Kröncke, I.; Bartholomä, A. Long-term changes in species composition of demersal fish and epibenthic species in the Jade area (German Wadden Sea/Southern North Sea) since 1972. *Estuar. Coast. Shelf Sci.,* **2016**, *181*, 284-293.
[http://dx.doi.org/10.1016/j.ecss.2016.08.047]

[56]   Deutsch, C.; Ferrel, A.; Seibel, B.; Pörtner, H.O.; Huey, R.B. Ecophysiology. Climate change tightens a metabolic constraint on marine habitats. *Science,* **2015**, *348*(6239), 1132-1135.
[http://dx.doi.org/10.1126/science.aaa1605] [PMID: 26045435]

[57]   Wolf, S. *Extinction. It's Not Just for Polar Bears. Report of the Center for Biological Diversity and Care for the Wild International,* **2010**. http://www.biologicaldiversity.org/ programs/climate _law_institute/ the_arctic_meltdown/pdfs/ ArcticExtinctionReport_Final.pdf

[58]   Bednaršek, N.; Feely, R.A.; Reum, J.C. *Limacina helicina* shell dissolution as an indicator of declining habitat suitability owing to ocean acidification in the California Current Ecosystem. *P. Roy. Soc. Lond. B Bio.,* **1785**, *2015*, 281.

[59]   Doubleday, Z.A.; Prowse, T.A.; Arkhipkin, A.; Pierce, G.J.; Semmens, J.; Steer, M.; Leporati, S.C.; Lourenço, S.; Quetglas, A.; Sauer, W.; Gillanders, B.M. Global proliferation of cephalopods. *Curr. Biol.,* **2016**, *26*(10), R406-R407.
[http://dx.doi.org/10.1016/j.cub.2016.04.002] [PMID: 27218844]

[60]   Gynther, I.; Waller, N.; Leung, L.K *Confirmation of the extinction of the Bramble Cay melomys MELOMYS RUBICOLA on Bramble Cay, Torres Strait: results and conclusions from a comprehensive survey in August-September 2014,* **2016**. https://www.ehp.qld.gov.au/wildlife/threatened-species/documents/bramble-cay-melomys-survey-report.pdf

[61]   Pistevos, J.C.; Nagelkerken, I.; Rossi, T.; Olmos, M.; Connell, S.D. Ocean acidification and global warming impair shark hunting behaviour and growth. *Sci. Rep.,* **2015**, *5*, 16293.
[http://dx.doi.org/10.1038/srep16293] [PMID: 26559327]

[62]   Oliver, J.K.; Berkelmans, R.; Eakin, C.M. Coral bleaching: patterns, processes, causes and consequences. van Oppen, M.J.H.; Lough, J.M. Eds. In: *Ecological Studies*; Springer: Heidelberg, Berlin, **2009**; Vol. 205, pp. 21-39.

[63]   Eakin, M.; Lough, J.M.; Heron, S. Coral bleaching: patterns, processes, causes and consequences. van Oppen, M.J.H.; Lough, J.M. Eds. In: *Ecological Studies*; Springer: Heidelberg, Berlin, **2009**; Vol. 205, pp. 41-67.

[64]   McClanahan, T.R.; Weil, E.; Cortés, J. Coral bleaching: patterns, processes, causes and consequences. van Oppen, M.J.H.; Lough, J.M. Eds. In: *Ecological Studies*; Springer: Heidelberg, Berlin, **2009**; Vol. 205, pp. 121-138.

[65]   Pratchett, M.S.; Wilson, S.K.; Graham, N.A. Coral bleaching: patterns, processes, causes and consequences, .van Oppen, M.J.H.; Lough, J.M. Eds. In: *Ecological Studies*; Springer: Heidelberg, Berlin, **2009**; Vol. 205, pp. 139-158.

[66]   Kuffner, I.B.; Lidz, B.H.; Hudson, J.H. A Century of Ocean Warming on Florida Keys Coral Reefs: Historic *in situ* Observations. *Estuaries Coasts,* **2015**, *38*, 1085-1096.
[http://dx.doi.org/10.1007/s12237-014-9875-5]

[67]   Hoegh-Guldberg, O. Climate change, coral bleaching and the future of the world's coral reefs. *Mar. Freshw. Res.,* **1999**, *50*, 839-866.
[http://dx.doi.org/10.1071/MF99078]

[68]   Schoepf, V.; Stat, M.; Falter, J.L.; McCulloch, M.T. Limits to the thermal tolerance of corals adapted to a highly fluctuating, naturally extreme temperature environment. *Sci. Rep.,* **2015**, *5*, 17639.
[http://dx.doi.org/10.1038/srep17639] [PMID: 26627576]

[69]   Burke, L.; Reytar, K.; Spalding, M. *Reefs at Risk Revisited, Key Findings,* **2011**.

http://www.wri.org/sites/default/files/reefs_at_risk_key_findings.pdf

[70] Burt, J.A.; Al-Khalifa, K.; Khalaf, E.; Alshuwaikh, B.; Abdulwahab, A. The continuing decline of coral reefs in Bahrain. *Mar. Pollut. Bull.,* **2013**, *72*(2), 357-363.
[http://dx.doi.org/10.1016/j.marpolbul.2012.08.022] [PMID: 22980773]

[71] Carpenter, K.E.; Abrar, M.; Aeby, G.; Aronson, R.B.; Banks, S.; Bruckner, A.; Chiriboga, A.; Cortés, J.; Delbeek, J.C.; Devantier, L.; Edgar, G.J.; Edwards, A.J.; Fenner, D.; Guzmán, H.M.; Hoeksema, B.W.; Hodgson, G.; Johan, O.; Licuanan, W.Y.; Livingstone, S.R.; Lovell, E.R.; Moore, J.A.; Obura, D.O.; Ochavillo, D.; Polidoro, B.A.; Precht, W.F.; Quibilan, M.C.; Reboton, C.; Richards, Z.T.; Rogers, A.D.; Sanciangco, J.; Sheppard, A.; Sheppard, C.; Smith, J.; Stuart, S.; Turak, E.; Veron, J.E.; Wallace, C.; Weil, E.; Wood, E. One-third of reef-building corals face elevated extinction risk from climate change and local impacts. *Science,* **2008**, *321*(5888), 560-563.
[http://dx.doi.org/10.1126/science.1159196] [PMID: 18653892]

[72] Albright, R.; Caldeira, L.; Hosfelt, J.; Kwiatkowski, L.; Maclaren, J.K.; Mason, B.M.; Nebuchina, Y.; Ninokawa, A.; Pongratz, J.; Ricke, K.L.; Rivlin, T.; Schneider, K.; Sesboüé, M.; Shamberger, K.; Silverman, J.; Wolfe, K.; Zhu, K.; Caldeira, K. Reversal of ocean acidification enhances net coral reef calcification. *Nature,* **2016**, *531*(7594), 362-365.
[http://dx.doi.org/10.1038/nature17155] [PMID: 26909578]

[73] Del Monaco, C.; Hay, M.E.; Gartrell, P.; Mumby, P.J.; Diaz-Pulido, G. Effects of ocean acidification on the potency of macroalgal allelopathy to a common coral. *Sci. Rep.,* **2017**, *7*, 41053.
[http://dx.doi.org/10.1038/srep41053] [PMID: 28145458]

[74] Schoepf, V.; Grottoli, A.G.; Levas, S.J.; Aschaffenburg, M.D.; Baumann, J.H.; Matsui, Y.; Warner, M.E. Annual coral bleaching and the long-term recovery capacity of coral. *Proc. Biol. Sci.,* **2015**, *282*(1819), 20151887.
[http://dx.doi.org/10.1098/rspb.2015.1887] [PMID: 26582020]

[75] Hutchins, D.A.; Walworth, N.G.; Webb, E.A.; Saito, M.A.; Moran, D.; McIlvin, M.R.; Gale, J.; Fu, F.X. Irreversibly increased nitrogen fixation in Trichodesmium experimentally adapted to elevated carbon dioxide. *Nat. Commun.,* **2015**, *6*, 8155.
[http://dx.doi.org/10.1038/ncomms9155] [PMID: 26327191]

[76] Cronin, H.A.; Cohen, J.H.; Berge, J.; Johnsen, G.; Moline, M.A. Bioluminescence as an ecological factor during high Arctic polar night. *Sci. Rep.,* **2016**, *6*, 36374.
[http://dx.doi.org/10.1038/srep36374] [PMID: 27805028]

[77] Molinos, J.G.; Halpern, B.S.; Schoeman, D.S. Climate velocity and the future global redistribution of marine biodiversity. *Nat. Clim. Chang.,* **2016**, *6*, 83-88.

[78] Kortsch, S.; Primicerio, R.; Fossheim, M.; Dolgov, A.V.; Aschan, M. Climate change alters the structure of arctic marine food webs due to poleward shifts of boreal generalists. *Proc. Biol. Sci.,* **2015**, *282*(1814), 20151546.
[http://dx.doi.org/10.1098/rspb.2015.1546] [PMID: 26336179]

[79] Irvine, J.; Loe, L.E.; Ropstad, E. *Small is beautiful: are reindeer 'shrinking' because of climate warming?,* **2016**. https://eventmobi.com/bes2016/agenda/207583/1074793

[80] Mason, T.H.; Stephens, P.A.; Apollonio, M.; Willis, S.G. Predicting potential responses to future climate in an alpine ungulate: interspecific interactions exceed climate effects. *Glob. Change Biol.,* **2014**, *20*(12), 3872-3882.
[http://dx.doi.org/10.1111/gcb.12641] [PMID: 24957266]

[81] Mason, T.H.; Apollonio, M.; Chirichella, R. Environmental change and long-term body mass declines in an alpine mammal. *Front. Zool.,* **2014**, *11*, 69.
[http://dx.doi.org/10.1186/s12983-014-0069-6]

[82] Scott, R.; Marsh, R.; Hays, G.C. Ontogeny of long distance migration. *Ecology,* **2014**, *95*(10), 2840-2850.
[http://dx.doi.org/10.1890/13-2164.1]

[83] Maxwell, C.D. *Restoration and Recovery of an Industrial Region - Progress in Restoring the Smelter-Damaged Landscape Near Sudbury, Canada*; Gunn, J.M., Ed.; Springer: New York, **1995**, pp. 219-231.
[http://dx.doi.org/10.1007/978-1-4612-2520-1_17]

[84] Secord, A.L.; Patnode, K.A.; Carter, C.; Redman, E.; Gefell, D.J.; Major, A.R.; Sparks, D.W. Contaminants of Emerging Concern in Bats from the Northeastern United States. *Arch. Environ. Contam. Toxicol.*, **2015**, *69*(4), 411-421.
[http://dx.doi.org/10.1007/s00244-015-0196-x] [PMID: 26245185]

[85] Gibbons, D.; Morrissey, C.; Mineau, P. A review of the direct and indirect effects of neonicotinoids and fipronil on vertebrate wildlife. *Environ. Sci. Pollut. Res. Int.*, **2015**, *22*(1), 103-118.
[http://dx.doi.org/10.1007/s11356-014-3180-5] [PMID: 24938819]

[86] Inger, R.; Gregory, R.; Duffy, J.P.; Stott, I.; Voříšek, P.; Gaston, K.J. Common European birds are declining rapidly while less abundant species' numbers are rising. *Ecol. Lett.*, **2015**, *18*(1), 28-36.
[http://dx.doi.org/10.1111/ele.12387] [PMID: 25363472]

[87] Hoppe, P.P.; Safer, A.; Amaral-Rogers, V.; Bonmatin, J.M.; Goulson, D.; Menzel, R.; Baer, B. Effects of a neonicotinoid pesticide on honey bee colonies: a response to the field study by Pilling *et al.* (2013). *Environ. Sci. Eur.*, **2015**, *27*(1), 28-31.
[http://dx.doi.org/10.1186/s12302-015-0060-7] [PMID: 27752429]

[88] Williamson, S.M.; Willis, S.J.; Wright, G.A. Exposure to neonicotinoids influences the motor function of adult worker honeybees. *Ecotoxicology*, **2014**, *23*(8), 1409-1418.
[http://dx.doi.org/10.1007/s10646-014-1283-x] [PMID: 25011924]

[89] Beketov, M.A.; Kefford, B.J.; Schäfer, R.B.; Liess, M. Pesticides reduce regional biodiversity of stream invertebrates. *Proc. Natl. Acad. Sci. USA*, **2013**, *110*(27), 11039-11043.
[http://dx.doi.org/10.1073/pnas.1305618110] [PMID: 23776226]

[90] Jepson, P.D.; Deaville, R.; Barber, J.L.; Aguilar, À.; Borrell, A.; Murphy, S.; Barry, J.; Brownlow, A.; Barnett, J.; Berrow, S.; Cunningham, A.A.; Davison, N.J.; Ten Doeschate, M.; Esteban, R.; Ferreira, M.; Foote, A.D.; Genov, T.; Giménez, J.; Loveridge, J.; Llavona, Á.; Martin, V.; Maxwell, D.L.; Papachlimitzou, A.; Penrose, R.; Perkins, M.W.; Smith, B.; de Stephanis, R.; Tregenza, N.; Verborgh, P.; Fernandez, A.; Law, R.J. PCB pollution continues to impact populations of orcas and other dolphins in European waters. *Sci. Rep.*, **2016**, *6*, 18573.
[http://dx.doi.org/10.1038/srep18573] [PMID: 26766430]

[91] Krkošek, M.; Ford, J.S.; Morton, A.; Lele, S.; Myers, R.A.; Lewis, M.A. Declining wild salmon populations in relation to parasites from farm salmon. *Science*, **2007**, *318*(5857), 1772-1775.
[http://dx.doi.org/10.1126/science.1148744] [PMID: 18079401]

[92] Senate Committee on Fisheries and Oceans. *An ocean of opportunities: Aquaculture in Canada*, **2015**.
http://www.parl.gc.ca/Content/SEN/Committee/412/pofo/rep/rep12jul15Vol3-e.pdf

[93] Zhang, D.-C.; Brouchkov, A.; Griva, G.; Schinner, F.; Margesin, R. Isolation and characterization of bacteria from ancient siberian permafrost sediment. *Biology (Basel)*, **2013**, *2*(1), 85-106.
[http://dx.doi.org/10.3390/biology2010085] [PMID: 24832653]

[94] Legendre, M.; Lartigue, A.; Bertaux, L.; Jeudy, S.; Bartoli, J.; Lescot, M.; Alempic, J.M.; Ramus, C.; Bruley, C.; Labadie, K.; Shmakova, L.; Rivkina, E.; Couté, Y.; Abergel, C.; Claverie, J.M. In-depth study of *Mollivirus sibericum*, a new 30,000-y-old giant virus infecting *Acanthamoeba*. *Proc. Natl. Acad. Sci. USA*, **2015**, *112*(38), E5327-E5335.
[http://dx.doi.org/10.1073/pnas.1510795112] [PMID: 26351664]

[95] Evengård, B. Vulnerable populations: health of humans and animals in a changed landscape. *Int. J. Circumpolar Health*, **2013**, *72* Suppl. 1, 58-60.

[96] Mitra, A.; Chatterjee, C.; Mandal, F.B. Synthetic chemical pesticides and their effects on birds. *Res. J. Environ. Toxicol.*, **2011**, *5*, 81-96.

[http://dx.doi.org/10.3923/rjet.2011.81.96]

[97] Kurle, C.M.; Bakker, V.J.; Copeland, H.; Burnett, J.; Jones Scherbinski, J.; Brandt, J.; Finkelstein, M.E. Terrestrial scavenging of marine mammals: Cross-ecosystem contaminant transfer and potential risks to endangered california condors (Gymnogyps Californianus). *Environ. Sci. Technol.,* **2016**, *50*(17), 9114-9123.
[http://dx.doi.org/10.1021/acs.est.6b01990] [PMID: 27434394]

[98] Kletou, D.; Hall-Spencer, J.M.; Kleitou, P. A lionfish (Pterois miles) invasion has begun in the Mediterranean Sea. *Mar. Biodivers. Rec.,* **2016**, *9*, 46.
[http://dx.doi.org/10.1186/s41200-016-0065-y]

[99] Roche, D.G.; Torchin, M.E. Established population of the North American Harris mud crab, Rhithropanopeus harrisii (Gould 1841) (Crustacea: Brachyura: Xanthidae) in the Panama Canal. *Aquat. Invasions,* **2007**, *2*(3), 155-161.
[http://dx.doi.org/10.3391/ai.2007.2.3.1]

[100] Gophen, M. Ecological devastation in Lake Victoria: Part B: Plankton and fish communities. *Open J. Ecol.,* **2015**, *5*, 315-325.
[http://dx.doi.org/10.4236/oje.2015.57026]

[101] Orr, S.; Pittock, J.; Chapagain, A. Dams on the Mekong River: Lost fish protein and the implications for land and water resources. *Glob. Environ. Change,* **2012**, *22*(4), 925-932.
[http://dx.doi.org/10.1016/j.gloenvcha.2012.06.002]

[102] Pavlovskaya, L.P. *Fishery in the lower Amu-Darya under the impact of irrigated agriculture,* http://www.fao.org/docrep/V9529E/v9529E04.htm

[103] Stirling, I.; Derocher, A.E. Effects of climate warming on polar bears: a review of the evidence. *Glob. Change Biol.,* **2012**, *18*(9), 2694-2706.
[http://dx.doi.org/10.1111/j.1365-2486.2012.02753.x] [PMID: 24501049]

[104] Villa, S.; Migliorati, S.; Monti, G.S. Risk of POP mixtures on the Arctic food chain. *Environ. Toxicol. Chem.,* **2016**, 1-12.
[PMID: 28054401]

[105] Miller, W.; Schuster, S.C.; Welch, A.J.; Ratan, A.; Bedoya-Reina, O.C.; Zhao, F.; Kim, H.L.; Burhans, R.C.; Drautz, D.I.; Wittekindt, N.E.; Tomsho, L.P.; Ibarra-Laclette, E.; Herrera-Estrella, L.; Peacock, E.; Farley, S.; Sage, G.K.; Rode, K.; Obbard, M.; Montiel, R.; Bachmann, L.; Ingólfsson, O.; Aars, J.; Mailund, T.; Wiig, O.; Talbot, S.L.; Lindqvist, C. Polar and brown bear genomes reveal ancient admixture and demographic footprints of past climate change. *Proc. Natl. Acad. Sci. USA,* **2012**, *109*(36), E2382-E2390.
[http://dx.doi.org/10.1073/pnas.1210506109] [PMID: 22826254]

[106] Stuart, S.N.; Chanson, J.S.; Cox, N.A.; Young, B.E.; Rodrigues, A.S.; Fischman, D.L.; Waller, R.W. Status and trends of amphibian declines and extinctions worldwide. *Science,* **2004**, *306*(5702), 1783-1786.
[http://dx.doi.org/10.1126/science.1103538] [PMID: 15486254]

[107] Pacifici, M.; Visconti, P.; Butcher, S.H. Species' traits influenced their response to recent climate change. *Nat. Clim. Chang.,* in press

[108] Australian Government. *Threatened Species Strategy,* https://www.environment.gov.au/ system/files/resources/51b0e2d4-50ae- 49b5-8317-081c6afb3117/files/ts-strategy.pdf

[109] Intergovernmental Platform of Biodiversity and Ecosystem Services. http://www.ipbes.net/index.php

[110] Rabb, G.B.; Saunders, C.D. The future of zoos and aquariums: conservation and caring. *Int. Zoo Year b.,* **2005**, *39*(1), 1-26.
[http://dx.doi.org/10.1111/j.1748-1090.2005.tb00001.x]

[111] Wigle, P.; Humphrey, T.; Daly, J. *Zoonotic diseases, human health and farm animal welfare,* **2013**. https://www.ciwf.org.uk/media/3756123/Zoonotic-diseases-human-health-and-farm-animal-welfare-

16-page-report.pdf

[112] Heinrich Böll Foundation and Friends of the Earth Europe. *Meat Atlas - Facts and Figures about the Animals we eat,* **2014**. https://www.foeeurope.org/sites/default/files/publications/foee_hbf_meatatlas_jan2014.pdf

[113] World Economic Forum. *Toward Transparency and Best Practices for Deep Seabed Mining. An initial multistakeholder dialogue,* **2016**. http://dosi-project.org/wp-content/ uploads/2015/08/ Toward_ Transparency_ Best_Practices_Deep_ Seabed_Mining_Bellagio_report_2016_0501.pdf

[114] Hume, B.; D'Angelo, C.; Burt, J.; Baker, A.C.; Riegl, B.; Wiedenmann, J. Corals from the Persian/Arabian Gulf as models for thermotolerant reef-builders: prevalence of clade C3 *Symbiodinium*, host fluorescence and *ex situ* temperature tolerance. *Mar. Pollut. Bull.,* **2013**, *72*(2), 313-322.
[http://dx.doi.org/10.1016/j.marpolbul.2012.11.032] [PMID: 23352079]

[115] Al-Dahash, L.M.; Mahmoud, H.M. Harboring oil-degrading bacteria: a potential mechanism of adaptation and survival in corals inhabiting oil-contaminated reefs. *Mar. Pollut. Bull.,* **2013**, *72*(2), 364-374.
[http://dx.doi.org/10.1016/j.marpolbul.2012.08.029] [PMID: 23014479]

CHAPTER 17

Human Beings: Benign Effects of Transformations

Abstract: The positive effects of industrialisation are summarised as follows: more reliabile sufficient food production, advancement in life sciences (dietetics, hygienics, medicine knowledge and technique, family planning, pharmacology, biology, biotechnology) and public health care (vaccination, immunisation), resulting in slow rise of average life expectancy and well being, and less poverty; decline of burden of communicable diseases; development of continent-wide electric grids and improvement of heating, cooking, and cold storage systems; improvement of physical and virtual mobility, as well as speed of information transfer; replacement of dangerous, exhausting, and monotonous work by machines or robots; improvement of occupational health; reduction of duration of the workweek and immense growth in economic performance; slow increase of general education, specialised knowledge, and of the world intellectual property index; increase of the amount of published scientific papers per annum; general IQ gains; increase of chip performance; progress in risk management, quality of prognoses, forecasts, and of early warning systems, thus preventing damages to life and property; improved performance of artificial photosynthesis and of solar cells; and immense progress in basic research concerning *e.g.* genomics, the subatomic world, material science, geosciences, aero- and astronautics, and cosmology.

Keywords: Dietetics, Digitalisation, Disparities, Education, Electrical revolution, Family planning, Food security, Genetical engineering, Hygienics, Intellectual development, Internet, Life expectancy, Medicine, Mobility, Occupational health, Pharmacology, Risk management, Robotisation, Technological development, Wealth.

The positive effects of industrialisation on human beings are the improvement in the quality of life and life expectancy, promoted *via* the realisation of ideas, innovations, inventions, and detections by eminent authorities in science and technology. A few of these were already mentioned in the third chapter. Of prime importance is the progress in the reliability of sufficient food production, realised by the application of artificial fertilisers and biocides, specialised machines, development of agricultural science and technology, as well as cultivation of crop plants and farm animals. Advancement in the science of dietetics, *e.g.* detection of essential vitamins, has resulted in improved quality of nutrition. Improvements were also made in medicinal science and technology, as well as in pharmaceuticals. Public health care measures, like immunisations and protective vaccinations against communicable diseases including smallpox, poliomyelitis, measles, pertussis, *etc.*, raised considerably the average health status of human population.

Since 1971, genetic engineering has revolutionised human medicine and pharmacology, as well as industrial agriculture and environmental remediation by designing transgene bacteria, plants, and animals (genetic and molecular pharming). Applications are: targeted creation of biopharmaceuticals *via* bacteria, plant, or animal pharming (production of *e.g.* Insulin, Humira, Herceptin); personalised tumour therapy (genetherapy); transgene crop plants characterised by higher yields and resiliences to climate change (*e.g.* drought), herbicides, and parasites; transgene plants and bacteria created for the generation of biofuels, decomposition of toxic waste, remediation of dump sites, organic wastes, contaminated terrain, waste water and waste gas, as well as purification of drinking water, *etc.* [1 - 3]. A transgenic line of mosquito *Aedes aegypti* (OX513A) was released to decrease the population size of that mosquito species and to suppress transfer of viral diseases like Zika, dengue, yellow fever, *etc.* [4, 5].

As a consequence of progress in medicine and pharmaceutics, average life expectancy[1], which in 1760 amounted to ca. 37 years in England [6], ran up in 2013 to ca. 71.5 years globally (country minimum 45 years, maximum 85 years) [7]. Although large regional disparities exist, global health development has improved: compared to 2004, surgical volume increased by 38% to 312.9 Million in 2012 [8]. Demoscopic assessments revealed that the age structure of the world population developed from a pyramid shape in 1950 to a bell shape in 2015. This reflects a transition to less poverty, less infant mortality, decline in birth rate, better health support, and rise of longevity, although enormous regional and structural disparities exist [9, 10]. Connected to rising average level of global development, life expectancy rose, the burden of communicable diseases declined, and also child and maternal mortality [11]. On 06/29/2015, the 100 millionth chemical substance was registered, designed to treat acute myeloid leukaemia [12].

Since the first implementation of a 2000 V/1.1 kW electric tele transmission line over 57 km from the village of Miesbach to the city of Munich in 1882 by engineer Oscar von Miller[2], electric supply grids have spread over the globe and yielded a bulk output of 3.2 TW in 2015. Technical progress during industrialisation also brought about heating systems in buildings and physical mobility has developed to a high degree. Ovens used for space heating and cooking purpose, which were fuelled with wood, coal, and petroleum, were replaced in ca. 1950 by more comfortable central heating systems and electric stoves, thus reducing indoor and, in part, outdoor aerosol emissions. Refrigerators, cold storage systems, and progress in cryo-engineering has simplified long-term storing of food. Since the invention of the low pressure steam engine, animal power (by horse, donkey, ox) was substituted by a series of fast-developing motor systems, fuelled at first with wood or coal and later with refined chemical derivatives of petroleum or natural gas, and finally with electricity. This development contributed enormously to the range expansion of individuals, travelling much more comfortably, trade range, logistical volume, and exchange speed of more and more diverse products. Since the 1970s, cumulated CO_2-emission intensity of produced unit of material stocks declined, due to technical development and improved efficiency, considerably by 48% to 11 kg C per material stock [13].

Dangerous, exhausting, and monotonous work has been locally replaced by machines[3]. Occupational health has improved on average due to enactment of numerous legislative regulations (*e.g.* ban of manufacturing of all kinds of asbestos, ban on tobacco smoking in all workstations), leading to a decrease in cases of injuries and of illnesses [14]. Engaged social movements sheltered working people from exploitation of their capacity to work: In 1825/1830 one workweek lasted in Germany/USA on average 82/69 hours and in 2000 one workweek lasted 35/36 hours [15 - 17]. In that way, the general public gained more time resources and power to promote cultural progress and to fulfil the pursuit of literacy, learning, and know-how. This trend originated in the era of enlightenment, was enhanced by the simplified dissemination of written text[4] and by the abolishment - at the end of the 18th century - of the Latin language as a common official medium in the humanities, natural science, and writing. General school attendance (in order to restrain analphabetism, raise general education, and specialised knowledge), the secondary school, the vocational school, the middle school *etc.*, were introduced in Europe in the 18th and 19th centuries. Between 1801 and 2012, far more universities - ca. 450 - were established in Europe than in medieval scholastic and early modern times (from 1088 to 1796: ca. 270) [18 - 21].

In the meantime, humankind has explored and temporarily contacted or inhabited even most adverse ecosystems. The availability of industrially produced artificial oxygen made possible to climb the highest summit on Earth (8844.43 ± 0.21 m asl) on 05/29/1953. Further technological development and the construction of deep submergence vehicles enabled access of the Mariana Trench on 03/26/2012, one of the deepest marine sites of Earth's surface (-10,984 ± 25 m). In January 2013, accretion ice containing relicts of bacteria, archaea, and eukaryota was recovered from Lake Vostok, the largest known subglacial lake in Antarctica, covered by ca. 3.6 km ice and isolated for at least 15 Ma from global geo-biosphere development. The world's deepest mine is situated in South Africa: Its shaft has been sunken to more than 4 km depth, where the temperature of ambient rock measures 60 °C. On 10/24/2014, a spacesuit protected parachutist survived a stratosphere jump (41,419 m). Since 1956, scientists have been permanently present at Amundsen-Scott South Pole Station. Since the first expedition crew has arrived on International Space Station in 11/02/2000, scientists have continuously occupied the orbiting laboratory to carry out experiments under microgravity conditions.

Automotives, airplanes, lokomotives, and ships, fuelled by cheap energy carriers, have contributed to rising physical mobility of individuals. Personal computers and internet added to virtual mobility and immediate availability of information. Between 1950 and 2015, global growth of economic performance and export grew by 900% and 3750% respectively [22]. The mass of freight physically transported globally in 2010 amounted to ca. 10.6 GT [23]. Falling production prices and an average growth rate of 45% between 1975 and 2015 resulted in an installed global solar photovoltaic capacity of 230 GW and - according to the experience curve law and learning rates - a significant decrease of environmental footprints of energy carrier applications and greenhouse gas emissions during production of poly- and monocrystalline-based photovoltaic systems. A break-even between environmental detriments and benefits is expected to occur between 1997 and 2018 [24].

According to the World Intellectual Property Indicator, more than 8.7 million patent and trademark applications, as well as industrial designs were registered in 2014 [25]. The world's technological capacity to transmit, store, and compute information has grown exponentially and has approached the magnitude of ca. 10^{23} bits present in human DNA [26]. Between 1975 and 2012, an approximately constant rate of doubling of computer chip performance every two years persisted (Moore's law[5]), which then slowed down to a 2.5 year pace at the passage from 22 nm processors to 14 nm processors [27]. A bibliometric analysis of total scientific output since the year 1650, cited in 38 million publications from 1980 to 2012, add up to ca. 755 million references. Growth rates of scientific literature amount to 1% in the 18th century, 2-3% in 1930, and 8-9% in 2012. From this, it has been concluded that, at the beginning of the 21st century, the doubling time of scientific knowledge is estimated to be 9 years [28]. In late 2014, ca. 34,550 professional scientific journals were listed, in which ca. 2.5 million scholarly peer-reviewed papers were collectively published [29]. The human brain project, initiated in October 2013 and committed to speeding up European industry, promotes and networks knowledge development in neuroinformatics, neurorobotics, neuromorphic computing, medical informatics, brain simulation, and brain-related medicine [30]. A psychometric intelligence metastudy, consisting of 271 independent samples including ca. 4 million participants from 31 countries and covering the years between 1909 and 2013, revealed nearly continuous global IQ gains in generational populations of, on average, 3 IQ points per decade, with generally stronger gains for adults and generally slightly smaller gains since ca. 1970. These facts, which are mainly based on more affluence, bettered health, education, nutrition, and higher GDP, partly substantiate the Flynn Effect, which states that each generation has developed higher full-scale intelligence (consisting of crystallised, fluid, and spatial intelligence) than the preceding one. This effect is also called rising population intelligence or generational IQ gain [31][6]. Similar positive trends were obtained during monitoring personality traits like *e.g.* self-confidence and sociability, ascertained for 419,523 Finnish military conscripts born between 1962 and 1976 [32].

Progress has been made in improving risk-management and hazard maps, early warning systems, and prognoses concerning catastrophes like storms, hurricanes, tornados, wildfires, droughts, floods, tsunamis, snow avalanches, and landslides to prevent damage, as well as losses of property and lives. However, despite immense efforts in active tectonics research, it is at present not possible to predict exactly earthquake occurrence probability, as well as time, hypocentre, magnitude and damage-potential of such hazards [33, 34]; the same applies to volcanic eruptions.

Two examples indicating the technical progress: Biotechnical conversion of the grreenhouse gas CO_2 into organic multicarbon compounds (artificial photosynthesis) has been invented by *in vitro* construction of a carbon fixation pathway similar to the Calvin cycle acting in *e.g.* plant cells. It contains partially engineered enzymes metabolising CO_2 faster and more efficiently, and will be developed as future source for random organic products like plastics, biodiesel, pharmaceuticals, *etc.* [35]. Promising progress in performance (22.1%) of hybrid organometallic halide perovskite solar cells has been achieved [36].

Sophisticated technical apparatuses - including land- and satellite-based - were constructed to observe distant, faint objects in deep space (Hubble and ESO telescopes), to analyse the chemical composition of planets, gas clouds, and stars, as well as the origination and population history of galaxies, their distribution, and architecture. 357 million distinct objects of an estimated 2 trillion present in the observable universe have been mapped and spectroscopically analysed [37, 38]. These observations will help to establish a revised standard model of cosmology [39]. Exploration of the forces and particles present in subatomic world occurs *via* collisional experiments with hadrons, leptons, *etc.* in synchro/cyclotrons and linear particle accelerators. Data and new findings from both spheres are expected to result in a Grand Unified Theory: the model unifying gravity with the electromagnetic, weak, and strong interactions [40].

More points concerning positive developments in recent human progress can be studied in detail on websites of several eminent historians and analysts of developments in science, economy, sociology, humanities, technology, health and well-being: *e.g.* Gregory Clark (economist and research associate of the Centre for Poverty Research), Angus Deaton (economist and researcher of determinants of health in poor and rich countries and of measuring poverty and inequality), Charles Kenny (economist and researcher of the relations between economic growth and happiness, of global health development, and of the impact of digitalisation on society), Steve Radelet (expert in foreign aid and research in the effects of trade between rich and poor countries), Steven Pinker (experimental psychologist and expert in science of cognition and development of social relations), Hans Rosling (medical doctor and statistician; expert in public health and relations between economic development, agriculture, poverty, and health; and health advisor at WHO and UNICEF), Max Roser (geologist, economist, and philosopher specialised in global trends of living conditions, education, health, violence, poverty, and income inequality) and Marian Tupy (policy analyst specialised in investigation of the effects of globalisation, global well-being, and national finances).

NOTES

[1] Progress in hygienics, medicine (*e.g.* mass vaccination) and public health care beginning during industrialisation, and the invention of artificial soil fertilisers led to the modern demographic transition. As it slowly began in ca. 1750 in Europe, average death rate began to decrease and average life expectancy started to increase. Because the birth rates remained constant in Europe for the time being, its population growth began to accelerate at an exponential rate, so that doubling to 2 billion was accomplished within 120 years in 1920. Due to the rising level of education, spread of family planning, and women's rights, fertilisation decreased in Europe below the replacement level of 2.3 after 1960. The populations outside Europe began to grow beginning ca. 1950 much faster because of established medicinal knowledge and locally higher fertility. Regress of fertility has been present since 1980 and is mostly caused by better education and decline in infant- and child-mortality. This occurs at very differing rates according to the population momentum: its power depends on the specific age structure of the population. Average global fertility decreased from 4.8 in 1960 to 2.5 in 2010 [41].

[2] He informed the public about the advantages of tele transmitted electricity by

illuminating electric bulbs and by powering the water pump of an artificial cascade presented in an exposition in the Crystal Palace of Munich.

[3] Economic reasons, which fostered the trend to replace ever more employees with machines, imply negative consequences like unemployment, sociopolitical instabilities, more consumption of resources (metals, energy carriers), implied dependencies, and the rising risk of resource depletion.

[4] Invention of letterpress printing by J. Gutenberg in 1450 and of the paper machine by L. Robert in 1799.

[5] A better concept is Moore's observation/projection/forecast.

[6] Nota bene: Changes in IQ gains were observed to be correlated with historical events: higher gains occurred between World War I and II, as well as post World War II; lower gains were measured during World War II [31].

CONFLICT OF INTEREST

The author (editor) declares no conflict of interest, financial or otherwise.

ACKNOWLEDGEMENTS

Declare none.

REFERENCES

[1]   Hadzimichalis, N. *Genetic Engineering: The Past, Present, and Future,* http://futurehumanevolution. com/genetic-engineering-the-past-present-and-future

[2]   Wang, L.K.; Ivanov, V.; Tay, J-H., Eds. *Environmental Biotechnology, Series Title: Handbook of Environmental Technology*; **2010**, Vol. 10.
        [http://dx.doi.org/10.1007/978-1-60327-140-0]

[3]   Primrose, S.B.; Twyman, R. Principles of Gene Manipulation and Genomics, **2006**.

[4]   FDA. *Final Environmental Assessment for Genetically Engineered Mosquito,* **2016**. http://www.fda. gov/AnimalVeterinary/NewsEvents/CVMUpdates/ucm490246.htm

[5]   Waltz, E. GM mosquitoes fire first salvo against Zika virus. *Nat. Biotechnol.,* **2016**, *34*(3), 221-222.
        [http://dx.doi.org/10.1038/nbt0316-221] [PMID: 26963535]

[6]   Galor, O.; Moav, O. *Natural Selection and the Evolution of Life Expectancy,* **2005**, http://sticerd.lse.ac.uk/seminarpapers/dg09102006.pdf

[7]   The World Bank. *The life expectancy at birth, total (years),* http://data.worldbank.org/indicator/ SP.DYN.LE00.IN/countries/1W?display=graph

[8]   Weiser, T.G.; Haynes, A.B.; Molina, G.; Lipsitz, S.R.; Esquivel, M.M.; Uribe-Leitz, T.; Fu, R.; Azad, T.; Chao, T.E.; Berry, W.R.; Gawande, A.A. Size and distribution of the global volume of surgery in

2012. *Bull. World Health Organ.,* **2016**, *94*(3), 201-209F.
[http://dx.doi.org/10.2471/BLT.15.159293] [PMID: 26966331]

[9]   World Life Expectancy. *World population pyramide,* http://www.worldlifeexpectancy.com/world-population-pyramid

[10]   Fried, L. *Routledge Handbook of Global Public Health*; Parker, R.; Sommer, M., Eds.; Taylor & Francis, Routledge: London, New York, **2011**, pp. 208-226.

[11]   IHME. *Rethinking development and health: Findings from the Global Burden of Disease Study,* http://www.healthdata.org/sites/default/files/files/images/news_release/2016/IHME_GBD2015.pdf **2016**.

[12]   Chemical Abstracts Service of the American Chemical Society. *CAS Assigns the 100 Millionth CAS Registry Number to a Substance Designed to Treat Acute Myeloid Leukemia,* http://www.cas.org/news/media-releases/100-millionth-substance

[13]   Krausmann, F.; Wiedenhofer, D.; Lauk, C.; Haas, W.; Tanikawa, H.; Fishman, T.; Miatto, A.; Schandl, H.; Haberl, H. Global socioeconomic material stocks rise 23-fold over the 20th century and require half of annual resource use. *Proc. Natl. Acad. Sci. USA,* **2017**, *114*(8), 1880-1885.
[http://dx.doi.org/10.1073/pnas.1613773114] [PMID: 28167761]

[14]   WHO. *Global Plan of Action on Workers' Health (2008-2017), Global Country Survey 2008/2009, Geneva,* **2013**. http://www.who.int/occupational_health/who_workers_health_web.pdf?ua=1

[15]   Strawe, C. *Arbeitszeit, Sozialzeit, Freizeit - Ein Beitrag zur Überwindung der Arbeitslosigkeit, Institut für Soziale Dreigliederung,* **1994**. http://www.dreigliederung.de/essays/1994-12-001.html

[16]   Whaples, R. *Hours of Work in U.S. History,* **2001**. http://eh.net/encyclopedia/hours- of-work-in-u-s-history/

[17]   Lee, S.; McCann, D.; Messenger, J.C. *Working Time Around the World; Routledge Studies in the Modern World Economy*; Taylor & Francis: London, New York, **2007**.

[18]   Ruegg, W., Ed. A History of the University in Europe, Universities in the Nineteenth and Early Twentieth Centuries (1800–1945). Cambridge University Press, **2004**, Vol. III, .
[http://dx.doi.org/10.1017/CBO9780511496868]

[19]   Ruegg, W., Ed. A History of the University in Europe, Universities Since 1945. Cambridge University Press, **2011**, Vol. IV, .

[20]   De Ridder-Symoens, H., Ed. A History of the University in Europe, Universities in Early Modern Europe (1500–1800). Cambridge University Press, **1996**, Vol. II, .

[21]   De Ridder-Symoens, H., Ed. A History of the University in Europe, Universities in the Middle Ages. Cambridge University Press, **1992**, Vol. I, .

[22]   World Trade Organization. **2016**. https://www.wto.org/english/res_e/publications_e/publ_by_subject_e.htm

[23]   Sustainable Europe Research Institut. *Distributing materials through global physical trade,* **2015**. http://www.materialflows.net/fileadmin/docs/materialflows.net/factsheets/matflow_FS7_2015.pdf

[24]   Louwen, A.; van Sark, W.G.; Faaij, A.P.; Schropp, R.E. Re-assessment of net energy production and greenhouse gas emissions avoidance after 40 years of photovoltaics development. *Nat. Commun.,* **2016**, *7*, 13728.
[http://dx.doi.org/10.1038/ncomms13728] [PMID: 27922591]

[25]   World Intellectual Property Organization. http://www.wipo.int/ipstats/en/wipi/index.html#patents

[26]   Hilbert, M.; López, P. The world's technological capacity to store, communicate, and compute information. *Science,* **2011**, *332*(6025), 60-65.
[http://dx.doi.org/10.1126/science.1200970] [PMID: 21310967]

[27]   Niccolai, J. *Intel pushes 10nm chip-making process to 2017, slowing Moore's Law, Infoworld,* **2015**.

http://www.infoworld.com/article/2949153/hardware/intel-pushes-10nm-chipmaking-process-to-2017-slowing-moores-law.html

[28]   Bornmann, L.; Mutz, R. Growth rates of modern science: A bibliometric analysis based on the number of publications and cited references. *J. Assoc. Inf. Sci. Technol.,* **2015**, *66*(11), 2215-2222. [http://dx.doi.org/10.1002/asi.23329]

[29]   Ware, M.; Mabe, M. *The STM Report, An overview of scientific and scholarly journal publishing,* **2015**. http://www.stm-assoc.org/2015_02_20_STM_Report_2015.pdf

[30]   *European Union, Human brain project, Geneva,* **2016**. https://www.humanbrainproject.eu/de/2016-overview

[31]   Pietschnig, J.; Voracek, M. One century of global IQ gains: a formal meta-analysis of the Flynn Effect (1909–2013). *Perspect. Psychol. Sci.,* **2015**, *10*(3), 282-306. [http://dx.doi.org/10.1177/1745691615577701] [PMID: 25987509]

[32]   Jokela, M.; Pekkarinen, T.; Sarvimäki, M.; Terviö, M.; Uusitalo, R. Secular rise in economically valuable personality traits. *Proc. Natl. Acad. Sci. USA,* **2017**, *114*(25), 6527-6532. [http://dx.doi.org/10.1073/pnas.1609994114] [PMID: 28584092]

[33]   Munich, Re. *Welcome to natural hazards site,* **2016**. https://www.munichre.com/touch/naturalhazards/en/homepage/index.html

[34]   Sobolev, G.A. Seismicity dynamics and earthquake predictability. *Nat. Hazards Earth Syst. Sci.,* **2011**, *11*, 445-458. [http://dx.doi.org/10.5194/nhess-11-445-2011]

[35]   Schwander, T.; Schada von Borzyskowski, L.; Burgener, S.; Cortina, N.S.; Erb, T.J. A synthetic pathway for the fixation of carbon dioxide *in vitro. Science,* **2016**, *354*(6314), 900-904. [http://dx.doi.org/10.1126/science.aah5237] [PMID: 27856910]

[36]   Yang, M.; Zhang, T.; Schulz, P.; Li, Z.; Li, G.; Kim, D.H.; Guo, N.; Berry, J.J.; Zhu, K.; Zhao, Y. Facile fabrication of large-grain $CH_3NH_3PbI_{3-x}Br_x$ films for high-efficiency solar cells *via* CH_3NH_3Br-selective Ostwald ripening. *Nat. Commun.,* **2016**, *7*, 12305. [http://dx.doi.org/10.1038/ncomms12305] [PMID: 27477212]

[37]   Abazajian, K.N.; Adelman-McCarthy, J.K.; Agüeros, M.A. The seventh data release of the Sloan Digital Sky Survey. *Astrophys. J. Suppl. Ser.,* **2009**, *182*(2) [http://dx.doi.org/10.1088/0067-0049/182/2/543]

[38]   Conselice, C.J.; Wilkinson, A.; Duncan, K. The evolution of galaxy number density at Z < 8 and its implications. *Astrophys. J.,* https://arxiv.org/pdf/1607.03909v2.pdf

[39]   Tiret, O.; Combes, F. Evolution of spiral galaxies in modified gravity. *Astron. Astrophys.,* **2007**, *464*, 517-528. [http://dx.doi.org/10.1051/0004-6361:20066446]

[40]   Hawking, S.W. *The Theory of Everything: The Origin and Fate of the Universe*; Phoenix Books: Beverly Hills, California, USA, **2006**.

[41]   Van Bavel, J. The world population explosion: causes, backgrounds and -projections for the future. *Facts Views Vis. ObGyn.,* **2013**, *5*(4), 281-291. [PMID: 24753956]

CHAPTER 18

Human Beings: Adverse Effects of Transformations

Abstract: World population growth, concentration in megacities, development of the technosphere and numerous novel chemical substances, industry, and agriculture have exposed humans to new kinds of health threats. Insufficient sewerage systems in megacities resulted in cholera epidemics. London smog caused impairment of the respiratory tract and rickets. Los Angeles smog brought about the presence of ground level toxic ozone in urban aerosol plumes. Mortality increased significantly in cities during summer heat waves. Permanent availability of artificial light has entailed unhealthy night and shift work. Rapid increase in global traffic resulted in annual losses of more than 1 million lives. Excess artificial noise causes 61,000 disability adjusted life years annually. Global annual fatalities due to air pollution run up to 5.5 millions. Clean-up of industrial brownfields impose financial burdens on public budgets. Health costs of Hg released into the environment are estimated at a minimum of 23,000 euros/kg. A positive correlation exists between health impairment and CO_2 emissions. Financial pressure is rising to organise constructions to shelter coastal cities against sea level rise. Growth of cropland has not kept pace with population growth since 1960. Ailments arise because of mass-consumption of cheap foodstuffs, luxury food, and stimulants. Globally, 1.8 billion persons lack access to good quality drinking water. Detected detrimental effects of certified chemicals and pharmaceuticals resulted in numerous health impairments. Global distillation has deteriorated the health status of the Arctic population. Health burdens arise due to industrial and nuclear havaries, and above ground nuclear tests. Intentional application of industrially produced warfare has caused ca. 145 million fatalities since 1800.

Keywords: Accidents, Adaptation, Civilised ailments, Concentration, Contamination, Food security, Global distillation, Health costs, Heat islands, Industrial havaries, Medicament side-effects, Noise, Overkill, Pollution, Population growth, Population resettlement, Social acceleration, Water scarcity, Waste, Weapons.

The immense gains in human health (see chapter 17) and technical development (see chapter 3) have come a very high price in deterioration of nature's ecological systems (see chapters 6-16). Because of the fact that degree of health depends on the health of ecosystems, continued degradation of nature due to economic growth will annihilate health gains made during the 20th century [1]. In the following, several examples will detail this.

Health problems arise from the growth and concentration of industry, economy, and population in cities. This will exemplarily be shown in the cases of the megacities London and Los Angeles:

The presence of a primitive sanitation system, dysfunctional sewers, and antiquated cesspits have caused urgent hygienic and health problems, as in London - world centre of industry and international trade at that time - experienced its fastest expansion and population growth between 1801 (959,300 residents) and 1891 (5,572,012 residents). Contamination of River Thames with filth and quality degradation of drinking water with pollutants from industrial and municipal organic wastes caused four severe cholera epidemics between 1831 and 1866, resulting in ca. 37,100 fatalities. As intolerable smells mounted in the hot and dry summer months July and August 1858, rising public fear of *miasmas*[1] pressured city authorities to decide to create a novel intercepting sewer system network - sustained by pumping stations -, which collected and reliably contained all the effluents and discharged them downriver out of the bounds of the city. Since the ca. 2000 km drainage system was built and started to operate in 1875, drinking water quality slowly improved, the waters of River Thames cleared and communicable diseases, as well as odours, began to disappear. The effects of this measure led to the acceptance that cholera is a water-borne disease transferred by bacteria [2, 3].

Between 1890-1920 scientific evidence became available that rickets, a children's disease already known to result from malnutrition, spread predominantly in large industrial towns, where sunlight is often dimmed or even blocked by polluted air and smog, preventing photochemical reactions in human skin necessary to synthesise vitamin D, which stimulates calcium and phosphate metabolism in the bodies of children [4].

Severe public health impairment occurred during the Big Smog[2] episode in London, which lasted from December 5-9 1952 and which was the worst ever recorded there. It required special atmospheric conditions and continuous emissions of atmospheric pollutants in the city. Massive *in situ* accumulation of these resulted from a temperature inversion phase with little or absent wind. Stagnant cold air and fog was trapped below a lid of warmer air, preventing convection and exchange of air masses. Pollutants were emitted by vehicles, coal-fired power stations (daily emitting 1 kT black carbon, 2 kT CO_2, 370 T SO_2, 140 T HCl, and 14 T fluorine-containing compounds), and domestic stoves, fuelled with cheap, low-grade sulphurous coal. The values of black carbon aerosol concentration exceeded 4000 $\mu g/m^3$ and had an average of ca. 1600$\mu g/m^3$ during this episode. Smog even penetrated indoor areas. Ca. 4000 persons (predominantly very young and elderly ones) died prematurely and about 25,000 persons got ill from impaired respiratory tracts (bronchopneumonia, purulent bronchitis). One month later, a second phase, coinciding with influenza, caused further 8000 fatalities [5 - 7].

The causal connection between air pollution and elevated mortality was recognised in the 1930s in a small industrial town in Belgium [5].

Since ca. 1945, rising amounts of the so-called "Los Angeles smog" became a major health concern in cities that were characterised by ample industrial volatile emissions and a high density of many motorised vehicles. Warm and dry air, as well as intense solar irradiation, present in cities located in the subtropical dry climate zone, fosters photochemical reactions in urban plumes between the primary pollutants (precursors) CO, volatile organic compounds (from incompletely combusted hydrocarbons, chemical solvents, windshield washer fluids *etc.*), and NO_x (produced during combustion), after a series of intermediate reactions, the formation of secondary atmospheric pollutants, *e.g.* peroxyacetylnitrate and ground-level (tropospheric) ozone[3] [8, 9]. Exposure to these pollutants causes eye irritation, probably cancer, respiratory diseases, and birth defects [10, 11].

Chemical composition of air pollution components changes with season, air temperature, moisture, and emitter behaviour. The last is shown as follows: high resolution mass-spectrometer measurements of diurnal concentration variations of primary and secondary organic submicron aerosols in the 2010 winter urban plume of Fresno, California were carried out. The species identified were related to different sources: traffic, cooking, and biomass burning. It was found that these organic pollutants have contributed - besides nitrate, sulphate, ammonium, and chloride aerosols - 67 masspercent to the $PM_{<1}$ fraction of aerosols. Measured diurnal cycles of the three types of organic aerosols were correlated positively with temporal emission profiles of local sources. Residential wood combustion during winter time turned out to diminish air quality and was a significant source of aerosol containing polycyclic aromatic hydrocarbons. Concentrations of organic aerosols varied within the 14 day monitoring period (with an average 85% relative humidity and average 9.8 °C; fog and rainfall) between 0.06-161 $\mu g/m^3$ (average 7.9 $\mu g/m^3$) [12].

In more recent times, long-term cohort studies on health impairment by air pollution have found increasing evidence that there does not exist a threshold of adverse health effects concerning the range of concentration of pollutants, that air pollution also has impacts on human health unrelated to the respiratory tract, and that multifactorial disease causation explains how air pollution can affect mortality at low levels [5].

In 2012, ca. 3.7 million premature deaths due to stroke, cardiovascular, and respiratory diseases were caused by polluted outdoor air containing PM_{10}[4], NO_x and ground level ozone [13, 14]. However, monitoring of air quality in 188 countries between 1990 and 2013 concerning the health consequences of poor air quality caused by emissions from power plants; industrial manufacturing; vehicle exhaust; burning coal, wood, dung, and waste, revealed that ca. 5.5 million premature deaths occurred globally in 2013 and that 85% of the world population live in areas where the WHO air quality guidelines[5] are exceeded[6]. Thus, air pollution has become the fourth highest risk factor for death and by far the first risk factor for diseases [15]. Research in biokinetics and toxic effects of inhaled µ-sized particles and nanoparticles revealed the following complex reactions in the pulmonary system, which were obtained from *in vivo* animal tests with canines and simians - exposed to engineered, biopersistent, and radiolabeled nanoparticles - as well as from cohort studies in occupational settings (*e.g.* industrial production of nanomaterials), and which were derived from correlations between citizens exposed to urban plumes and more frequently diagnosed diseases after hospitalisation.

- Long-term retention time of inhaled particles increases substantially in conducting airways with decreasing particle size;
- Inhaled particles are relocated in several ways after deposition on the alveolar epithelium:
  1. They are removed from the alveolar space *via* mucocilians of the bronchioles and are excreted or swallowed;
  2. They are absorbed by alveolar macrophages and cleared as in 1) or they are translocated by them into the interstitial space entering the systemic circulation:
     2a) They can pass into blood capillaries or into lymphatic vessels.
     2b) They relocate onto the alveolar epithelium and get cleared by ciliated airways.
     2c) The smallest particle fraction translocates directly across the alveolar-capillary wall into the interstitial space and enters the system circulation, where they redeposite and accumulate *e.g.* in distal organs.

The toxicological mechanisms triggered by inhalation of µ-sized particles and nanoparticles include oxidative stress, pulmonary and systemic inflammation, genotoxicity, changes in fibrinogen and prothrombin level, platelet activation, von Willebrand factor induction (causing defective blood coagulation), reduced heart rate variability, increased blood pressure, lipid peroxidation products, vasomotor dysfunction, disturbed lipid metabolism, and inflammation in the central nerve system. Health effects, which occur after transgression of the tolerable dose, include rehospitalisation with myocardial infarction, acute asthma, increased systolic blood pressure, ischaemic stroke, impaired lung function, allergic inflammation, myocardial ischaemia, arrhythmia, lung cancer, bronchitis, deep vein thrombosis, cognitive and behavioural change, neurophathy and neurogenerative diseases, low birth weight, pre-term birth and small gestational age [16, 17]. Inhaled combustion-related nanoparticles and $PM_{2.5}$ were detected to have deposited into functional units of the brains of young urban residents, being affected by brain inflammation, breakdown of nasal, olfactory and blood-brain barriers, volumetric and metabolic brain changes, attention and short-term memory deficits, and Alzheimer's and Parkinson's diseases. The airborne and iron-rich combustion-derived particles are considered to be a novel path into Alzheimer pathogenesis [18]. Some further examples:

Diesel exhaust particles were found to stimulate allergic disorders like asthma and to alter airway immune responses, because these pollutants act as an adjuvant for immature dendritic cell maturation [19]; this effect is probably augmented by obesity [20]. The biological activity and potential health burdens of mass-emitted ultra-fine particulate matter (< 10 nm), which is produced by new diesel combustion systems, is at present under study and may bear mutagenic, cytotoxic and inflammatory potentials [21].

Particulate platinum, emitted *e.g.* from motor vehicle catalytic converters or from industrial plants, is known to react in small quantities to soluble halite complexes. So it cannot be excluded that exposure to these secondary pollutants do not contribute to asthmatic diseases [22].

Biomonitoring and health risk assessment of local human populations exposed to particulate releases from smelters - *e.g.* in Sudbury (Canada) - and refineries and zinc plants revealed that the dust and aerosol immissions (PM_{10}) containing As, Cd, Cr, Ni, Pb, and SO_2 often exceed health-based guidelines. Derived exposure potency indexes[7] and other data point to elevated risk of pulmonary cancer, as well as adverse cardiorespiratory and neurobehavioural symptoms [23]. Another assessment found that the exposure to these toxic substances also occurs *via* skin absorption and eating of local food grown on contaminated soil [24].

Further health burdens from fine particulate matter emissions consist in low birth weight at fullterm birth and in elevated risks of preterm births: data from 183 countries concerning population-weighted, annual average ambient anthropogenic $PM_{2.5}$ concentration and frequency of preterm births revealed that estimated 2.7 million (*i.e.* 18% of the total preterm births globally) preterm births in 2010 were probably caused by maternal exposure to fine particulate matter, if a low concentration cut off is set at 10 μgm^{-3} in ambient air [25].

Health impairments in European countries in 2013 caused by transboundary atmospheric pollution from the emissions of 280 lignite and hard coal fired power stations, which receive 10 billion euros in subsidies per year from European governments, are: more than 22,900 premature deaths, ca. 21,000 hospital admission, 6,575,800 work days lost, 23,502,800 restricted activity days, and up to 62.3 billion euros in health costs. The pollutants consist in primary (soot) and secondary (NH_4NO_3, $(NH_4)_2SO_4$) aerosol, $PM_{2.5}$, ground level ozone, mercury, *etc.* [26]. It is supposed that there is also some evidence that exposure to polluted air is positively correlated with cognitive decline [27].

While tolerable µPM exposure dose during the entire life time has already been determined, the discussion about nanoparticle exposure is still going on [16, 28].

According to an analysis of the Lancet Commission, global environmental pollution - toxic air, water, soil, and workplace - killed ca. 9.0 million persons in the year 2015. Ca. 92% of these fatalities occurred in low- and middle-income countries. The annual costs of welfare losses and pollution-related diseases are estimated at ca. 4.6 trillion USD/a, equivalent to 6.2% of the world GDP in the year 2015 [29].

Negative effects of summer heat waves are more pronounced in urban areas because of the origination of heat islands: by machine-based heat production in buildings, traffic, and industry, and by heat absorption and storage in asphalt and concrete, causing diminished cooling of air during night. Heat is caught in between numerous buildings, which, in addition, delay exchange with cooler air masses. Excess premature mortality during the European heat wave in 2003 - of estimated 22,080 people - struck predominantly those persons of the age-group above 74 years and located on the lowermost socioeconomic level [30].

Atmospheric warming makes necessary technical measures to compensate for adaptive limits: these actions raise the financial burdens to run air conditioning at working places and in dwellings, hospitals, foodstuff deposits, automotivs, server farms, *etc.* during summer time; but the CO_2 emissions from the energy carriers spent for cooling contribute to warming.

Densification of cities and the resulting loss of green space has been shown to be positively correlated with problematic health outcomes in cities like air pollution, noise, heat, lack of natural settings, *etc.*, which degrade quality of life [31].

Safety aspects recommend also reliable calculations of stability of built infrastracture long-term exposed to corrosive fluids, acid rain and other wet immissions (see chapter 13 atmosphere). But exact predictions of dissolution kinetics of natural and/or artificial solid matters appeared to be complex and problematic, because in addition to variable extrinsic factors (temperature, pH, inhibitors, electrolytes) also variable intrinsic factors exist like microscopic kink sites, dissolved or sorbed nanoparticles, and colloids on the reacting crystal surface. These factors control the development of surface roughness and the distribution of surface energy. Both determine differing dissolution rate spectra on the surface, resulting in a variability of the mineral surface reactivity over the time [32].

The invention of electric light in the late 19th century, subsequent industrial production of lamps, lanterns, headlights, *etc.*, and their integration into everyday life, made artificial light permanently available and activities, as well as work, independent from natural day night cycles.

Neurasthenia spread during the "electric revolution" in ca. 1900, which brought about artificial illumination and the telephone, led to adaptive societal problems to this technological and cultural change. Humans were exposed to incessant floods of stimuli and work intensification, causing restlessness and distress [33]. Those societal adaptive stresses are comparable to those (*e.g.* burn out, depression) caused by the current digital revolution.

Since the beginning of industrialisation, the amount of around-the-clock, shift, and night work grew considerably because of

- The physical and chemical properties of components installed in machines, arrangements, and apparatuses, which compel permanent monitoring and control, *e.g.* blaste furnace, nuclear power plant, server farm;
- The growing size of plants and enterprises, their rising number of components, and the growing number of international connections and sub-contractors;
- Trade rivalry and focus on profit-making.

But long-term night workers are at elevated risk of lower health statuses because the absence of natural light diminishes the strength of their immune systems and deteriorates their vitamin D statuses. A cohort study proved that night workers have poorer dietary habits and metabolic profiles, as well as higher cardiovascular risks than day workers [34]. The reason lies in long-term circadian misalignment [35].

Multitasking, technically accelerated pace of life, permanent personal availability, and distraction by floods of news, information, and advertisement cause a fractured mind, diminishes attention, destroys one's own focus, often disturbs one's peace of mind, and impedes the finding of self-identity [36 - 38]. This lifestyle has also caused insufficient sleep (< 6 hours) and recreation for significant parts of populations in the USA (18%), Canada (6%), GB (16%),

Germany (9%), and Japan (16%), resulting in increases in mortality up to 13%, reduced workplace productivity and annual economic damage of up to 680 billion USD [39].

Intense exposure to growing amounts of electromagnetic radiation is caused by numerous industrial technical mass applications (power lines, TV, PC, mobile phones, base stations, *etc.*). This development evoked the question whether this kind of radiation is detrimental to human health. In 2010, ca. 4.6-5.0 billion mobile phone subscriptions existed. The Interphone Study, carried out between 2000-2010 in 13 countries, included 5000 participants. It revealed that, although uncertainties remain, mobile phone use does not cause brain tumours [40]. However, another working group came to the conclusion that radiofrequency electromagnetic fields, including those used in wireless communication, are possibly carcinogenic to humans [41]. Radiowave induced thermal damage to brain tissue may appear after six minutes, although large uncertainties exist [42].

Light pollution and artificial sky glow has degraded - in addition to smog and waste heat pollution - the transparency of air masses above cities and surroundings [43]. Thus, representations of cosmic objects viewed in city observatories have been blurred and the results of investigations got imprecise. Therefore, the sites of modern observatories have been chosen in remote high mountain ranges in the subtropical arid zone of the Chilean Altiplano, in the tropical low humid zone of Mauna Kea summit (Hawaii), or in the nival arid zone of eastern Antarctica, where the best atmospheric transparency over a broad spectral range is present [44]. In addition, observatories orbit in the exosphere.

The average relation between number of motorised vehicles and persons present in Germany developed from 1:134 in 1930 [45] to 1:1.546 in 2014 (see below). According to data from 2010 to 2014, the number of registered motorised vehicles present in 179 countries exceeded 1.776 billions[8]. Although traffic incidents decreased considerably in westen Europe and the USA due to numerous legislative and educational, as well as technical prevention measures, the annual global death toll is ca. 1.25 millions. Tens of millions of persons were injured or permanently disabled in road traffic [46] and accident costs amounting to billions of dollars arose. Additional detriments were caused by traffic jams. For example, ca. 694,000 jams resulting in a cumulative length of 1.3 million km occurred 2016 in Germany, amounting to 419,000 useless hours [47], many ten thousands of litres of fuel spent uselessly and economic losses of some millions of euros. Another point is wasting time and energy carriers during search for parking space in cities. A cost-benefit study conducted in Copenhagen considering externalities (accidents, climate change, health, travel time) arising from urban car driving revealed that each driven kilometer incures 50 eurocent to society [48]. However, according to the urban mobility expert M. Colville-Anderson, this value totals to 85 eurocent.

Despite many security and preventative measures, innumerable accidents have occurred during the controlling and operating of machines. The global occupation-related mortality, caused by work accidents and work-related diseases, was estimated to be 2.3 million fatalities. The economic costs amount to be an average of 4% of GDP [49].

Since the beginning of industrialisation, noise emitted from gas-firing motors, developed *e.g.* for application in vehicles, aircraft, and industry, has become a rising health concern, because the development of fully effective sound absorption techniques remained unsolved. In addition, noise is emitted from many different sources - *e.g.* cyclic/continuous explosive combustion of gas in pistons/jet turbine engines, accidents, rolling wheels, sirens, ignited explosives, whistling air streams, supersonic bangs. Artificial noise is positively correlated with urban and industrial concentration. Noise pollution affects the biosphere on land and below the water surface. It occurs predominantly on roadways, in cities, during concerts and mass events equipped with amplifiers and loudspeakers, during car races, near wind turbines and transformer houses, at building and mining sites, at airports and below flying lanes, at harbours and on sea routes, at railway stations and along railroads, during rocket lift-offs, during sonar activities below the water surface, and during offshore drilling and mining. During night time in residential areas, noise intensity must be < 35 dB (A) to guarantee a good quality of sleep. 40% of Europeans are exposed to noise intensities above the absolute threshold of pain by noise, which is 50 dB (A). Excess noise causes hearing impairment and hypertension [50]. Health effects of environmental noise in Europe was assessed from urban areas of > 250,000 residents and the burden of diseases quantified in terms of disability adjusted life years. The most important are: ischaemic heart disease (61,000 years), cognitive decline of children (45,000 years), sleep disturbance (903,000 years), tinnitus (22,000 years), and annoyance (654,000 years) [51].

Increased mobility induced the higher presence of fair-skinned persons in low latitude regions while increased levels of UVB, caused by stratospheric ozone loss due to industrial production of chlorofluorocarbons and N_2O, resulted in the loss of ca. 1.5 million disability adjusted life years and 60,000 premature deaths in the year 2000; the greatest of these diseases were cortical cataract and cutaneous malignant melanoma [52].

The overall magnitude of the burden of non-communicable diseases and injuries has risen [53].

Globally rising life expectancy and ageing of industrialised societies destabilise financial balances of budgets of civil servants and company pensions, as well as those of social insurances. In addition, higher life expectancy, retirement of the baby boomer generation, decline in birth rates in Europe and the USA since ca. 1965, ongoing substitution of employees by faster and cheaper robots and machines, and the growth of the low wage sectors generate precarious working and viability conditions, as well as old-age poverty.

Global atmospheric warming affects life in terms of rising risks, climate sensitive diseases, mortality rates, number of deaths, and disability adjusted life years. According to a human life cycle impact assessment, the degree of damage (*e.g.* malaria, cholera, dengue, diarrhoea, cardiovascular and respiratory disease, malnutrition, and coastal and inland flooding), depending on socioeconomic situations, is directly correlated to CO_2 emissions [54 - 56].

Recent climate change has had long-term influences on marine prokaryote communities that *e.g.* sea surface warming by 1.5 °C between 1958-2011 in the North Atlantic and the North Sea was positively correlated with the increase in abundance of pathogens like *Vibrio*, as well as occurrences of environmentally acquired *Vibrio* infections in waterborn food (*e.g.* oyster) [57].

Atmospheric warming will also diminish transport safety along aviation corridors, *e.g.* the transatlantic cruising routes situated at the boundary between troposphere and stratosphere: a climate modelling study indicates strenghening of vertical wind shears - as response to warming - within the atmospheric jet streams, thus increasing the occurrence probability of shear instabilities, which cause significantly more events of clear-air turbulences [58].

50% of 32 examined studies reported that anthropogenic climate change probably increases frequency and/or intensity of events and developments detrimental to humans: heat waves, droughts, wildfires, winterstorms, heavy rains, and tropical cyclones, as well as the shrinkage of Antarctic shelf ice extent and elevated sea surface temperatures [59].

The total annual costs related to the health impairment caused by the detrimental effects of artificial reactive nitrogen in the environments of Europe were estimated to be 70-320 billion euros. The effects include NO_x and secondary aerosols from ammonia (air pollution), global warming, and stratospheric ozone depletion by N_2O, ground and surface water degradation by NO_3^- *etc.* [60].

The overall economic and ecological damages caused by pesticide application has been estimated in the USA. 12 billion USD due to pesticide impact on human health; including: 1.1 billion disability-adjusted life years: 2.5 billion USD, defeating resistances of pests: 1.5 billion, crop losses by contamination by pesticides: 1.4 billion, wildlife losses: 2.2 billion, and treatment of contaminated surface and ground water: 2 billion [61].

Occupational and dietary exposure to metabolic products of the agriculturally applied commercial formulation of glyphosate is problematic because it proved to be cytotoxic: concentrations of it far below agricultural recommendations cause *in vitro* oxidative stress and apoptosis to human hepatic cells, which are sensitive to dietary pollutants [62]. In Europe, the safety approval of glyphosate was extended until December 2017 in order to have a complete review of this herbicide by the European Chemical Agency about its carcinogenicity to humans [63].

According to integrated assessment model calculations, the expenditures for the most effective measures that can be taken within 5 a (*e.g.* reduction of fertilisers containing N and P) in the agricultural areas of the Mississippi-Atchafalaya Basin to restrict the size of the hypoxic zone in the Northern Gulf of Mexico to 5000 km^2, is estimated to be 2.7 billion USD/a [64].

CH_3Hg containing wastewater from an industrial acetaldehyde production site on the southwestern coast of Kyushu (Japan) was discarded into seawater between 1932-1968 causing epidemic poisoning of the local population, because of ingestion of fish containing bioaccumulated methylmercury. The victims suffer from neuropathology and cerebral diseases among other. By 1995, of 2252 officially recognised victims, 1052 died of Minamata disease [65, 66].

Many thousand tonnes of Hg, which causes *e.g.* neurological defects, heart disease, and loss of IQ points, are annually released into the environment (see previous chapters). Related health costs aggregate - depending on the value of the effect-threshold - between 22,937-52,129 euros/kg Hg [67].

Sea level has risen by ca. 21.0 cm on average between 1880 and 2009 [68]. This fact has excited expensive plans to shelter coastal urban communities against more frequent detrimental effects of floods, inundation, storms, hurricanes, *etc*. Ongoing rise of sea level will result in elevated risks and adaptation stress to coastal residents [69].

Security of food availability is very probably at risk, because between 1960 and 2006 cropland area increased by ca. 11%, but world population nearly doubled. This results in a per capita cropland area decrease on average by 1% per annum, totalling 44% in 2006 [70]. On the other hand, since 1960 world food production grew faster than world population due to improved agricultural techniques and the falling of many real resource prices, so that from 1961 to 2009 the amount of food products increased on average by 41% per capita [71]. But uncertainties remain:

e.g. resource depletion: continued soil fertility: resilience of crop plants against heat: drought and pests: biodiversity loss: deforestation: global warming and environmental pollution [72]. Another point is that under elevated atmospheric CO_2 concentrations crops generate higher yields, but with less proteins and other micronutrients (Fe, Mg, Ca, Zn) [73].

Rising disequilibrium between the growth of world population and corresponding growth of the amount of available drinking water meant that ca. 1.8 billion people in 2016 lacked access to good quality drinking water. A further 158 million people from Africa, southeastern and southern Asia use surface waters (rivers, lakes, ponds) as drinking water due to lack of tap water. This development causes elevated suffering, mortality, and straining of health care systems [74].

Inappropriate wastewater treatment and disposal management by anti-infective drug manufacturing industries caused the origination of multidrug-resistant pathogens in ambient aquatic environments near Hyderabad (India), thus disseminating antimicrobial resistance [75].

Because China's per capita agricultural land is half of the world average and the per capita potable water is one fourth of the world average, food safety and health risks are of growing concern in that land. Concerns include water scarcity, pesticide over-application, chemical pollution, and heavy metal contamination of agricultural land, the latter caused by irrigation with untreated industrial and mine waste waters. This resulted in pollutant-containing food, detrimental to the health of consumers and creating problems like cancer villages [76]. 280 million Chinese residents are exposed to drinking water containing critical levels of inorganic and organic contaminants [77].

Insufficient physical activity in combination with increasing mass availability and often cheap, industrially produced foodstuffs, luxury food, and stimulants (characterised by addictive properties) like sugar, meat, tobacco, alcohol, drugs, *etc.* have caused civilised ailments on epidemic scale since ca. 1950 including obesity, adiposities, diabetes mellitus type 2, gout, cancer, hypertension, stroke, cardiovascular and pulmonary diseases, arteriosclerosis, osteoarthritis, dementia, hepatitis, and cirrhosis. Globally 6.6 million persons die annually of smoking and from passive smoke, and 3.3 million persons die globally per annum from harmful use of alcohol. Economic insecurity, drug and alcohol poisoning, diabetes, suicides, and liver cirrhosis caused a reversal in the long-term significant decline in mortality rate: from 1979 to 1998 the rate of decline was 1.8%/a in USA concerning the fraction of the age group 45-54; between 1999 and 2013 the mortality rate for middle-aged non-Hispanic men and women reversed and rose by 0.5%/a. Ca. 0.5 million fatalities could have been avoided in that period. This fatal development concerned particularly the least educated subset. The economic burdens caused by obesity evaluated in Australia (1989-1990), France (1992), the Netherlands (1981-1989), and the USA (1994) came up to 48.7 billion USD [78 - 81]. Epidemiologic studies in 195 countries revealed that in 2015 107.7 million children and 603.7 million adults were obese, and that a high body mass index accounted to 4.0 million deaths and 120 million disability adjusted life years globally [82]. Early childhood overweight and obesity, resulting from poor diet, inadequate physical activity and excessive sedentary behaviour, is of concern for 42 million pre-schoolers globally; this development has raised further healthcare expenditures [83]. A two-year study, which considered clinical data of 94,647 persons and environmental data, revealed that elderly persons (> 60 a) having type 2 diabetes mellitus are at elevated risk of cerebral and myocardial infarction, if exposed at temperatures above 25 °C to higher levels (> $25\mu g/m^3$) of $PM_{2.5}$ [84]. This association points to increased fatal vulnerability to degraded environmental conditions in the case of epidemic civilised ailments.

Exposure to antimicrobials by consuming the meat of livestock treated with veterinary drugs has resulted in a significant public health threat because of the development of drug-resistant pathogens [85].

A rising number of industrial products containing artificial endocrine disrupting chemicals[9] (at present ca. 800) have been identified, which impair the role of hormones in the development of the human body (as well as in animals) and can cause adverse health effects, *e.g.* cancer, birth defects, obesity, diabetes, and neurotoxic effects. They enter the human body *via* ingestion, inhalation, and skin absorption. Exposures to endocrine disrupting chemicals during foetal or infant states causes increased susceptibility to diseases in later life. Some of these substances are effective as persistent organic pollutants and others bioaccumulate in the food chain. Differing views about the chemical toxicity of these substances were settled and an international consensus about their identification was confirmed in order to create a basis for regulatory criteria concerning endocrine disrupting chemicals in the European Union [86, 87]. Official evaluations based on toxicity screenings and tests conducted by experts resulted in the ban of some of these chemicals. A positive correlation between child-obesity and prenatal exposure to perfluoroalkyl substances - applied as outer coating on products to make them more stain-resistant, waterproof and nonstick - was ascertained [88, 89].

Side-effects of some certified pharmaceutical products detrimental to human health have occurred. In the case of Contergan, a non-prescriptive, popular sleeping and tranquilising agent containing thalidomide[10], embryonic damage, and birth defects were caused to an estimated 10,000 children between 1957 and 1961 [90, 91].

The American Chemical Society listed 265 toxic environmental chemicals, to which humans are exposed [88]. The Lancet Commission stated that ca. 140,000 chemicals and pesticids were synthesised since 1950, and 5000 of them were applied in greatest volumes, resulting in global dispersion. Of these high-production chemicals, however, less than 50% underwent thorough testing for safety and toxicity [29].

Intentional utilisation of motor waste gases and several T of industrially produced Zyklon B occurred in six Nazi extermination camps between 1941 and 1945. At least 1 million persons were industrially mass-murdered in technically organised, efficient, and economically rationalised manner by coercive exposure to carbon monoxide or hydrocyanic acid[11] in gas chambers. Their dead bodies were annihilated in the cremation furnaces attached to these death factories [92]. Industrially produced poison gases were later applied during hostile military operations in Iran (1987), Iraq (1988), and Syria (2013-2017). The very first hostile application of chlorine gas occurred during World War I on 04/22/1915 at Ypern (Belgium).

Strategic bombing of the Japanese cities Hiroshima (08/06/1945, 15 kT explosive force, ^{235}uranium) and Nagasaki (08/09/1945, 21 kT explosive force, ^{239}plutonium) by US airforces caused at least 80,000 and 35,000 immediately kills. The death count, predominantly due to radiation illness, rose in Hiroshima to 155,000 (end of 1946) and in Nagasaki to 80,000 (end of 1945) [93]. A 45 year (1950-1995) longevity follow-up of a cohort of 120,200 A-bomb survivors from Hiroshima and Nagasaki, regarding biological and health consequences of exposure to low radiation doses, revealed that life expectancy decreased with dose increase at a rate of ca. 1-3 years/Gy and that medium loss of longevity among all survivors was estimated to ca. 4 months. Shortening of life expectancy has been estimated to be 1-2%/Gy [94].

135,000 persons that lived within a radius of 30 km around Chernobyl were quickly evacuated after the nuclear havary on 04/26/1986. The estimated internal doses accumulated in their thyroid glands, because of incorporation of ^{131}J *via e.g.* contaminated milk, was on average ca. 490 mGy and the average effective external and internal whole body dose caused by other radionuclides amounted to 35 mSv [95].

247,000 clean-up workers and liquidators were exposed to an average external dose of 150 mSv. Fallout derived radiation exposure of the population outside the 30 km zone resulted from soil and food contaminated with ^{134,137}Cs. An estimated

6.1 million persons in the former USSR were exposed to ^{137}Cs soil activity above 37 kBq/m^2 [95].

Subsequent monitoring of 1.6 million persons aged 1-18 living in highly contaminated regions revealed that ca. 100,000 were affected by thyroid gland doses of > 300mGy. Estimation of the collective dose (from radionuclides except iodine) received between 1986 and 2005 by ca. 6.3 million persons inhabiting highly contaminated areas, comes up to 58,640 Sv. Their average effective dose in 1986 was 88 mSv. Health effects include significantly elevated rates of acute radiation syndrome, skin burns, leukaemia, cancer of the thyroid gland, other forms of cancer, cardiovascular diseases, and cataracts. These consequences have been observed in many cohort studies [95].

The havary at the Fukushima nuclear power plant on 03/11/2011 made necessary the immediate evacuation of ca. 200,000 residents living in the vicinity of the site and in potentially affected regions. Additional mortalities resulting from cancer cases caused by radiation exposure are expected to be ca. 1000 persons. Expenditures necessary for clean-up of the site, decontamination of soil, and compensation payments to local residents are estimated at 250 billion USD [96].

Assessment of the degree of contamination with ^{134}Cs and ^{137}Cs in human placentas revealed that women living within 290 km of the Fukushima Nuclear Power Plant had levels up to 0.742 Bq/kg of ^{134}Cs and up to 0.922 Bq/kg of ^{137}Cs, and that these exposures were lower than those measured in Japan and Canada in the 1960s during the numerous above ground nuclear tests and lower than those detected in Italy 1986-1987 after havary of the Chernobyl nuclear power plant [97].

On 12/03/1984, a gas leak in a pesticide plant[12] in Bhopal (India) caused the release of ca. 40 T of methylisocyanate and reactive products within two hours, to which ca. 500,000 inhabitants were exposed because no emergency plans had been set up prior. 3400 died immediately and further 15,000-20,000 persons during the following twenty years. Ca. 102,000 persons were permanently disabled. Early health effects of the toxic gas mixture included chemosis, pulmonary edema, persistent diarrhoea, anxiety, and impaired audio and visual memory. Late health effects included corneal opacity, chronic conjunctivitis, decreased lung function, increased pregnancy loss and infant mortality, increased chromosomal abnormalities, and impaired associative learning [98].

At least 350 major industrial disasters, which have caused in sum some 100,000 fatalities and hundreds of billions in costs, have occurred in the defence, energy, food, manufacturing, and other sectors since 1860 [99, 100].

Assessment, clean-up, and development costs of contaminated industrial sites (brownfields) positioned in urban contexts have put huge financial burdens to public budgets [101].

Epidemiological studies, conducted between 1971 and 2005 and based on a base line cohort of 36,461 residents of the Arctic population in Finland, on cancer risk following radiation exposure caused by global fallout from above ground nuclear tests indicated an increased cancer incidence associated with radiation exposure received during childhood (< age 15 years). Cancer risk estimations (n = 3073) were statistically compared with the observed numbers of cancer cases (n = 2630) based on whole body measurements of bioaccumulated ^{137}Cs [102].

The Arctic Monitoring Assessment Program, conducted since 1991 in order to investigate the health effects of aeolian import and deposition of persistent organic pollutants and toxic chemicals into boreal ecosystems, revealed that indigenous people are exposed to acidic Arctic haze, elevated levels of UVB due to depletion of stratospheric ozone, and to contaminated water and food, in which toxins and radioactive matter have accumulated [103, 104].

The radioactivity caused by total global deposition of selected radionuclides[13] emitted during above ground nuclear explosions was estimated to 1.8×10^{20} Bq. The average effective external dose between 1953 and 2000 per capita in the USA from global fallout came up to ca. 0.74 mGy and the effective internal dose to 0.4 mSv, giving rise to the induction of thyroid cancer and leukaemia in the population [105].

Four years of monitoring of prenatal exposure to unconventional natural gas development activity, involving an inverse distance squared model that incorporated distance to the mother's home, as well as dates and durations of well pad development, drilling, hydraulic fracturing, and production volume found that a positive correlation exists between unconventional natural gas development activity and preterm birth [106]. Chemicals applied and emitted during hydraulic fracking operations are negatively impacting human health and have the potential to contaminate groundwater [107].

42 of the 50 biggest open municipal waste dumpsites are situated either within populated urban areas or are positioned within 2 km of them. Residents are therefore impacted by contaminated water and polluted air due to the release of toxic gases, dust, odours, and soot, the latter generated by open combustion of waste. Health impairments, especially of waste recyclers, include respiratory ailments due to inhalation of toxic aerosol, polycyclic aromatic hydrocarbon vapours, phthalate, dioxins, furans, *etc.*; blood lead levels above the toxic level of 10 µg/decilitre; muscular atrophy; allergic dermatitis; diarrhoeal disease; size and

staining abnormalities of red blood cells caused by heavy metal poisoning [108]. Uncontrolled slide events of ill-managed municipal solid waste dumpsites occurred due to sudden biogas release and/or rainwater induced slope failure. 2.7 million m³ of waste were mobilised by a failure, destroying the lives of 147 people living in the vicinity of a large dumpsite in West Java Province [109]. The Koshe rubbish dump landslide on 03/11/2017 (Addis Abeba, Ethiopia) killed more than 113 persons and was probably caused by construction works for a biogas plant on top of the waste dump site [110].

Since World War II, spread of open cast mining in the Rhenish Lignite Mining Region, covering an area of 2500 km² in 2012, caused the loss of 103 villages and the resettlement of 47,600 residents having lived there [111]. Problematic measures like that have occurred since the beginning of industrialisation in many other places globally. Many more displacements and exproprietations were carried out because of other large-scale building measures like superhighways, airports and dams (see below).

Implementation of one of the world's largest hydropower plants, the Three Gorges Dam[14] (22.5 GW capacity) in China, made necessary the relocation of 1.3 million residents. Since the beginning of the filling of the reservoir in 2003, flood control, increased shipping, and the generation of electricity was enabled, but also elevated risks arose concerning rock- and landslides, as well as reservoir-induced seismicity, which are both related to fluctuating water levels and changing water pressure. These have caused some tens of fatalities [112].

The construction of several large dams on the Mekong River for flood protection, irrigation, and hydroelectricity also made necessary population resettlement in economically underdeveloped regions. Inundation-caused loss of agricultural land, used for crop production for subsidence, and diminished fishcatch caused reduction of economic bases and of food security [113, 114].

Settling, territorial development, and economic expansion of European emigrants in the USA, Canada and Australia in the 18th and 19th centuries resulted in displacement of First Nations and aborigines from their homelands.

Since 1960, the Aral Sea environmental crisis, caused by inappropriate water abstraction of its affluents, has resulted in water level decline, hypersalinisation and desiccation, and caused the loss of thousands of fishing-related jobs and the exodus of ca. 100,000 people living at the former coasts. Due to bioaccumulation and exposure to sandstorm-transported contaminated dust, scoured from the dried-up sea floor polluted with agrochemicals (DDT, HCH, Toxophene, PCBs, Phosalone), caused considerable decline of the health status of the local population [115].

Discrimination of triggered seismicity, caused by the reactivation of tectonic faults, from natural seismicity is difficult. Documented and probable cases including magnitudes between 1 and 7.9 have been reported since 1929, which have caused thousands of fatalities [116, 117].

Exploratory drilling for natural gas at Sidoarjo (Java) brought about the eruption of a mud volcano on 05/29/2006. Since then, a continuous outflow of gas and mud mixed with clasts has covered ca. 10 km^2, having caused the displacement of ca. 40,000 persons, and an estimated 2.7 billion USD damage [118].

Excess ground water abstraction from an aquifer below Mexico City (ca. 20 million inhabitants) caused decrease of interstitial water pressure at the base of the overlying aquitard, consisting of lakustrine, highly compressible clay and silt deposits. Upward diffusion of negative pore pressure resulted in compaction, differential surface subsidence, and considerable damage to urban infrastructure by ground sagging. Geodetic monitoring (2003-2007) *via* satellite radar interferometry revealed vertical terrain subsidence rates up to 38.7 cm/a [119].

With expansion of settlements into regions with steep slopes and temporarily plentiful rainfall, the number of fatalities caused by landslides has increased and amounted to 32,322 people between 2004 and 2010. Settling in earthquake zones and close to flood-prone areas entailed ca. 11,500 fatalities between 1990 and 2006 [120 - 122]. Landslides caused by thawing permafrost soil resulted in infrastructure damages and losses of life.

The fate of many industrial monocities is explained here in the case of Asbestos (Canada): 11 phases of lateral extensions of the Jeffrey Mine between 1928 and 1980 into the adjacent area of the City of Asbestos, which developed from a small mining town to an industrial centre, effected that ca. 54% of the town's centre was converted from residential and commercial use by 1967 by bills of expropriation into industrial use, in order to foster open cast chrysotile mining and to keep operators' positions. Applications of this raw material were construction and automobile components, sheeting for fighter jets and commercial airplanes, fabrics to cover furniture, *etc.* Since 1970 however, growing awareness of significant negative health effects caused by inhalation of the mineral dust and the related closing of global asbestos markets, resulted in economic collapse of the mine company in 1983 and the loss of the exclusive and precarious economic foundation of Asbestos City [123]. Globally, an estimated 107,000 people annually die from asbestos-related diseases (asbestosis, mesothelioma, lung cancer). In 1987, the International Agency for Research on Cancer evaluated all forms of asbestos as carcinogenic to humans [124].

Many toxic substances like lead (lead tetraethyl as an anti knock agent in gasoline) [125], copper, mercury, sodium cyanide (for extracting Au), U, Pu, *etc.* were tolerated because of their appreciated physical or chemical properties.

The evolution in and trend of risky resource consumption have been calculated for 20 critical and abundant metallic mineral raw materials through the year 2100. The deposits, necessary to cover future demand must contain 5-10 times the amount of material from known deposits and that not yet discovered deposits probably do not exist in sufficient size and quantity, and on accessible terrain [126].

Technical development, combined with large-scale industrial production, of ever more precisely operating warfare material, is characterised by augmented mobility, accuracy of aim, velocity of projectiles, as well as multiplied destructive power, and resulted in rising numbers of killed and war-disabled persons per unit time[15]. The arms race spurred the invention of the tank in 1916 [127] and that of the warplane in 1918 [128].

In the meantime, the overall global destructive power of war material[16] of industrial-military complexes has multiplied and resulted in mass destruction and overkill potential, which peaked during the Cold War. The degree of aim of long-distance weapons has developed to perfection due to laser- and GPS-based techniques. The wars carried out globally in the 19th and 20th centuries caused the death of at least 145 million persons [129] and many more war-disabled persons.

Since the onset of industrialisation the per capita killing potential of weapons in conflict situations has multiplied enormously.

Industrialisation and technological progress has resulted in simplification of killing. The export and trade of ammunition and weapons caused globalised availability of these optimised and potent technologies, resulting in rising death tolls in the case of wars, terrorist attacks, and crimes. Arms sales in 2014, as derived from collected information from the global top 100 weapons producing companies, resulted in expenditures of ca. 401 billion USD [130].

Estimated global military expenditures increased tenfold from 1.74×10^{11} USD in 1949 to 1.73×10^{12} USD in 2015. During the final phase of the Cold War, military expenditures reached their maximum at 1.55×10^{12} USD in 1988. The relative maximum of military expenditures, related to the GDPs of the referring countries, was reached in 1991 at 4.2% [131].

This development and enhanced trade with warfare material enabled fast and long lasting killing rates: *e.g.* murdering of five persons/minute over 100 days - a killing rate five times faster than in Auschwitz - has occurred during the genocide in Ruanda in 1994, according to UN-majór-general Roméo Dallaire.

Post-traumatic stress disorder, which was first interpreted as shell shock or combat fatigue, is now seen by neuropathologists as traumatic brain injury. It is caused by ultrasonic pressure waves propagating from detonating TNT or dynamite, applied for the first time in 1902 by the Reichsgerman military, resulting in the microscarring of brain tissue. These micro lesions cause dysfunction of the brain that very probably result in depression, panic attacks, sleeplessness, other psychic disorders, and amnesia [132].

It was estimated that in early 2016 the total nuclear weapon inventory consisted in Russia of 7300 and in the USA of 4670 long-range strategic, as well as shorter-range tactical warheads [133, 134].

Analyses of events concerning armed conflicts and climate-related natural calamities between 1980 and 2010 found that the global coincidence rate is only 9%, but that it increases up to 23% in ethnically divided or societally fissured states. This points to the fact that the disruptive character of heat waves or droughts makes the outbreak of armed conflicts more probable and that anthropogenic climate change will contribute to the frequency of these co-occurrences [135].

According to the Gun Violence Archive, 13,286 citizens suffered death and 26,819 suffered injuries by firearms 2015 in the USA because legislation permits private ownership of firearms. Between 1968 and 2011, ca. 1.4 million citizens were shot during private conflicts [136]. 19,392 persons committed suicide by firearm in 2009 in the USA [137]. An estimated 310 million privately owned firearms existed in 2009 in the USA [138].

The huge amount of weapons and the overkill potential is probably a threat because of the legacy of human conspecific violence, which developed temporarily to a high degree in ancestral lines of primates and hominids and consists of a phylogenetic and a cultural (social and territorial) component. But, despite the remarks above, the present average rate of human lethal violence has declined to 0.1% [139].

Another point is the problem of dual-use of economic goods: *e.g.* technical devices can be abused for military, terroristic and suppressive purposes as well as to store and screen huge masses of data for control, surveillance and persecution in totalitarian states and dictatorships, which destroy the free society and civil rights [140 - 142].

The last point of this chapter will discuss one example of new technological applications, which can't be fuelled and run sustainably with mineral raw materials and energy carriers: *i.e.* the very high ecological footprint of light-duty batteries for e-cars, consisting, among other, of the metals lithium and cobalt. The calculation resulted in an estimate of 175 kg CO_2eq./kWh emissions on average [143], which originate during the industrial production of one Li-ion battery. Referring to a battery capacity of only 30 kWh, the emissions total to ca. 5.3 T CO_2eq. Assuming that 2.64 kg CO_2 is produced during combustion of 1 L diesel in automotives [144], a standard passenger vehicle must cover a horizontal distance of 29.4×10^3 km to generate the amount of CO_2 necessary for the industrial production of that Li-ion battery.

Lithium/cobalt from mining amounted in 2014 globally to ca. 36,000/130,222 T. The world reserves of the metals lithium/cobalt are estimated to 13.5/7.1 MT and the world resources to 39.5/145 MT [145, 146]. Caused by the forced transition to future electromobility, an annual demand of 800,000 T of Li can be expected by 2040 and its supply range would shrink to 17 a (reserves) and to 50 a (resources) [147]. Lithium/cobalt mineral commodities are characterised by moderate to high market concentration (Herfindahl-Hirschman-Index: 2669/3599) and the supply with Co is characterised by high country risk [148]. In consequence, the very high ecological rucksacks of Li-ion batteries and the criticality of the metals Li and Co exclude Li-ion batteries from representing sustainable technology [149], unless these devices can be scaled up or the metals Li and Co can be substituted or the resources/reserves in these metals increase [147].

NOTES

[1] Gases and vapours from organic wastes were at that time thought to be poisonous, so that their inhalation was thought to cause communicable diseases like cholera.

[2] This term, combining the notions smoke and fog, was coined ca. 1893 in England. The physician H. A. des Voeux used that term 1905 in his paper "Fog and Smoke", presented during a meeting of the Public Health Congress [https://en.wikipedia.org/wiki/Smog].

[3] Simplified formations of peroxyacetylnitrate: $C_XH_Y + O_2 + NO_2 + h\upsilon = CH_3COOONO$ and of ozone: $CO + 2O_2 + h\upsilon = CO_2 + O_3$. The origination of photochemical smog was detected in the 1950s by biochemist Arie Jan Haagen-Smit [150]. Atmospheric chemist Phillip Leighton found 1961 that tropospheric

ozone production interdepends on intensity of solar irradiation and concentration of NO [151].

[4] Particulate matter measuring < 10.0 μm of maximum diameter.

[5] World Health Organisation (WHO) air quality guidelines set daily particulate matter at 25 micrograms per cubic meter air.

[6] Between 1800 and 1920, no statistics exist concerning fatalities caused by air pollution, when the average concentration of PM was estimated to $400 \mu g/m^3$ in European cities [5].

[7] It is defined as the quotient between the annual average concentration of the toxic substance in ambient air and the dose, that induces a 5% increase in incidence or mortality due to relevant tumours, derived from toxicological studies.

[8] Top ten 2014: USA: 265,043,362, China: 250,138,212, India: 159,490,578, Indonesia: 104,211,132, Japan: 91,377,312, Brazil: 81,600,729, Germany: 52,391,000, Italy: 51,269,218, Russia: 50,616,136, France: 42,792,103 [45].

[9] Commercial applications of: dichloro-diphenyl-trichloroethane (DDT): pesticide; polychlorinated biphenyls (PCB): industrial coolants and lubricants; polybrominated diphenyl ethers (PBDE): flame retardants; phthalates: cosmetics, medical equipment, floorings, soft toys; bisphenol A (BPA): paper coatings, plastic components; organotins: boat varnish, anti-fouling paint; *etc.*

[10] The active agent thalidomide, chemically synthesised from the aromatic hydrocarbon species 1.2-benzenedicarboxylic acid, caused embryopathies by suppressing morphogenesis of limbs, tubular bones, and organs. However it turned out that it also suppresses leprosy, Crohn's disease, and several forms of cancer [152].

[11] Industrial production of hydrocyanic acid by dehydration of formamide at 300 °C. The latter is chemically synthesised from ammonia and carbon monoxide.

[12] Because of dysfunctional technical devices and underdeveloped safety systems, ca. 1000 litres of water came into contact with ca. 40 T of impure

methylisocyanate, resulting in a violent exothermic chemical reaction forming CO_2, methylamine, dimethylamine, di- and trimethylisocyanurate, as well as the thermal decomposition product of methylisocyanate: hydrocyanic acid (HCN) [98].

[13] Radionuclides taken into account: $^{239,240}Pu$, ^{137}Cs, ^{90}Sr, ^{106}Ru, ^{144}Ce, ^{95}Zr, ^{89}Sr, ^{103}Ru, ^{95}Nb, ^{141}Ce, ^{140}Ba, ^{131}J, ^{3}H, and ^{14}C.

[14] Amount of materials used were 27.2×10^6 m³ concrete, and 3.54×10^5 T of steel. Amount of gravel, sand, and rock removed by excavating and blasting: 102.6×10^6 m³ [153].

[15] It took about three years (149-146 BC) for the Roman military to conquer and destroy the city of Carthage; it took less than three hours on 04/26/1937 to destroy Guernica by carpet bombing, carried out by the Legion Condor of the Reichsgerman air forces; it took only seconds on 08/06/1945 to melt down and vaporise Hiroshima's centre by dropping one uranium bomb by the US air force.

[16] The destructive power of war material fabricated from metal (iron and copper) has already been described by Georgius Agricola (1494-1555): *".. for it is employed in making swords, javelins, spears, pikes, arrows - weapons by which men are wounded, and which cause slaughter, robbery, and wars. These things so moved the wrath of Pliny that he wrote: „Iron is used not only in hand to hand fighting, but also to form the winged missiles of war, sometimes for hurling engines, sometimes for lances, some times even for arrows. I look upon it as the most deadly fruit of human ingenuity. For to bring Death to men more quickly we have given wings to iron and taught it to fly." The spear, the arrow from the bow, or the bolt from the catapult and other engines can be driven into the body of only one man, while the iron cannon-ball fired through the air, can go through the bodies of many men, and there is no marble or stone object so hard that it cannot be shattered by the force and shock. Therefore it levels the highest towers to the ground, shatters and destroys the strongest walls. Certainly the ballistas which throw stones, the battering rams and other ancient war engines for making breaches in walls of fortresses and hurling down strongholds, seem to have little power in comparison with our present cannon..."* [154].

CONFLICT OF INTEREST

The author (editor) declares no conflict of interest, financial or otherwise.

ACKNOWLEDGEMENTS

Declare none.

REFERENCES

[1] Commission of the Rockefeller Foundation. *Safeguarding human health in the Anthropocene epoch: report of The Rockefeller Foundation-Lancet Commission on planetary health,* **2015**. http://www.thelancet.com/commissions/planetary-health

[2] Halliday, S. Death and miasma in Victorian London: an obstinate belief. *BMJ,* **2001**, *323*(7327), 1469-1471.
[http://dx.doi.org/10.1136/bmj.323.7327.1469] [PMID: 11751359]

[3] Halliday, S. *The Great Stink of London.,* **2013**. http://www.thehistorypress.co.uk/publication/the-great-stink-of-london/9780750925808/

[4] Gibbs, D. Rickets and the crippled child: an historical perspective. *J. R. Soc. Med.,* **1994**, *87*(12), 729-732.
[PMID: 7503834]

[5] Anderson, H.R. Air pollution and mortality: A history. *Atmos. Environ.,* **2009**, *43*, 142-152.
[http://dx.doi.org/10.1016/j.atmosenv.2008.09.026]

[6] Bell, M.L.; Davis, D.L.; Fletcher, T. A retrospective assessment of mortality from the London smog episode of 1952: the role of influenza and pollution. *Environ. Health Perspect.,* **2004**, *112*(1), 6-8.
[http://dx.doi.org/10.1289/ehp.6539] [PMID: 14698923]

[7] Met Office. *The Great Smog of 1952,* **2015**. http://www.metoffice.gov.uk/learning/learn-about-the-weather/weather-phenomena/case-studies/great-smog

[8] LaFranchi, B.W.; Wolfe, G.M.; Thornton, J.A. Closing the peroxy acetyl nitrate budget: observations of acylperoxy nitrates (PAN, PPN, and MPAN) during BEARPEX 2007. *Atmos. Chem. Phys.,* **2009**, *9*, 7623-7641.
[http://dx.doi.org/10.5194/acp-9-7623-2009]

[9] Lei, W.; de Foy, B.; Zavala, M. Characterizing ozone production in the Mexico City Metropolitan Area: a case study using a chemical transport model. *Atmos. Chem. Phys.,* **2007**, *7*, 1347-1366.
[http://dx.doi.org/10.5194/acp-7-1347-2007]

[10] Vyskocil, A.; Viau, C.; Lamy, S. Peroxyacetyl nitrate: review of toxicity. *Hum. Exp. Toxicol.,* **1998**, *17*(4), 212-220.
[http://dx.doi.org/10.1177/096032719801700403] [PMID: 9617633]

[11] Padula, A.M.; Mortimer, K.; Hubbard, A.; Lurmann, F.; Jerrett, M.; Tager, I.B. Exposure to traffic-related air pollution during pregnancy and term low birth weight: estimation of causal associations in a semiparametric model. *Am. J. Epidemiol.,* **2012**, *176*(9), 815-824.
[http://dx.doi.org/10.1093/aje/kws148] [PMID: 23045474]

[12] Ge, X.; Setyan, A.; Sun, Y. Primary and secondary organic aerosols in Fresno, California during wintertime: Results from high resolution aerosol mass spectrometry. *J. Geophys. Res.,* **2012**, *117*, D19301.
[http://dx.doi.org/10.1029/2012JD018026]

[13] WHO. *Ambient (outdoor) air quality and health, Fact sheet N°313,* **2014**. http://www.who.int/mediacentre/factsheets/fs313/en/

[14] Lelieveld, J.; Evans, J.S.; Fnais, M.; Giannadaki, D.; Pozzer, A. The contribution of outdoor air pollution sources to premature mortality on a global scale. *Nature,* **2015**, *525*(7569), 367-371.
[http://dx.doi.org/10.1038/nature15371] [PMID: 26381985]

[15] The University of British Columbia. *Poor air quality kills 5.5 million worldwide annually,* **2016**.

http://news.ubc.ca/2016/02/12/poor-air-quality-kills-5-5-million-worldwide-annually/

[16]   Kreyling, W.G.; Semmler-Behnke, M.; Takenaka, S.; Möller, W. Differences in the biokinetics of inhaled nano- *versus* micrometer-sized particles. *Acc. Chem. Res., 2013, 46*(3), 714-722.
       [http://dx.doi.org/10.1021/ar300043r] [PMID: 22980029]

[17]   Stone, V.; Miller, M.R.; Clift, M.J.; Elder, A.; Mills, N.L.; Møller, P.; Schins, R.P.; Vogel, U.; Kreyling, W.G.; Jensen, K.A.; Kuhlbusch, T.A.; Schwarze, P.E.; Hoet, P.; Pietroiusti, A.; De Vizcaya-Ruiz, A.; Baeza-Squiban, A.; Tran, C.L.; Cassee, F.R. Nanomaterials *vs* ambient ultrafine particles: an opportunity to exchange toxicology knowledge. *Environ. Health Perspect., 2016*.
       [http://dx.doi.org/10.1289/EHP424] [PMID: 27814243]

[18]   González-Maciel, A.; Reynoso-Robles, R.; Torres-Jardón, R. Combustion-derived nanoparticles in key brain target cells and organelles in young urbanites: Culprit Hidden in Plain Sight in Alzheimer's Disease Development. *J. Alzheimer's Disease, 2017*. in print

[19]   Bleck, B.; Tse, D.B.; Jaspers, I.; Curotto de Lafaille, M.A.; Reibman, J. Diesel exhaust particle-exposed human bronchial epithelial cells induce dendritic cell maturation. *J. Immunol., 2006, 176*(12), 7431-7437.
       [http://dx.doi.org/10.4049/jimmunol.176.12.7431] [PMID: 16751388]

[20]   Moon, K-Y.; Park, M-K.; Leikauf, G.D.; Park, C.S.; Jang, A.S. Diesel exhaust particle-induced airway responses are augmented in obese rats. *Int. J. Toxicol., 2014, 33*(1), 21-28.
       [http://dx.doi.org/10.1177/1091581813518355] [PMID: 24536021]

[21]   Pedata, P.; Stoeger, T.; Zimmermann, R.; Peters, A.; Oberdörster, G.; D'Anna, A. "Are we forgetting the smallest, sub 10 nm combustion generated particles?". *Part. Fibre Toxicol., 2015, 12*, 34.
       [http://dx.doi.org/10.1186/s12989-015-0107-3] [PMID: 26521024]

[22]   WHO. In: *Air Quality Guidelines, second ed.; - WHO Regional Office for Europe: Copenhagen, Denmark, 2000*. http://www.euro.who.int/__data/assets/pdf_file/0015/123081/AQG2ndEd_6_11 Platinum.PDF

[23]   Newhook, R.; Hirtle, H.; Byrne, K.; Meek, M.E. Releases from copper smelters and refineries and zinc plants in Canada: human health exposure and risk characterization. *Sci. Total Environ., 2003, 301*(1-3), 23-41.
       [http://dx.doi.org/10.1016/S0048-9697(02)00229-2] [PMID: 12493182]

[24]   Khatter, K. *Sudbury Human Health Risk Assessment Briefing, 2008*. http://play.psych.mun. ca/~mont/pdfs/Sudbury%20Human%20Health%20Risk%20Assessment.pdf

[25]   Malley, C.S.; Kuylenstierna, J.C.; Vallack, H.W.; Henze, D.K.; Blencowe, H.; Ashmore, M.R. Preterm birth associated with maternal fine particulate matter exposure: a global, regional and national assessment. *Environ. Int., 2017, 101*, 173-182.
       [http://dx.doi.org/10.1016/j.envint.2017.01.023] [PMID: 28196630]

[26]   Jones, D.; Huscher, J.; Myllyvirta, L. *Europe's dark cloud - How coal-burning countries are making their neighbours sick, CAN Europe, HEAL, Sandbag, WWF, 2016*. http://www.caneurope. org/docman/position-papers-and-research/coal-2/2924-report-europe-s-dark-cloud-how-coal-burning-countries-are-making-their-neighbours-sick/file

[27]   Peters, R.; Peters, J.; Booth, A.; Mudway, I. Is air pollution associated with increased risk of cognitive decline? A systematic review. *Age Ageing, 2015, 44*(5), 755-760.
       [http://dx.doi.org/10.1093/ageing/afv087] [PMID: 26188335]

[28]   US-EPA. *Particulate Matter (PM) Standards Revision, 2006*; Vol. 2006, US Government, **2006**.

[29]   Landrigan, P.J.; Fuller, R.; Acosta, N.J. The Lancet Commission on pollution and health. *Lancet,* in press
       [http://dx.doi.org/10.1016/S0140-6736(17)32345-0]

[30]   Kosatsky, K. *The 2003 European heat waves, 2005*. http://www.eurosurveillance.org/ViewArticle. aspx?ArticleId=552

[http://dx.doi.org/10.2807/esm.10.07.00552-en]

[31]   WHO Europe. *Urban green spaces and health,* **2016.** http://www.euro.who.int/__data/assets/pdf_file/0005/321971/Urban-green-spaces-and-health-review-evidence.pdf?ua=1

[32]   Fischer, C.; Kurganskaya, I.; Schäfer, T. Variability of crystal surface reactivity: What do we know? *Appl. Geochem.,* **2014,** *43,* 132-157.
[http://dx.doi.org/10.1016/j.apgeochem.2014.02.002]

[33]   Radkau, J. *Das Zeitalter der Nervosität: Deutschland zwischen Bismarck und Hitler,* **1998.** https://www.hanser-literaturverlage.de/buch/das-zeitalter-der-nervositaet/978-3-446-25355-1/

[34]   Lasfargues, G.; Vol, S.; Cacès, E.; Le Clésiau, H.; Lecomte, P.; Tichet, J. Relations among night work, dietary habits, biological measure, and health status. *Int. J. Behav. Med.,* **1996,** *3*(2), 123-134.
[http://dx.doi.org/10.1207/s15327558ijbm0302_3] [PMID: 16250759]

[35]   Baron, K.G.; Reid, K.J. Circadian misalignment and health. *Int. Rev. Psychiatry,* **2014,** *26*(2), 139-154.
[http://dx.doi.org/10.3109/09540261.2014.911149] [PMID: 24892891]

[36]   Crawford, M. *The World Beyond Your Head: On Becoming an Individual in an Age of Distraction*; Farrar, Straus and Giroux: New York, **2015.**

[37]   Konersmann, R. *Die Unruhe der Welt*; Fischer Verlag: Frankfurt a. M., **2015.**

[38]   Rosa, H. *Social Acceleration. A New Theory of Modernity*; Columbia University Press: New York, **2015.**

[39]   Hafner, M.; Stepanek, M.; Taylor, J. *Why Sleep Matters - The Economic Costs of Insufficient Sleep. A Cross-Country Comparative Analysis*; RAND Corporation: Santa Monica, Calif., **2016.**
[http://dx.doi.org/10.7249/RR1791]

[40]   Swerdlow, A.J.; Feychting, M.; Green, A.C.; Leeka Kheifets, L.K.; Savitz, D.A. International Commission for Non-Ionizing Radiation Protection Standing Committee on Epidemiology. Mobile phones, brain tumors, and the interphone study: where are we now? *Environ. Health Perspect.,* **2011,** *119*(11), 1534-1538.
[http://dx.doi.org/10.1289/ehp.1103693] [PMID: 22171384]

[41]   International Agency for Research on Cancer. *WHO. IARC classifies radiofrequency electromagnetic fields as possibly carcinogenic to humans,* **2011.** http://www.iarc.fr/en/media-centre/pr/2011/pdfs/pr208_E.pdf

[42]   Sienkiewicz, Z.; van Rongen, E.; Croft, R.; Ziegelberger, G.; Veyret, B. A closer look at the threshold of thermal damage: Workshop report by an ICNIRP task group. *Health Phys.,* **2016,** *111*(3), 300-306.
[http://dx.doi.org/10.1097/HP.0000000000000539] [PMID: 27472755]

[43]   Cinzano, P.; Falchi, F.; Elvidge, C.D. The first world atlas of the artificial night sky brightness. *Mon. Not. R. Astron. Soc.,* **2001,** *328*(3), 689-707.
[http://dx.doi.org/10.1046/j.1365-8711.2001.04882.x]

[44]   Bustos, R.; Rubio, M.; Otárola, A. Parque Astronómico de Atacama: An ideal site for millimeter, submillimeter, and mid-infrared astronomy. *Publ. Astron. Soc. Pac.,* **2014,** *126,* 1126-1132.
[http://dx.doi.org/10.1086/679330]

[45]   Wenzlaff, A. "Made in Germany" - 125 Jahre Automobil. *Münchner Statistik,* **2011,** *4,* 17-36.

[46]   WHO. *Registered vehicles - data by country. Global Health Observatory data repository,* **2016.** http://apps.who.int/gho/data/view.main.51210

[47]   Allgemeiner Deutscher Automobilclub. *Jeden Tag mehr als 1900 Staus,* **2017.** https://www.adac.de/infotestrat/adac-im-einsatz/motorwelt/staubilanz_2016.aspx

[48]   Gössling, S.; Choi, A.S. Transport transitions in Copenhagen: Comparing the cost of cars and bicycles. *Ecol. Econ.,* **2015,** *113,* 106-113.
[http://dx.doi.org/10.1016/j.ecolecon.2015.03.006]

[49]   Nenonen, N.; Saarela, K.L.; Takala, J. *Global estimates of occupational accidents and work-related illnesses 2014,* **2014**. https://www.wsh-institute.sg/files/wshi/upload/cms/file/Global%20Estimates%20of%20Occupational%20Accidents%20and%20Work-related%20Illness%202014.pdf

[50]   WHO. *Noise, data and statistics,* **2016**. http://www.euro.who.int/en/health-topics/ environment-an--health/noise/data-and-statistics

[51]   Fritschi, L.; Brown, A.L.; Kim, R. *Burden of disease from environmental noise - Quantification of healthy life years lost in Europe, World Health Organization, Europe,* **2011**. http://www.euro.who.int/__data/assets/pdf_file/0008/136466/e94888.pdf?ua=1

[52]   Lucas, R.; McMichael, T.; Smith, W. *Environmental Burden of Disease Series*; Prüss-Üstün, A.; Zeeb, H.; Mathers, C., Eds.; WHO: Geneva, **2006**.

[53]   Institute for Health Metrics and Evaluation. *Rethinking development and health: Findings from the Global Burden of Disease Study,* **2016**. http://www.healthdata.org/sites/default/files/files/images/news_release/2016/IHME_GBD2015.pdf

[54]   Tang, L.; Ii, R.; Tokimatsu, K. Development of human health damage factors related to CO_2 emissions by considering future socioeconomic scenarios. *Int. J. Life Cycle Assess.,* **2015**, 1-12.

[55]   International Research Institute for Climate and Society. http://iri.columbia.edu/ our-expertise/publi--health/

[56]   National Intelligence Council. *Implications for US National Security of Anticipated Climate Change,* **2016**. https://www.dni.gov/files/documents/Newsroom/Reports%20and%20Pubs/Implications_for_US_National_Security_of_Anticipated_Climate_Change.pdf

[57]   Vezzulli, L.; Grande, C.; Reid, P.C.; Hélaouët, P.; Edwards, M.; Höfle, M.G.; Brettar, I.; Colwell, R.R.; Pruzzo, C. Climate influence on Vibrio and associated human diseases during the past half-century in the coastal North Atlantic. *Proc. Natl. Acad. Sci. USA,* **2016**, *113*(34), E5062-E5071. [http://dx.doi.org/10.1073/pnas.1609157113] [PMID: 27503882]

[58]   Williams, P.D. Increased light, moderate, and severe clear-air turbulence in response to climate change. *Adv. Atmos. Sci.,* **2017**, *34*(5), 576-586. [http://dx.doi.org/10.1007/s00376-017-6268-2]

[59]   Herring, S.C.; Hoerling, M.P.; Kossin, J.P. *Explaining extreme events of 2014 - From A Climate Perspective. Special Supplement to the Bulletin of the American Meteorological Society,* **2015**. http://journals.ametsoc.org/doi/pdf/10.1175/BAMS-D-15-00157.1

[60]   Brink, C.; van Grinsven, H. *The European Nitrogen Assessment*; Sutton, M.A.; Howard, C.M.; Erisman, J.W., Eds.; Cambridge University Press: EU, **2011**.

[61]   Pimentel, D.; Burgess, M. *Integrated Pest Management, Pesticide Problems*; Pimentel, D.; Peshin, R., Eds.; Springer: Heidelberg, New York, Dordrecht, London, **2014**, Vol. 3, pp. 47-73. [http://dx.doi.org/10.1007/978-94-007-7796-5_2]

[62]   Chaufan, G.; Coalova, I.; Ríos de Molina, M.C. Glyphosate commercial formulation causes cytotoxicity, oxidative effects, and apoptosis on human cells: differences with its active ingredient. *Int. J. Toxicol.,* **2014**, *33*(1), 29-38. [http://dx.doi.org/10.1177/1091581813517906] [PMID: 24434723]

[63]   Stokstad, E. *European Commission gives controversial weed killer a last-minute reprieve,* **2016**. [http://dx.doi.org/10.1126/science.aag0630]

[64]   Rabotyagov, S.S.; Campbell, T.D.; White, M.; Arnold, J.G.; Atwood, J.; Norfleet, M.L.; Kling, C.L.; Gassman, P.W.; Valcu, A.; Richardson, J.; Turner, R.E.; Rabalais, N.N. Cost-effective targeting of conservation investments to reduce the northern Gulf of Mexico hypoxic zone. *Proc. Natl. Acad. Sci. USA,* **2014**, *111*(52), 18530-18535. [http://dx.doi.org/10.1073/pnas.1405837111] [PMID: 25512489]

[65]   Harada, M. Minamata disease: methylmercury poisoning in Japan caused by environmental pollution. *Crit. Rev. Toxicol.,* **1995**, *25*(1), 1-24.
[http://dx.doi.org/10.3109/10408449509089885] [PMID: 7734058]

[66]   Aschner, M.; Aschner, J.L.; Kimelberg, H.K. *The Vulnerable Brain and Environmental Risks; Toxins in Food*; Isaacson, R.L.; Jensen, K.F., Eds.; Springer: New York, **1992**, Vol. 2, pp. 3-17.
[http://dx.doi.org/10.1007/978-1-4615-3330-6_1]

[67]   Nedellec, V.; Rabl, A. Costs of health damage from atmospheric emissions of toxic metals: Part 2 - Analysis for Mercury and Lead. *Risk Anal.,* **2016**, *36*(11), 2096-2104.
[http://dx.doi.org/10.1111/risa.12598] [PMID: 26992113]

[68]   Church, J.A.; White, N.J. Sea-Level Rise from the Late 19th to the Early 21st Century. *Surv. Geophys.,* **2011**, *32*, 585.
[http://dx.doi.org/10.1007/s10712-011-9119-1]

[69]   Jevrejeva, S.; Jackson, L.P.; Riva, R.E.; Grinsted, A.; Moore, J.C. Coastal sea level rise with warming above 2 °C. *Proc. Natl. Acad. Sci. USA,* **2016**, *113*(47), 13342-13347.
[http://dx.doi.org/10.1073/pnas.1605312113] [PMID: 27821743]

[70]   Wilkinson, B.H.; McElroy, B.J. The impact of humans on continental erosion and sedimentation. *Geol. Soc. Am. Bull.,* **2006**, *119*(1-2), 140-156.
[http://dx.doi.org/10.1130/B25899.1]

[71]   Lam, D. How the world survived the population bomb: lessons from 50 years of extraordinary demographic history. *Demography,* **2011**, *48*(4), 1231-1262.
[http://dx.doi.org/10.1007/s13524-011-0070-z] [PMID: 22005884]

[72]   Becker, S. Has the world really survived the population bomb? (Commentary on "how the world survived the population bomb: lessons from 50 years of extraordinary demographic history"). *Demography,* **2013**, *50*(6), 2173-2181.
[http://dx.doi.org/10.1007/s13524-013-0236-y] [PMID: 23955197]

[73]   Medek, D.E.; Schwartz, J.; Myers, S.S. Estimated effects of future atmospheric CO_2 concentrations on protein intake and the risk of protein deficiency by country and region. *Environ. Health Perspect.,* **2017**, *125*(8), 087002. http://www.dx.doi.org/10.1289/EHP41
[http://dx.doi.org/10.1289/EHP41] [PMID: 28885977]

[74]   Stockholm World Water Week. *Pollution or Prosperity?,* **2016**. http://programme.worldwaterweek. org/event/5601

[75]   Lübbert, C.; Baars, C.; Dayakar, A. Environmental pollution with antimicrobial agents from bulk drug manufacturing industries in Hyderabad, South India, is associated with dissemination of extended-spectrum beta-lactamase and carbapenemase-producing pathogens. *Infection,* **2017**. in press

[76]   Lu, Y.; Song, S.; Wang, R.; Liu, Z.; Meng, J.; Sweetman, A.J.; Jenkins, A.; Ferrier, R.C.; Li, H.; Luo, W.; Wang, T. Impacts of soil and water pollution on food safety and health risks in China. *Environ. Int.,* **2015**, *77*, 5-15.
[http://dx.doi.org/10.1016/j.envint.2014.12.010] [PMID: 25603422]

[77]   Yi, J. *More Than 80 Percent of China's Groundwater Polluted,* **2016**. http://en.tuidang.org/ news/environment/2016/04/more-than-80-percent-of-chinas-groundwater-polluted.html

[78]   WHO. *Obesity,* **2004**. http://www.who.int/nutrition/publications/obesity/WHO_TRS_894/en/

[79]   WHO. *Tobacco,* **2015**. http://www.who.int/tobacco/global_report/2015/en/

[80]   WHO. *Global Information System on Alcohol and Health,* **2016**. http://www.who.int/gho/alcohol/en/

[81]   Case, A.; Deaton, A. Rising morbidity and mortality in midlife among white non-Hispanic Americans in the 21st century. *Proc. Natl. Acad. Sci. USA,* **2015**, *112*(49), 15078-15083.
[http://dx.doi.org/10.1073/pnas.1518393112] [PMID: 26575631]

[82] Murray, C.J.; Afshin, A. Forouzanfar; M. H. Health Effects of Overweight and Obesity in 195 Countries over 25 Years. *N. Engl. J. Med.,* **2017**.
[http://dx.doi.org/10.1056/NEJMoa1614362]

[83] Brown, V.; Moodie, M.; Baur, L.; Wen, L.M.; Hayes, A. The high cost of obesity in Australian pre-schoolers. *Aust. N. Z. J. Public Health,* **2017**, *41*(3), 323-324.
[http://dx.doi.org/10.1111/1753-6405.12628] [PMID: 28110505]

[84] Hoshino, T.; Hoshino, A.; Nishino, J. Assessment of associations between ischaemic attacks in patients with type 2 diabetes mellitus and air concentrations of particulate matter <2.5 μm. *J. Int. Med. Res.,* **2016**, *44*(3), 639-655.
[http://dx.doi.org/10.1177/0300060516631702] [PMID: 27020595]

[85] Van Boeckel, T.P.; Brower, C.; Gilbert, M.; Grenfell, B.T.; Levin, S.A.; Robinson, T.P.; Teillant, A.; Laxminarayan, R. Global trends in antimicrobial use in food animals. *Proc. Natl. Acad. Sci. USA,* **2015**, *112*(18), 5649-5654.
[http://dx.doi.org/10.1073/pnas.1503141112] [PMID: 25792457]

[86] Solecki, R.; Kortenkamp, A.; Bergman, Å. *Scientific principles for the identification of endocrine disrupting chemicals - a consensus statement - Outcome of an international expert meeting organized by the German Federal Institute for Risk Assessment (BfR),* **2016**. http://www.bfr.bund.de/ cm/349/scientific-principles-for-the-identification-of-endocrine-disrupting-chemicals-a-consensus-statement.pdf

[87] Bergman, Å.; Heindel, J.J.; Jobling, S. *State of the science of endocrine disrupting chemicals 2012,* **2013**. http://www.unep.org/pdf/WHO_HSE_PHE_IHE_2013.1_eng.pdf

[88] US Department of Health & Human Services, Center of Disease Control and Prevention. *National Report on Human Exposure to Environmental Chemicals, Updated Tables,* **2015**. http://www.cdc.gov/ biomonitoring/pdf/FourthReport_UpdatedTables_Feb2015.pdf

[89] Braun, J.M.; Chen, A.; Romano, M.E.; Calafat, A.M.; Webster, G.M.; Yolton, K.; Lanphear, B.P. Prenatal perfluoroalkyl substance exposure and child adiposity at 8 years of age: The HOME study. *Obesity (Silver Spring),* **2016**, *24*(1), 231-237.
[http://dx.doi.org/10.1002/oby.21258] [PMID: 26554535]

[90] Mcbride, W.G. Thalidomide and congenital abnormalities. *Lancet,* **1961**, *278*(7216), 1358.
[http://dx.doi.org/10.1016/S0140-6736(61)90927-8]

[91] Sjöström, H.; Nilsson, R. *Thalidomide and the Power of the Drug Companies*; Penguin Books: UK, **1972**.

[92] Levi, P. *So war Auschwitz. Zeugnisse 1945-1986. Mit L. de Benedetti. Scarpa, D*; Levi, F., Ed.; Hanser: München, **2017**.

[93] US Strategic Bombing Survey. *Chapter II. The Effects of the Atomic Bombings of Hiroshima and Nagasaki*; General effects, 1. Casualties: Washington, **1946**.

[94] Cologne, J.B.; Preston, D.L. Longevity of atomic-bomb survivors. *Lancet,* **2000**, *356*(9226), 303-307.
[http://dx.doi.org/10.1016/S0140-6736(00)02506-X] [PMID: 11071186]

[95] DAtF Deutsches Atomforum e. *V. Der Reaktorunfall in Tschernobyl - Unfallursachen, Unfallfolgen und deren Bewältigung; Sicherung und Entsorgung des Kernkraftwerks Tschernobyl, Berlin,* **2011**. http://www.kernenergie.de/kernenergie-wAssets/docs/service/025reaktorunfall_tschernobyl2011.pdf

[96] Koo, Y-H.; Yang, Y-S.; Song, K-W. Radioactivity release from the Fukushima accident and its consequences: A review. *Prog. Nucl. Energy,* **2014**, *74*, 61-70.
[http://dx.doi.org/10.1016/j.pnucene.2014.02.013]

[97] Suzuki, M.; Terada, H.; Unno, N.; Yamaguchi, I.; Kunugita, N.; Minakami, H. Radioactive cesium (^{134}Cs and ^{137}Cs) content in human placenta after the Fukushima nuclear power plant accident. *J. Obstet. Gynaecol. Res.,* **2013**, *39*(9), 1406-1410.

[http://dx.doi.org/10.1111/jog.12071] [PMID: 23815637]

[98] Broughton, E. The Bhopal disaster and its aftermath: a review. *Environ. Health,* **2005**, *4*(1), 6.
[http://dx.doi.org/10.1186/1476-069X-4-6] [PMID: 15882472]

[99] Wikipedia. *List of industrial disasters,* **2017**. https://en.wikipedia.org/wiki/List_of _industrial_disasters

[100] Mihailidou, E.K.; Antoniadis, K.D.; Assael, M.J. The 319 major industrial accidents since 1917. *IReChE,* **2012**, *4*(6), 529-540.

[101] Beaulieu, M. *Post-mining Landscapes. Reclamation, Ecology, Nature Preservation and Socio-Economy in Practice*; Xylander, W.E.R. Peckiana, **2004**.

[102] Kurttio, P.; Pukkala, E.; Ilus, T.; Rahola, T.; Auvinen, A. Radiation doses from global fallout and cancer incidence among reindeer herders and Sami in Northern Finland. *Occup. Environ. Med.,* **2010**, *67*(11), 737-743.
[http://dx.doi.org/10.1136/oem.2009.048652] [PMID: 20798008]

[103] Stone, D.P. *The Changing Arctic Environment - The Arctic Messenger*; Cambridge University Press: Cambridge, **2015**.
[http://dx.doi.org/10.1017/CBO9781316146705]

[104] Kallenborn, R., Ed. *Implications and Consequences of Anthropogenic Pollution in Polar Environments*; Springer: Berlin, Heidelberg, **2016**.
[http://dx.doi.org/10.1007/978-3-642-12315-3]

[105] CDC Radiation, Centers for Disease Control and Prevention. Feasibility Study of Weapons Test Fallout, Volume 1, Technical Report, Chapter 2, *Fallout from nuclear weapons,* **2014**. http://www.cdc.gov/nceh/radiation/fallout/feasibilitystudy/technical_vol_1_chapter_2.pdf

[106] Casey, J.A.; Savitz, D.A.; Rasmussen, S.G.; Ogburn, E.L.; Pollak, J.; Mercer, D.G.; Schwartz, B.S. Unconventional Natural Gas Development and Birth Outcomes in Pennsylvania, USA. *Epidemiology,* **2016**, *27*(2), 163-172.
[PMID: 26426945]

[107] Reap, E. The risk of hydraulic fracturing on public health in the UK and the UK's fracking legislation. *Environ. Sci. Eur.,* **2015**, *27*(1), 27.
[http://dx.doi.org/10.1186/s12302-015-0059-0] [PMID: 27752428]

[108] D-Waste. *Waste Atlas, Global Waste Generation Clock,* **2016**. http://www.atlas.d-waste.com/

[109] Lavigne, F.; Wassmer, P.; Gomez, C. The 21 February 2005, catastrophic waste avalanche at Leuwigajah dumpsite, Bandung, Indonesia. *Geoenviron. Disasters,* **2014**, *1*, 10.
[http://dx.doi.org/10.1186/s40677-014-0010-5]

[110] Africa Review. *Ethiopia rubbish dump landslide death toll soars to 115,* **2017**. http://www.africareview.com/news/Ethiopia-rubbish-dump-landslide-death-toll-/979180-3851950-107ybg0z/

[111] Flor, V. Mining in European history and its impacts on environment an human societies *Proc. for the 2nd Mining in European History Conference of the FZ HiMAT, Session III, Societal Interaction and Ecology,* **2013**, , pp. 133-138.

[112] Hvistendahl, M. *China's Three Gorges Dam: An Environmental Catastrophe?,* **2008**. https://www.scientificamerican.com/article/chinas-three-gorges-dam-disaster/

[113] Tilt, B.; Gerkey, D. Dams and population displacement on China's Upper Mekong River: Implications for social capital and social-ecological resilience. *Glob. Environ. Change,* **2016**, *36*, 153-162.
[http://dx.doi.org/10.1016/j.gloenvcha.2015.11.008]

[114] Orr, S.; Pittock, J.; Chapagain, A. Dams on the Mekong River: Lost fish protein and the implications for land and water resources. *Glob. Environ. Change,* **2012**, *22*(4), 925-932.
[http://dx.doi.org/10.1016/j.gloenvcha.2012.06.002]

[115] Whish-Wilson, P. The Aral Sea environmental health crisis. *J. Rural Remote Environ. Health,* **2002**, *1*(2), 29-34.

[116] Davies, R.; Foulger, G.; Bindley, A. Induced seismicity and hydraulic fracturing for the recovery of hydrocarbons. *Mar. Pet. Geol.,* **2013**, *45*, 171-185.
[http://dx.doi.org/10.1016/j.marpetgeo.2013.03.016]

[117] Dahm, T.; Cesca, S.; Hainzl, S. Discrimination between induced, triggered, and natural earthquakes close to hydrocarbon reservoirs: A probabilistic approach based on the modeling of depletion-induced stress changes and seismological source parameters. *J. Geophys. Res.,* **2015**, *120*(4), 2491-2509.
[http://dx.doi.org/10.1002/2014JB011778]

[118] Tingay, M.R.; Rudolph, M.L.; Manga, M. Initiation of the Lusi mudflow disaster. *Nat. Geosci.,* **2015**, *8*(7), 493-494.
[http://dx.doi.org/10.1038/ngeo2472]

[119] López-Quiroz, P.; Doin, M-P.; Tupin, F. Time series analysis of Mexico City subsidence constrained by radar interferometry. *J. Appl. Geophys.,* **2009**, *69*, 1-15.
[http://dx.doi.org/10.1016/j.jappgeo.2009.02.006]

[120] Petley, D. Global patterns of loss of life from landslides. *Geology,* **2012**, *40*(10), 927-930.
[http://dx.doi.org/10.1130/G33217.1]

[121] Perkins, S. Death toll from landslides vastly underestimated. *Nature,* **2012**.
[http://dx.doi.org/10.1038/nature.2012.11140]

[122] Klose, H.; Highland, L.; Damm, B. *Landslide Science for a Safer Geoenvironment: Methods of Landslide Studies*; Sassa, K.; Canuti, P.; Yin, Y., Eds.; Vol. 2, Springer: Berlin, **2014**.

[123] Van Horssen, J. *A town called Asbestos - Environmental contamination, health and resilience in a resource community*; UBC Press: Vancouver, Toronto, **2016**. http://www.ubcpress.ca/search/title_book.asp?BookID=299174603

[124] CCS, Canadian Cancer Society. *Asbestos - Our position,* **2016**. http://www.cancer.ca/en/ get-involved/take-action/what-we-are-doing/asbestos/?region=on

[125] Kaufman, A.S.; Zhou, X.; Reynolds, M.R.; Kaufman, N.L.; Green, G.P.; Weiss, L.G. The possible societal impact of the decrease in U.S. blood lead levels on adult IQ. *Environ. Res.,* **2014**, *132*, 413-420.
[http://dx.doi.org/10.1016/j.envres.2014.04.015] [PMID: 24853978]

[126] Patiño Douce, A.E. Metallic Mineral Resources in the Twenty-First Century: II. Constraints on Future Supply. *Nat. Resour. Res.,* **2016**, *25*(1), 97-124.
[http://dx.doi.org/10.1007/s11053-015-9265-0]

[127] Bishop, C., Ed. *The encyclopedia of weapons of world war II*; Barnes & Noble: New York, **2002**.

[128] Filchner, G.; Gundler, B. *Deutsches Museum*; Flugwerft Schleißheim: München, **2005**.

[129] White, M. *Atrocitology: Humanity's 100 deadliest achievements*; Canongate Books Ltd.: UK, **2011**.

[130] Stockholm International Peace Research Institute. *Arms production,* **2016**. https://www.sipri.org/ research/armament-and-disarmament/arms-transfers-and-military-spending/arms-production

[131] Perlo-Freeman, S. *SIPRI extended milex database beta,* **2016**. https://www.sipri.org/databases/milex

[132] Shively, S.B.; Horkayne-Szakaly, I.; Jones, R.V.; Kelly, J.P.; Armstrong, R.C.; Perl, D.P. Characterisation of interface astroglial scarring in the human brain after blast exposure: a post-mortem case series. *Lancet Neurol.,* **2016**, *15*(9), 944-953.
[http://dx.doi.org/10.1016/S1474-4422(16)30057-6] [PMID: 27291520]

[133] Kristensen, H.M.; Norris, R.S. United States nuclear forces. *B. Atom. Sci.,* **2016**, *72*(2), 63-73.

[134] Kristensen, H.M.; Norris, R.S. Russian nuclear forces. *B. Atom. Sci.,* **2016**, *72*(3), 125-134.

[135] Schleussner, C-F.; Donges, J.F.; Donner, R.V.; Schellnhuber, H.J. Armed-conflict risks enhanced by climate-related disasters in ethnically fractionalized countries. *Proc. Natl. Acad. Sci. USA,* **2016**, *113*(33), 9216-9221.
[http://dx.doi.org/10.1073/pnas.1601611113] [PMID: 27457927]

[136] News, B.B. *Guns in the US: The statistics behind the violence,* **2016**. http://www.bbc.com/news/world-us-canada-34996604

[137] National Center for Health Statistics. *10 Leading Causes of Injury Deaths by Age Group Highlighting Violence-Related Injury Deaths, United States,* **2010**. http://www.cdc.gov/injury/wisqars/pdf/10LCID_Violence_Related_Injury_Deaths_2010-a.pdf

[138] Curry, T. *Gun control offers no cure-all in America,* **2009**. http://nbcpolitics.nbcnews.com/_news/2012/12/18/15977143-gun-control-offers-no-cure-all-in-america?lite

[139] Gómez, J.M.; Verdú, M.; González-Megías, A.; Méndez, M. The phylogenetic roots of human lethal violence. *Nature,* **2016**, *538*(7624), 233-237.
[http://dx.doi.org/10.1038/nature19758] [PMID: 27680701]

[140] Orwell, G. *Nineteen Eighty-Four*; Oceania: London, **1949**.

[141] Huxley, A. *Brave New World*; Chatto & Windus: UK, **1932**.

[142] Lundqvist, N. *The Unit*; Other Press LLC: New York, USA, **2006**.

[143] Romare, M.; Dahllöf, L. *The Life Cycle Energy Consumption and Greenhouse Gas Emissions from Lithium-Ion Batteries,* **2017**. http://www.ivl.se/download/18.5922281715bdaebede9559/14960462 18976/C243+The+life+cycle+energy+consumption+and+CO2+emissions+from+lithium+ion+batteries+.pdf

[144] Herminghaus, H. *CO₂ Rechner: CO₂ Emissionen eines PKW,* **2011**. http://www.co2-emissionen-vergleichen.de/verkehr/PKW/CO2-Emissionen-PKW.html

[145] Jaskula, B.W. *Lithium, Mineral Commodity Summaries. United States Geological Survey,* **2015**. https://minerals.usgs.gov/minerals/pubs/commodity/lithium/mcs-2015-lithi.pdf

[146] Shedd, K.B. **2016**. https://minerals.usgs.gov/minerals/pubs/commodity/cobalt/mcs-2016-cobal.pdf

[147] Hunt, T. *Is There Enough Lithium to Maintain the Growth of the Lithium-Ion Battery Market?,* **2015**. https://www.greentechmedia.com/articles/read/is-there-enough-lithium-to-maintain-the-grow-h-of-the-lithium-ion-battery-m

[148] Reichl, C.; Schatz, M.; Zsak, G. *Minerals Production,* **2016**. http://www.wmc.org.pl/sites/default/files/WMD2016.pdf

[149] Vikström, H.; Davidsson, S.; Höök, M. Lithium availability and future production outlooks. *Appl. Energy,* **2013**, *110*, 252-266.
[http://dx.doi.org/10.1016/j.apenergy.2013.04.005]

[150] Haagen-Smit, A.J. The air pollution problem in Los Angeles. *Eng. Sci.,* **1950**, *XIV*, 7-13.

[151] Calvert, J.G.; Orlando, J.J.; Stockwell, W.R. *The mechanisms of reactions influencing atmospheric ozone*; Oxford University Press: UK, **2015**.

[152] Calabrese, L.; Fleischer, A.B. Thalidomide: current and potential clinical applications. *Am. J. Med.,* **2000**, *108*(6), 487-495.
[http://dx.doi.org/10.1016/S0002-9343(99)00408-8] [PMID: 10781782]

[153] Wertz, R.R. *Special Report, Three Gorges Dam,* **2011**. http://www.ibiblio.org/chinesehistory/contents/07spe/specrep01.html#Quick%20Facts

[154] Hoover, H.C.; Hoover, L.H. *De re metallica,* **1950**. https://archive.org/stream/deremetallica50agri/deremetallica50agri_djvu.txt

Causes of the Environmental Crisis and Proposals for its Mitigation

Abstract: The transformations caused by industrialisation are ambivalent: Progress in technology, medicine, science, human well being, mobility, food security, *etc.* contrasts with the degradation of the environment and implied health impairments for human beings. The latter has occurred, because humans created many more sources than sinks and artificial material flows exceeded the natural ones; in that way, waste and toxic matter accumulated in the biosphere. Genius ideas of scientists and inventors, as well as treasures generated by nature inside the Earth were utilised and exploited economically. Innate behaviour, ideologies, traditional economic systems, and permanent growth marginalised ecological concerns. Insufficient technological diversity, population growth, and energy subsidies entailed dependencies and the risk of resource depletion. Recommendations to avoid further aggravation of the ecological crisis are: Dematerialisation; terminating the use of fossil energy carriers; ecological, social, and humanitarian concerns must be equally entitled beside economic concerns; realistic pricing of products and fair trade; shift to economic systems and human reproductive societies characterised by sufficiency; fostering of positive human creativity and intellect, as well as fine arts, aesthetics, and morals.

Keywords: Creativity, Dependencies, Dematerialisation, Ecology, Economy, Environmental degradation, Exploitation, Extraction, Fairness, Fine arts, Human development, Human spirit, Humanities, Ideologies, Industrialisation, Innate behaviour, Material flows, Morals, Risk society, Subsidies, Sufficiency.

Positive cultural and civil progress is on the way, if maintenance of the standards in democracy, morals, and humaneness; transparency; balance and pluralism; the pursuit to minimise pain and to maximise pleasure for a maximum of number of persons [1, 2]; and the preservation of nature, biodiversity, and sustainability are realised [3].

It is obvious that, despite persisting large disparities, the averaged quality and standard of living of mankind has made progress during industrialisation. However, the drawbacks and adverse effects of the transformations impairing global ecosystems and life were detailed in part IV. The problems of sustainability and resource depletion abound. The fact that healthiness and the functioning of ecosystem services have been recognised as essential for the survival of humans must result in appreciation of the worth of ecology as at least equivalent to that of economics.

By utilising fossil carbon and hydrocarbons at the beginning of industrialisation, man surmounted the physical, chemical, and energetic constraints of the only previously available recent energy carriers produced in the biosphere by photosynthesis, and in this way, also impaired the ecological equilibria established in natural material and energetic cycles. That is the balanced natural cycles of CO_2, NH_3, NO_x, CH_4, S, P, Fe, Cu, *etc.* were brought in disequilibrium, because many more sources than sinks were opened, so that no, or only a few, closed artificial loops/cycles exist and natural sinks have become overstrained. As a consequence, intentionally and unintentionally produced waste and exhaust gases accumulate as poisons in the environment. In addition, man has also overstrained flora and fauna, as well as himself, which had developed over geological times under natural constraints. Many of the species are unable to adapt to human made transformations in the ecospheres.

In the meantime, there exists major consensus in the scientific community that recent human agencies and effects have caused a global ecological crisis, which ultimately led to the concept of the Anthropocene. But there is less consensus that the knowledge and experience of the humanities should be fully included and consulted to assess and evaluate the political, social, and environmental consequences of measures and innovations in technology, science, and economics [4].

The following facts, influences, ideologies, and innate human behavioural dispositions, as well as the properties and momenta of implemented economic systems, have contributed to the present ecological crisis:

- Repeated de-deification of nature has impaired previous respectful interrelations with it. The spread of Christianity, which caused suppression/extinction of original natural religions of tribes and spread the idea of Dominium terrae[1], powered the human instinct of self-preservation and neglect of effects on the ecospheres. De-deification occurred a second time with the beginning of the Age of Enlightenment, during which rationalism, free will, freedom, and individualism spread and were later incorporated into the foundations of capitalistic societies.
  In contrast to the respectful understanding of nature, seen as infinite and mysterious realm during the Romantic Period, a conception spread during Industrial Age that machines could redeem mankind and that the self-purification potential of nature is infinite. A naïve faith in technology spread [5].

Example: *"The creation of the superhighway as magnificent monument, prodigy and work of art .... The industrial society took in that an oath on perpetual growth"* [6].

- Boundlessness: Plato's pleonexia: unlimited accumulation of matter, affluence, and fortune[2]. Popular winged words: *"I want it all, and I want it now"* (Brian May & Freddie Mercury, Queen 1989, album The Miracle). *"Greed is good"* (Gordon Gekko in the film "Wall Street 1987"). It is problematic that global distribution of wealth levels developed towards an inequality maximum: In January 2016, the assets of the 62 richest persons, consisting predominantly of industrial and economic leaders, amounted to 1.76 trillion USD, which was equal to the same amount of property possessed by the poorest half of the world population combined [7].

- Exploitation: economic utilisation of workers and slaves, of treasures generated by nature inside the earth, and of genius ideas from the brilliant minds of inventors, scientists, and technicians. The presence of common natural resources and their exploitation by boundlessly growing economies are argued to be the main reasons for global environmental problems and climate change [3, 8].

- Insufficient comprehension of deep history [9]: the huge amount of time elapsed in earth history (4.5×10^9 a) and thereon evolved enormous complexity of nature and its ecosystems, resulting from human sense of time, human life cycles ($< 10^2$ a) and only a few hundred years of scientific development.

- A causal context has been recognised between humanity's supreme biological success during its evolutionary history and the global environmental demise: The archaic process of natural selection and adaptation favours actions that are immediately useful, profitable, and advantageous, regardless of long-term consequences. These problematic properties are constituents of the human genome and they determine tribal and group collaboration, but also irrational competition and hostilities that were once useful but now ruinously dysfunctional. However, human intellect, self-reflection, and perception enable us to deliberately oppose that inherited archaic behaviour and to struggle for types of behaviour providing sustainable advantage and benefit, because these properties are not established in the human genome [10].

- The principle of permanent economic growth.

- Superbia, arrogance, pride: *"eritis sicut deus scientes bonum et malum"* (Genesis 3:5). *"Our imagined self-importance, our delusion that we have privileged position in the universe.."* [11]. *"... cultivation of plants and taming and breeding of animals, controlled use of fire, invention of gas firing motor, discovery of nuclear fission etc.* gave the illusion of human's supremacy over nature, although man shares 98.5% of its genes with primates." [12].

- Egoism, indolence, laziness, habituation, irresponsibility, manipulability, indifference.

- The problems and risks of division of labour and processes, of dissipation of responsibilities, as well as of the vanishing of reality [13], alienation from work by heteronomy, and uprooting from nature.
- Ruinous economic competition, resulting in too cheap products [14].
- Competitive growth in cooperating societies.

The above-mentioned properties and behaviours resulted in the environmental crisis and provoked the concept of ecology and the idea of its importance for the well-being of humans. Therefore, parallel to the ideas of individualism and egoism, the idea of global interdependence is winning relevance, emphasising the infinitely complex web of life, matter, and energy, which developed over 4 Ga, but was excluded from the economy and trade originating in the late Stone Age [15]. The spread of the concept of ecology to the public, however, started not earlier than in 1866 thanks to Ernest Haeckel[3] [5].

A future problem is that new inventions, detections, and their technological applications open up infinitely further possibilities and higher magnitudes of deterministic economic modes of life. Continued utilisation of novel resources and options to win superiority, to fuel ever more traditional economic growth, and to fulfil new demands and pretensions, create the possibility of irreversible deterioration of environment or final depletion, which will inevitably result in further risks and supply/substitution problems. This challenge must be met by acting rationally and indeterministically and by setting up globally valid agreements comparable to these of the Nuclear Test Ban Treaty (1963), Non-Proliferation Treaty (1968), and Montreal Protocol (1987). Valuable proposals have been made, like taxation of resources to minimise externalities. Corporations must develop a strong concept of sustainability and internalisation of ecological, social, and cultural costs [16]. Dematerialisation [16], regulation of world finances, world economy, and the transformation of free enterprise economy into an eco-social market economy is recommended. The Carnoules Declaration (1994) [17] and the Paris Agreement (2015) under the United Nations Framework Convention on Climate Change must be realised.

The actual environmental situation can partly be compared with the time span shortly before the outbreak of epidemics in medieval Europe: at that time citizens unwittingly poisoned their wells with sewage disposed of untreated waste in the ambient ground. At present, a world population a hundred times larger disposes of *e.g.* its huge masses of waste gases, largely unfiltered, into the atmosphere and knowingly risks the consequences.

It must be possible for the majority of mankind to be able to exert the most decisive, indeterministic act: the free will to stop utilisation of fossil energy carriers, leaving them in place below the earth's surface. What has occurred over 250 years according to deterministic economic rules and generated rising dependencies, risks, and constraints, turned out to be fatal to environmental quality, health, as well as societal and political peace, and detracted from peacefully "*tilling and keeping the Garden of Eden*" (Genesis 2, 16).

The International Monetary Fund (IMF) calculated that the effective, direct, and indirect costs (including externalities) of global energy subsidies (resulting from the mining, extracting, and commercialising fossil energy carriers) rose from 2011 to 2015 by 1100 billion USD to 5300 billion USD. This is ca. 6.5% of global gross domestic product. The IMF therefore recommends an energy subsidy reform [18]. However, it has been noted that leading global banks divest from risky energy projects, fostering a shift in banks' lending strategies away from the high carbon towards a low carbon economy, thus following recommendations of environmentalists, who consider coal projects as particularly harmful to the ecosystems [19].

Coal, petroleum, natural gas and methane hydrate deposits consist of natural accumulations (*i.e.* common graves in humane sense) of partly redeposited organic relics of myriads of individuals of flora and fauna species, fossilised and preserved over geological times. Each creature, even its fossil (post mortem) state, owns per se a moralic worth and a right to exist unimpaired in those huge thanatocoenoses. But these accumulations were utilised as energy carriers, exploited for making money, combusted in rising quantities and functionalised to propel industrialisation. Due to these purposes, the creatures died a second time, because their relicts were transformed by combustion into inorganic matter. This kind of body stripping, grave desecration and grave robbery, which spread over the globe since 250 years, is psychologically and morally detrimental and desensibilising: it diminishes respect of other forms of life, of peace in death and - concerning the actors - shame and selfworth. In the meantime, the bad omens of this behaviour became true in form of rising ecological problems.

Mining and extraction of mineral raw materials, which are natural concentrates of matter that originated only once in earth history in limited quantity, is expensive, risky, and dangerous work, and the recoverable amounts are even more limited. So the responsibility of primary industry and economic geologists should be to not waste mineral commodities for the production of huge amounts of luxury goods and excess wealth, instead focussing on meeting merely the essential demands of humans. It is neither sensible nor sensitive to waste beneficiated matter, which was mined in concentrations of grams/T in the host rock.

The ecological crisis also resulted from insufficient diversity in technological development. Rising dependencies, criticalities of strategic metals, and risk of resource depletion occurred because of the one-sided, long-term practice in all sectors of industry, private households, and transport systems to use predominantly fossil energy carriers to operate machines, energy plants, robots, *etc.* It would have also been a better decision to promote, parallel to the gas firing motor, developments of alternative propulsion systems. Plato's allegory of the cave is clearly applicable here, because a vast majority sticks to the established current economic and technical system, confirmed by climate change deniers, often represented by conspiratorial and authoritarian conservative males [20, 21]. A metastudy of scientific consensus on human-made global warming reveals that more than 90% of climate scientists share this opinion in papers taking position on that theme and that the degree of consensus is positively correlated with scientific expertise [22].

Human creativity always enables us to proceed on safer paths; some possibilities among many are mentioned here:

- One promising solution to the problem that, in 2015, only ca. 2.7% (141 GW) of global grid-connected (5250 GW) energy could be stored, was the invention of a liquid organic hydrogen carrier system: Catalytic hydrogenation and dehydrogenation of dibenzyltoluol represents an innovative technological step towards secure and long-term chemical storage and concentration of fluctuating renewable energy of low-density and its conversion into a steady and base-load compatible availability-state [23].
- The application of alumosilicates of the Zeolite group in latent heat storage systems resulted in sufficient capacities to compensate for electricity peak loads in buildings [24] and in the capability to store energy of low density, *e.g.* industrial waste heat, *etc.*
- A possible utilisation of CO_2 to substitute petroleum-derived plastic (polyethylene terephthalate [PET]) *via* synthesis of *e.g.* plant-derived plastic (polyethylene furandicarboxylate [PEF]) was experimentally realised by CO_3^{2-}-promoted C-H carboxylation [25].
- Recycling of critical and environmentally problematic resources: Investigations concerning the recycling of phosphate, in the form of the chemical species magnesium ammonium phosphate, from domestic sewage sludge by gravity and magnetic separation techniques resulted in recoveries above 90% [26].
- Fostered by biochemist and food scientist P. O. Brown, biotechnological development of plant-based meat substitutes may become an appropriate alternative to industrial livestock farming, detrimental to climate, water and soil qualities, health and biodiversity [27].

The following recommendations are given:

- Obligatory school lessons in ecology. Compulsory evaluation of ecological ramifications in all technical, economic, and scientific studies, in which the full environmental impacts of the methods learned are explained in detail.
- More lectures in music and fine arts to augment and promote aesthetic perception.
- Realistic pricing of products, fair trade and fairness in general.
- Establishment of economic systems and reproductive human societies based on sufficiency, because quality matters more than quantity. Concentration always entails declines in quality, intellect, humaneness, as well as food and resource security.

Since the onset of human existence, violence has been perpetrated against nature, because of our constant aggression and the fulfilment of our bare necessities, demands and greed; human beings tolerate that and hazard the consequences. A small part of our violent behaviour is necessary in civilised and sensible forms of biological reproduction, hunting, gathering and creating food, mining, creation of buildings, streets, devices for transport and communication, as well as the acquisition of new knowledge, technical know-how, and goods. It is taken for granted that this form of necessary aggressive behaviour does not negatively influence our fellow human beings and descendants directly or indirectly, or short- or long-term. The large remaining part of the aggressive behaviour of human beings, which is based in rivalry and archaic impulses for power, greediness, and profit, and has resulted in permanent conflicts between our species and nature, is not necessary and must therefore be rejected immoral. In this way, human beings have created and tolerated constant problems concerning quality of life and the environment. The situation will improve only when human being cease abject violence, and consequently create wise and long-lasting positive relationships with our fellow human beings and our environment. There exists no reasonable alternative to life characterised by intelligent self-limitation as well as conscientiously minimised conflicts and violence. This means a fundamental change in cultural values, resulting in less self-violation of humankind, and impairment of nature.

Considering the urgent need to nourish more than 7.5 billion persons; the presence of more than 1.7 billion motor vehicles; more than 10,000 atomic missiles; more than 600 nuclear reactors; more than 50 billion food animals; more than 5000 slaughter houses; more than 140 million war-victims since 1800; *etc.*: from this it can be concluded that humankind has - in part - problems with morals, peaceableness, self-limitation, self-reflexion, as well as finding and realising reasonable decisions. Thus, a second Era of Enlightenment must occur, during which the above mentioned recommended changes in behaviour make their way.

NOTES

[1] *"Be fruitful, and multiply, and replenish the earth, and subdue it: and have dominion over the fish of the sea, and over the fowl of the air, and over every living thing that moveth upon the earth"*(Genesis 1,28 LUT).

[2] *"The angel who talked with me had a measuring rod of gold to measure the city, its gates and its walls. The city was laid out like a square, as long as it was wide. He measured the city with the rod and found it to be 12,000 stadia (ca. 2.2 km) in length, and as wide and high as it is long. ..... The wall was made of jasper, and the city of pure gold, as pure as glass. The foundations of the city walls were decorated with every kind of precious stone. The first foundation was jasper, the second sapphire, the third agate, the fourth emerald, the fifth onyx, the sixth ruby, the seventh chrysolite, the eighth beryl, the ninth topaz, the tenth turquoise, the eleventh jacinth, and the twelfth amethyst. The twelve gates were twelve pearls, each gate made of a single pearl. The great street of the city was of gold, as pure as transparent glass."* Revelation 21-15 to 21-21 (new international version) *http://biblehub.com/niv/revelation/21.htm*; retrieved 08/04/2015.

[3] He was influenced by the scholar Alexander von Humboldt, who earlier addressed in his opus magnum, entitled Cosmos, some ecological thoughts about the interdependent web of life and matter, and that human activities impinge nature: *e.g.* forest clearances because of agricultural expansion in Cuba and because of smelting silver in Mexico [28]. Nota bene: Ecologic thoughts have already been expressed by Georgius Agricola: *"..that the fields are devastated by mining operations, for which reason formerly Italians were warned by law that no one should dig the earth for metals and so injure their very fertile fields, their vineyards, and their olive groves. Also they argue that the woods and groves are cut down, for there is need of an endless amount of wood for timbers, machines, and the smelting of metals. And when the woods and groves are felled, then are exterminated the beasts and birds, very many of which furnish a pleasant and agreeable food for man. Further, when the ores are washed, the water which has been used poisons the brooks and streams, and either destroys the fish or drives them away. Therefore the inhabitants of these regions, on account of the devastation of their fields, woods, groves, brooks and rivers, find great difficulty in procuring the necessaries of life, and by reason of the destruction of the timber they are forced to greater expense in erecting buildings.."* [29].

CONFLICT OF INTEREST

The author (editor) declares no conflict of interest, financial or otherwise.

ACKNOWLEDGEMENTS

I am grateful to my wife Helga Pfoertner, who accompanied and sustained this work with never-ending patience and endurance. Thanks to the Deutsches Museum Munich for the information presented in the sections about environment, mining and extracting, as well as power engineering. Thanks to Dr. Paolo Sammuri (Rome), who read the pages and gave valuable comments. Many thanks to the scientific publishing houses and to the Bavarian State Library, which made all the information available online. Thanks to Mr. Alexander Wall (GB) for proofreading. Thanks to the anonymous reviewers, who helped to improve the text by giving comments and helpful advice. Thanks to teacher Gerald Zonsius, who in 1973, recommeded the 10th grade class of a secondary school in Upper Bavaria (Germany) to study the German edition of D. & D. Meadow's book "Limits to Growth".

REFERENCES

[1]   Freud, S. *Das Unbehagen in der Kultur*; Bayer, L.; Krone-Bayer, K.H., Eds.; ; Bayer, L.; Krone-Bayer, K.H., Eds.; Reclam: Stuttgart, **2012**.

[2]   Bentham, J. *An introduction to the principles of morals and legislation*; Dover Philosophical Classics: Mineola, New York, **2007**.

[3]   Meadows, D.; Randers, J.; Meadows, D. *The Limits to Growth: The 30 Year Update*; Chelsea Green Publishing Company: Claremont, NH, **2004**.

[4]   Hartmann, S. *Unpacking the Black Box: the need for Integrated Environmental Humanities (IEH)*, http://www.futureearth.org/blog/2015-jun-3/unpacking-black-box-need-integrated-environmental-humanities-ieh

[5]   Pepper, D. *Modern Environmentalism: An Introduction*; Routledge: London, New York, **1996**. [http://dx.doi.org/10.4324/9780203412244]

[6]   Schütz, E.; Gruber, E. *Mythos Reichsautobahn*; Weltbild Verlag: Augsburg, **2006**.

[7]   Oxfam. *62 people own the same as half the world, reveals Oxfam Davos report*, **2017**. https://www.oxfam.org/en/pressroom/pressreleases/2016-01-18/62-people-own-same-half-world-reveals-oxfam-davos-report

[8]   MacLellan, M. The Tragedy of Limitless Growth: Reinterpreting the Tragedy of the Commons for a Century of Climate Change. *Environ. Hum.*, **2015**, *7*, 41-58. [http://dx.doi.org/10.1215/22011919-3616326]

[9]   Conrad, S. *Geschichte der Welt 1750-1870: Wege zur modernen Welt 1750-1870*; Conrad, S.; Osterhammel, J.; Beck, C.H., Eds.; Munich, **2016**, Vol. 4, pp. 512-559. [http://dx.doi.org/10.17104/9783406641145]

[10]   De Duve, C.R. *Genetics of Original Sin: The Impact of Natural Selection on the Future of Humanity*; Yale University Press: New Haven, **2012**.

[11]   Sagan, C. *Cosmos*; Random House: New York, **1980**.

[12] Glickson, A.Y.; Groves, C. *Climate, Fire and Human Evolution. The Deep Time Dimensions of the Anthropocene*; Glickson, A.Y.; Groves, C., Eds.; Modern Approaches in Solid Earth Sciences Springer: Heidelberg, New York, **2016**, Vol. 10, pp. 177-188.
[http://dx.doi.org/10.1007/978-3-319-22512-8_6]

[13] Guggenberger, B. *Das Menschenrecht auf Irrtum: Anleitung zur Unvollkommenheit*; Hanser: München, **1987**.

[14] Carolan, M. *Cheaponomics. The High Cost of Low Prices*; Routledge: Oxford, UK, **2014**.

[15] Horan, R.D.; Bulte, E.; Shogren, J.F. How trade saved humanity from biological exclusion: an economic theory of Neanderthal extinction. *J. Econ. Behav. Organ.*, **2005**, *58*(1), 1-29.
[http://dx.doi.org/10.1016/j.jebo.2004.03.009]

[16] Sukhdev, P. *Corporation 2020: Transforming Business for Tomorrow's World*; Island Press: Washington, DC, **2012**.

[17] Schmidt-Bleek, F. *Grüne Lügen, 4. Aufl.; Ludwig*; Random House: München, **2014**.

[18] Coady, D.; Parry, I.; Sears, L. *How Large Are Global Energy Subsidies?*, **2015**.
http://www.imf.org/external/pubs/ft/wp/2015/wp15105.pdf
[http://dx.doi.org/10.5089/9781513532196.001]

[19] Corkery, M. *As coal's future darkens, banks curtail funding*; The New York Times International Weekly, **2016**.

[20] Jylhä, K.M.; Cantal, C.; Akrami, N. Denial of anthropogenic climate change: Social dominance orientation helps explain the conservative male effect in Brazil and Sweden. *Pers. Individ. Dif.*, **2016**, *98*, 184-187.
[http://dx.doi.org/10.1016/j.paid.2016.04.020]

[21] Lewandowsky, S.M.; Cook, J.; Lloyd, E. The 'Alice in Wonderland' mechanics of the rejection of (climate) science: simulating coherence by conspiracism. *Synthese*, **2016**, 1-22.

[22] Cook, J.; Oreskes, N.; Doran, P.T. Consensus on consensus: a synthesis of consensus estimates on human-caused global warming. *Environ. Res. Lett.*, **2016**, *11*, 048002.
[http://dx.doi.org/10.1088/1748-9326/11/4/048002]

[23] Hydrogenious Technologies. *The breakthrough of global energy storage*, http://www.hydrogenious.net/en/energy-storage/

[24] Johannes, K.; Kuznik, F.; Hubert, J.-L. Design and characterisation of a high powered energy dense zeolite thermal energy storage system for buildings. *Appl. Energy*, **2015**, *159*, 80-86.
[http://dx.doi.org/10.1016/j.apenergy.2015.08.109]

[25] Banerjee, A.; Dick, G.R.; Yoshino, T.; Kanan, M.W. Carbon dioxide utilization *via* carbonate-promoted C-H carboxylation. *Nature*, **2016**, *531*(7593), 215-219.
[http://dx.doi.org/10.1038/nature17185] [PMID: 26961655]

[26] Hirajima, T.; Hagino, T.; Kose, M. Recovery and upgrading of phosphorus from digested sewage sludge as MAP by physical separation techniques. *J. Environ. Prot. (Irvine Calif.)*, **2016**, *7*, 816-824.
[http://dx.doi.org/10.4236/jep.2016.76074]

[27] Fellet, M. A fresh take on fake meat. *ACS Cent. Sci*, **2015**, *1*(7), 347-349.
[http://dx.doi.org/10.1021/acscentsci.5b00307] [PMID: 27162992]

[28] Wulf, A. *The invention of nature. The adventures of Alexander von Humboldt - The lost hero of science*; John Murray: London, **2015**. https://www.hodder.co.uk/books/Sdetail.page?isbn=9781473637184

[29] Hoover, H.C.; Hoover, L.H. *De re metallica*, **1950**. https://archive.org/stream/deremetallica50agri/deremetallica50agri_djvu.txt

Abbreviations and Acronyms

a	annum (year)
asl	above sea level
AD	anno Domini
bn	billion
BOD	biological oxygen demand
Bq	Becquerel
bsl	below sea level
d	day
DOC	dissolved organic carbon
E	exa (10^{18})
g	gram
G	giga (10^{9})
Ga	billion years
GDP	gross domestic product
Gy	gray: $1m^2/s^2 = 1$ joule/kg: energy dose
kg	kilogram
km	kilometer
L	litre
LNG	liquified natural gas
m	meter
M	mega (10^{6})
Ma	million years
mGy	milligray
mSV	millisievert
p	partial pressure
P	peta (10^{15})
pM	pico (10^{-12}) molar
PM	particulate matter
ppb	parts per billion
ppm	parts per million
Sv	sievert: 1joule/kg: equivalent dose
T	metric tonnes

TWH terawatt hours

UV ultraviolet radiation

VOC volatile organic compounds

W watt

μ micro (10^{-6})

Used Encyclopedias and Online Resources

Der Brockhaus 5 volumes, 9th edition, Leipzig-Mannheim, 2000.

Der Brockhaus Philosophie second edition, Leipzig-Mannheim, 2009.

Lexikon der Chemie 3 volumes, Spektrum Akademischer Verlag, Heidelberg-Berlin, 1998.

Lexikon der Physik 6 volumes, Spektrum Akademischer Verlag, Heidelberg-Berlin, 1998.

Lexikon der Geowissenschaften 6 volumes, Spektrum Akademischer Verlag, Heidelberg-Berlin, 2000.

Lexikon der Biochemie 2 volumes, first edition, Elsevier Spektrum Akademischer Verlag, Heidelberg-Berlin, 1999/2000.

Internet: Several search engines, online-libraries of scientific publishing houses, Wikipedia, Google Scholar, Bavarian State Library online.

SUBJECT INDEX